To Frank with best wishes
Barry Taylor

The Fundamental Constants and Quantum Electrodynamics

The Fundamental Constants and Quantum Electrodynamics

B. N. Taylor
RCA Laboratories, Princeton, New Jersey

W. H. Parker
Department of Physics, University of California, Irvine, California

D. N. Langenberg
Department of Physics and Laboratory for Research on the Structure of Matter, University of Pennsylvania, Philadelphia, Pennsylvania

A Reviews of Modern Physics Monograph

Academic Press/New York/London/1969

ACADEMIC PRESS, INC.
111 Fifth Avenue, New York, New York 10003

United Kingdom Edition published by
ACADEMIC PRESS, INC. (LONDON) LTD.
Berkeley Square House, London W1X6BA

LIBRARY OF CONGRESS CATALOG CARD NUMBER: 76-101416

Originally printed in the
REVIEWS OF MODERN PHYSICS,
Volume 41, Number 2, July 1969

PRINTED IN THE UNITED STATES OF AMERICA

Acknowledgments

We should like to thank (Mrs.) Sheila Taylor for writing the several computer programs used throughout this work, and E. Maute for his help with the many numerical computations. We very gratefully acknowledge useful conversations and/or correspondence with the following individuals, many of whom kindly supplied us with information and results prior to publication: J. C. Baird, J. A. Bearden, E. Bergstrand, S. Bludman, H. S. Boyne, J. R. Brandenberger, S. J. Brodsky, R. A. Brown, J. Cameron, H. Capptuller, E. R. Cohen, A. H. Cook, B. L. Cosens, P. Crane, L. Csillag, R. D. Cutkosky, D. H. Douglass, Jr., J. W. M. DuMond, R. D. Deslattes, S. D. Drell, R. L. Driscoll, G. W. Erickson, J. E. Faller, F. J. M. Farley, T. Fulton, W. J. Hamer, J. A. Hammond, K. Hara, F. K. Harris, C. K. Iddings, R. R. Jacobs, S. L. Kaufman, T. Kinoshita, H. A. Kirkpatrick, J. W. Knowles, B. E. Lautrup, K. R. Lea, M. Leventhal, A. G. McNish, D. Mader, J. B. Marion, W. C. Martin, J. E. Mercereau, H. Metcalf, T. Myint, M. Narasimham, C. Nordling, C. H. Page, B. W. Petley, F. W. M. Pichanick, F. M. Pipkin, A. Rich, R. T. Robiscoe, F. L. Roesler, M. A. Ruderman, A. Sakuma, W. R. Shields, L. G. Smith, D. R. Tate, J. Terrien, A. M. Thompson, P. A. Thompson, J. S. Thomsen, K. R. Trigger, P. Vigoureux, T. E. Wells, W. H. Wing, F. Winkler, D. R. Yennie, and V. I. Zingerman.

Glossary of Symbols and Units

a_e Magnetic moment anomaly of the free electron: $a_e = \frac{1}{2}(g_s - 2)$. (When necessary to distinguish between the electron and positron, the symbols a_e^- and a_e^+ are used.)

a_μ Magnetic moment anomaly of the free muon: $a_\mu = \frac{1}{2}(g_\mu - 2)$

amu Atomic mass unit (unified scale, $^{12}C \equiv 12$)

Å* Angstrom-star x-ray unit defined by $\lambda(WK\alpha_1) \equiv 0.2090100$ Å*

A_{ABS} Absolute ampere: The ampere is that constant current which, if maintained in two straight parallel conductors of infinite length, of negligible circular cross section, and placed 1 m apart in vacuum, would produce between these conductors a force equal to 2×10^{-7} N/m of length.

A_{NBS} Ampere as maintained by the National Bureau of Standards (NBS)

BIPM Bureau International des Poids et Mesures

c Velocity of light

g Acceleration due to gravity

g_s g factor of the free electron: $g_s = 2\mu_e/\mu_B$. (When necessary to distinguish between the electron and positron, the symbols g_s^- and g_s^+ are used.)

g_μ g factor of the free muon

h Planck's constant

Hz Hertz (cycles per second)

e Electron charge

F Faraday constant

kg	Kilogram: The kilogram is the unit of mass; it is equal to the mass of the international prototype of the kilogram
kxu	kx-unit based on $\lambda(\mathrm{Cu}K\alpha_1) \equiv 1.537400$ kxu
K	Kelvin: The kelvin, unit of thermodynamic temperature, is the fraction 1/273.16 of the thermodynamic temperature of the triple point of water.
K	Ratio of A_{NBS}/A_{ABS}
m	Meter: The meter is the length equal to 1650763.73 wavelengths in vacuum of the radiation corresponding to the transition between the levels $2p_{10}$ and $5d_5$ of the krypton-86 atom.
m_μ	Muon rest mass
M_p	Proton rest mass
$M_p{}^*$	Proton rest mass in amu (unified scale, $^{12}\mathrm{C} \equiv 12$)
m_e	Electron rest mass
$m_e{}^*$	Electron rest mass in amu (unified scale, $^{12}\mathrm{C} \equiv 12$)
mgal	10^{-5} m/sec^2
N	Avogadro's number (unified scale, $^{12}\mathrm{C} \equiv 12$)
NBS	National Bureau of Standards
NPL	National Physical Laboratory (national standards laboratory of Great Britain)
ppm	Parts per million
r_i	Residual of a particular input datum in a least-squares adjustment
R	Birge ratio
RSS	Square root of the sum of the squares or root-sum-square
R_∞	Rydberg constant for infinite mass
sec	Second: The second is the duration of 9192631770 periods of the radiation corresponding to the transition between the two hyperfine levels of the ground state of the cesium-133 atom.

SI	Système International, the official name of the system of units based on the ampere, kilogram, meter, second, and candela.
$\mathcal{S}$	Lamb shift in hydrogenic atoms ($nP_{1/2}$–$nS_{1/2}$ interval)
T	Tesla (one tesla $= 10^4$ G)
V_{ABS}	Absolute volt
V_{NBS}	Volt as maintained by the National Bureau of Standards (NBS)
WQED	Without quantum electrodynamic theory
Z	Nuclear charge
α	Fine structure constant
α^{-1}	Inverse fine structure constant
γ_p	Gyromagnetic ratio of the free proton
γ_p'	Gyromagnetic ratio of protons in H_2O (spherical sample)
λ_C	Compton wavelength of the electron
ΔE	Fine-structure splitting in hydrogenic atoms ($nP_{1/2}$–$nP_{3/2}$ interval)
Λ	kx-unit-to-angstrom conversion factor
Λ^*	Å*-to-angstrom conversion factor
μ_e	Magnetic moment of the free electron
μ_B	Bohr magneton
μ_n	Nuclear magneton
μ_p	Magnetic moment of the free proton
μ_p'	Magnetic moment of protons in H_2O (spherical sample)
μ_μ	Magnetic moment of the free muon
μ_0	Permeability of free space ($4\pi \times 10^{-7}$ henry/m)
ν_{Hhfs}	Ground-state hyperfine splitting in hydrogen
ν_{Mhfs}	Ground-state hyperfine splitting in muonium
ν_{Phfs}	Ground-state hyperfine splitting in positronium
$\sigma(H_2O)$	Diamagnetic shielding correction for protons in H_2O (spherical sample)
σ_μ	Diamagnetic shielding correction for muons in H_2O (spherical sample)
χ^2	the statistic "chi squared"
χ_{H_2O}	Magnetic susceptibility of H_2O
ω_c	Proton cyclotron frequency

ω_e	Electron cyclotron frequency
ω_p	Proton spin flip or precession frequency
ω_p'	Spin flip or precession frequency for protons in H_2O (spherical sample)
ω_s	Electron spin flip or precession frequency
Ω_{ABS}	Absolute ohm
Ω_{NBS}	Ohm as maintained by the National Bureau of Standards (NBS)

[The definitions of the ampere, kilogram, meter, and second were taken from "Definitions of Basic SI Units," Metrologia **4,** 147 (1968).]

Contents

ACKNOWLEDGMENTS v

GLOSSARY OF SYMBOLS AND UNITS vii

I. Introduction

A. Importance of the Fundamental Physical Constants 1

B. Justification for a New Least-Squares Adjustment at This Time 2

C. Significance of the Output Values of a Least-Squares Adjustment of the Constants 6

D. Outline of Paper 8

II. Review of Experimental Data

A. Introduction

1. Auxiliary Constants, Stochastic Input Data, and Adjustable Constants 11
2. Conversion Factors as Adjustable Constants . . 12
3. Form of the Operational Equations Used in the Adjustments 14
4. Treatment of Error 18

B. Auxiliary Constants

1. Comparisons of As-Maintained Electrical Units . 21
2. Velocity of Light, c 26
3. Ratio of the Absolute Ohm to the NBS Ohm, $\Omega_{ABS}/\Omega_{NBS}$ 32
4. Acceleration Due to Gravity, g 35
5. Magnetic Moment of the Electron in Units of the Bohr Magneton, μ_e/μ_B 47

6. Magnetic Moment of the Proton in Units of the Bohr Magneton, μ_p/μ_B 50
7. Atomic Masses and Mass Ratios 56
8. Rydberg Constant for Infinite Mass, R_∞ . . . 58
9. Summary of the Auxiliary Constants 67

C. Stochastic Input Data

1. Josephson-Effect Value of e/h 68
2. Ratio of the NBS Ampere to the Absolute Ampere, $K \equiv A_{NBS}/A_{ABS}$ 71
3. Faraday Constant, F 77
4. Gyromagnetic Ratio of the Proton, γ_p 85
5. Magnetic Moment of the Pronton in Units of the Nuclear Magneton, μ_p/μ_n 112
6. X-Ray Experiments 129
(a) Measurements of hc/e 132
(b) Measurements of N 135
(c) Measurements of A 138
(d) Measurements of λ_C 140
7. Summary of the Stochastic Input Data . . . 142

III. Least-Squares Adjustment to Obtain Values of the Constants Without QED Theory

A. Preliminary Search for Discrepant Data

1. Inconsistencies among Data of the Same Kind . 153
2. Inconsistencies among Dissimilar Data 156

B. Least-Squares Search for Discrepant Data

1. Summary of Least-Squares Procedure 165
2. Analysis of Variance 169

C. Final Adjustment to Obtain Best Values of the Constants Without QED Theory 180

IV. Implications for Quantum Electrodynamics

A. g Factors

1. Free Electron and Positron 195
2. Free Muon 201

B. Ground-State Hyperfine Splittings

1. Atomic Hydrogen 211
2. Muonium 217
3. Positronium 226

C. Fine Structure of Hydrogenic Atoms

1. Lamb Shift in H and D, $n = 2$ 227
2. Fine Structure Splitting in H and D, $n = 2$ . . 243
3. Other Fine Structure Measurements 260

D. Comparison of Experimental Data via the Fine Structure Constant 263

V. Final Recommended Set of Fundamental Constants

A. Selection of Input Data 275
B. Final Adjustment and Recommended Constants . 291

VI. Summary and Conclusions

A. Conclusions Concerning Quantum Electrodynamics 299
B. Conclusions Concerning Superconductivity . . . 307
C. Weaknesses in the Present Adjustment, and Future Work 309
D. Further Fundamental-Constant Experiments Utilizing Macroscopic Quantum Phase Coherence in Superfluids 312
E. Recommendation for Reporting Results 319

VII. Notes Added in Proof 323

APPENDIX
Variance-Covariance Matrices 337

REFERENCES 343

Es irrt der Mensch, solang' er strebt.

Goethe

I. Introduction

A. Importance of the Fundamental Physical Constants

It has often been said that all of the apparently divergent branches of physics are really intimately related to one another. This unity of the various branches of physics is clearly emphasized by the far reaching implications of the new value of e/h obtained from the ac Josephson effect in superconductors (Parker, Langenberg, Denenstein, and Taylor, 1969). We shall show here that this solid-state physics experiment has important consequences for fields as far removed from the solid state as quantum electrodynamics (QED), high-energy physics, atomic physics, x rays, and our general knowledge of the fundamental physical constants. Our analysis is based on a complete least-squares adjustment of the fundamental physical constants. These constants are important links in the chain of physical theory which binds all of the diverse branches of physics together, and the careful study of their numerical values as obtained from various experiments in the different fields of physics can give significant information about the over-all consistency and correctness of the basic theories of physics themselves. Thus, as Cohen and DuMond (1965) have emphasized, measurements of the fundamental physical constants to ever greater levels of accuracy are important, not just because they "add another decimal point" and provide us with a more consistent set of constants to work with, but because they may lead to the discovery of a previously unknown inconsistency or the removal of a known inconsistency in our physical description of nature. We have taken this view as a guiding principle throughout the present study.

B. Justification for a New Least-Squares Adjustment at This Time

Quantum electrodynamics, which describes the interaction between electrons, muons, photons and external electromagnetic sources, is one of the most important and precise of our modern theories. Since the coupling constant or expansion parameter of the theory is the fine structure constant α, an accurate value of α is essential for comparing the theoretical predictions of QED with experiment. Heretofore, the most accurate values of α were obtained from experiment with the aid of theoretical equations containing significant contributions from QED. This made it difficult to compare QED theory and experiment unambiguously since the theory had to be evaluated using values of α derived from the experiments themselves. Such comparisons were therefore limited to the testing of internal consistency among various experiments of this type. Now, however, by combining the value of e/h obtained from the ac Josephson effect with the measured values of certain other constants, a highly accurate indirect value of α can be obtained without any essential use of QED theory. As a result, direct and unambiguous comparisons can be made between QED theory and experiment. In addition to the Josephson-effect value of e/h, the experimental input data used in deriving the indirect value of α include the Faraday constant, the gyromagnetic ratio of the proton, the magnetic moment of the proton in units of the nuclear magneton, the ratio of the ampere as maintained by the United States National Bureau of Standards to the absolute ampere, and certain accurately known auxiliary constants. It should be noted that these input data and the constants to be derived from them are independent of QED only in the practical sense that it is not essential to make use of any QED theory in the analysis. In a deeper philosophical sense it is clear that these input data *are* dependent on electrodynamic interactions which may be described by some QED theory of the future. For the present, however, it is precisely the absence of such a theory which permits us to treat these quantities simply as well-defined measurable parameters which can be interrelated without the use of existing QED theory. In order to differentiate the constants derived in this way from the "best" or recommended constants we derive later in the paper using, in part, data which must be analyzed with the help of QED theory, we shall adopt the notation WQED meaning "*without quantum electrodynamic theory.*"

One way of obtaining an indirect value of α from e/h is to use the following equation (to be derived in Sec. III.A.2):

$$\alpha^{-1}=C_1[(1/\gamma_p)(2e/h)]^{1/2}, \tag{1}$$

where C_1 is a combination of accurately known auxiliary constants (see Sec. II.A.1) and γ_p is the gyromagnetic ratio of the proton. An alternate expression (or route) for deriving α is

$$\alpha^{-1}=C_2\left(\frac{1}{F}\frac{1}{\mu_p/\mu_n}\frac{2e}{h}\right)^{1/2}, \tag{2}$$

where C_2 is again a combination of auxiliary constants, F is the Faraday constant, and μ_p/μ_n is the magnetic moment of the proton in units of the nuclear magneton. Still another route one can follow for deriving α is (see Sec. II.C.6d)

$$\alpha^{-1}=C_3(1/\lambda_C)^{1/2}, \tag{3}$$

where λ_C, the electron Compton wavelength, is the wavelength of the radiation emitted by an electron–positron pair when they annihilate (both assumed to be initially at rest).

The situation becomes more complex when conversion factors are taken into account. For example, λ_C is presently measured in terms of the arbitrary unit of length used in the field of x rays, the so-called x-unit, rather than in absolute units. Information on the x-unit-to-milliangstrom conversion factor Λ can be obtained by direct diffraction grating experiments, by measuring the short wavelength limit of the continuous x-ray spectrum, or from measurements of the density and crystal lattice spacing of a particular crystal species. These three experiments yield, respectively, Λ, $h/e\Lambda$, and $N\Lambda^3$ (N is Avogadro's number). Similarly, the Josephson-effect value of e/h as well as F, γ_p, and $h/e\Lambda$ are measured in terms of as-maintained electrical units (see Sec. II.A.2) rather than absolute electrical units, thereby requiring an as-maintained-unit-to-absolute-unit conversion factor. Knowledge of this factor follows from direct-current balance measurements as well as a comparison of high-field and low-field measurements of γ_p.

It is obvious from even this brief discussion that information concerning α can be obtained from a variety of sources, and that such information results in a highly overdetermined set of equations for deriving α. The most straightforward and consistent method for handling such an overdetermined set is by the method of least squares. [For a particularly clear and concise discussion of the method, see Bearden and Thomsen (1957). See also Cohen and DuMond (1965) and Cohen, Crowe, and DuMond (1957).] This technique permits the calculation of a "best" compromise value of α which approximately satisfies all of the relevant equations.

Another motivation for carrying out a new adjustment at this time is that the previously accepted values for the fundamental constants, those resulting from the 1963 adjustment of Cohen and DuMond (1965),* are found to change significantly as a result of the Josephson-effect value for e/h. This is because the new value of e/h implies that the value of α used in the 1963 adjustment is too small by about 20 ppm (Parker, Taylor, and Langenberg, 1967; Langenberg, Parker, and Taylor, 1968). As a result, the previously accepted values of e, h, and N are changed by approximately +60, +100, and −60 ppm, respectively. The implications are clear: a new and complete least-squares

* Other related papers by these authors have since appeared: DuMond (1966) and Cohen (1968; 1969; 1966).

analysis should be carried out in order to obtain a new set of best or recommended values for all of the constants.

Our final reason for believing that a new adjustment and critical reevaluation of the constants is presently called for is that, since 1963, several important experiments and theoretical calculations have been completed. The new experimental work, in addition to the Josephson-effect determination of e/h, includes: (1) a measurement of the fine structure splitting in atomic hydrogen by Metcalf, Brandenberger, and Baird (1968); (2) a measurement of the $2S_{1/2}$–$2P_{3/2}$ splitting in atomic hydrogen by Kaufman, Lamb, Lea, and Leventhal (1969a; 1969b) [see also Kaufman (1968) and Kaufman, Leventhal, and Lea (1968)]; (3) a measurement of μ_p/μ_B, the magnetic moment of the proton in units of the Bohr magneton, by Myint, Kleppner, Ramsey, and Robinson (1966); (4) measurements of μ_p/μ_n, the magnetic moment of the proton in units of the nuclear magneton, by Mamyrin and Frantsuzov (1968; 1965; 1964) and by Petley and Morris (1968a; 1967); (5) a measurement of the hyperfine splitting in muonium in low as well as high magnetic fields by Hughes and his coworkers (Thompson, Amato, Crane, Hughes, Mobley, zu Putlitz, and Rothberg, 1969; Thompson, Amato, Hughes, Mobley, and Rothberg, 1967; Hughes, 1967; 1966; Cleland, Bailey, Eckhause, Hughes, Mobley, Prepost, and Rothberg, 1964); (6) a recalculation of μ_e/μ_B, the magnetic moment of the electron in units of the Bohr magneton, by Rich (1968a, 1968b, 1968c) from the experimental data of Wilkinson and Crane; (7) the recalculation of the Lamb shift by Robiscoe (1968) from his own data; (8) a new measurement of the Lamb shift in deuterium by Cosens (1968); and (9) a new measurement of the muon g factor at CERN (Bailey, Bartl, von Bochmann, Brown, Farley, Jöstlein, Picasso, and Williams, 1968).

The new theoretical work includes: (1) a comprehensive calculation of the second-order QED corrections to the hyperfine splitting (hfs) in atomic hydrogen by

Brodsky and Erickson (1966); (2) a critical analysis of the proton polarizability contribution to the hfs in hydrogen by Iddings (1965), by Drell and Sullivan (1967), and by Guérin (1967a; 1967b); (3) the calculation of the sixth-order contribution to the anomalous magnetic moment of the electron by Drell and Pagels (1965) and by Parsons (1968); and (4) an exact calculation of the fourth-order correction to the Lamb shift by Soto (1966). These developments (and others) shed new light on the over-all consistency of QED and our knowledge of the fundamental physical constants.

C. Significance of the Output Values of a Least-Squares Adjustment of the Constants

We should like to emphasize at the very outset that the numbers resulting from a least-squares adjustment of the fundamental physical constants must be taken with a grain of salt. This is not to say that such an exercise is without merit. On the contrary, a least-squares adjustment (LSA) is one of the few ways in which the over-all consistency of physical theory can be systematically investigated. Moreover, it provides a consistent set of constants at a particular epoch which can be used by all workers requiring them. However, it must be realized that the final values of the constants resulting from an LSA are generally obtained from a highly expurgated group of experimental input data. Most of the inconsistent or "bad" data are discarded (usually with a rationalization involving improper experimental procedure). This censorship results in output values which are so consistent that they would appear inviolate. Such confidence is not justified! For example, consider the fine structure constant. In Fig. 1 we show the values of α^{-1} (and its one-standard-deviation uncertainty) which have resulted from various adjustments carried out since 1950, including the output value of the present work which is used as a reference. The important point to note is that changes in α^{-1} significantly larger than the assigned uncertainties have taken place over the years. One of the main reasons for such

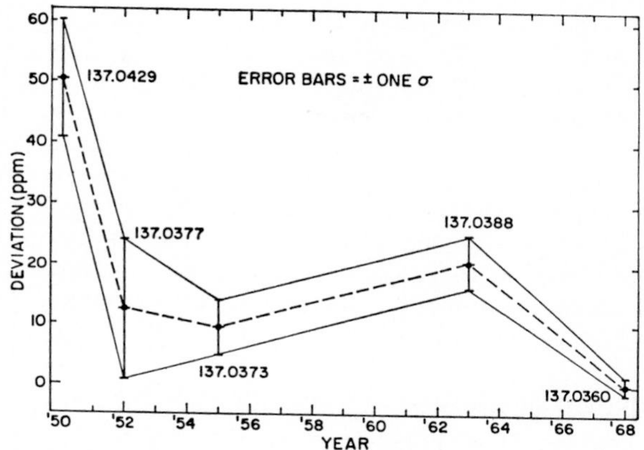

FIG. 1. Plot showing how our knowledge of α^{-1} and its one-standard-deviation uncertainty has changed over the last 18 years. The final recommended value of the present adjustment is used as a reference. [Value for 1963, Cohen and DuMond (1965); value for 1955, Cohen, DuMond, Layton, and Rollett (1955); value for 1952, DuMond and Cohen (1953); value for 1950, DuMond and Cohen (1952; 1951).] For similar comparisons of other constants, see Fig. 8.

large changes in the constants is the intimate relationships which exist among them. A significant change in one will generally have important effects on the others, to wit, the present situation in which the ac Josephson-effect value of e/h implies changes in the output values of the 1963 least-squares adjustment which exceed their assigned one-standard-deviation uncertainties by several times.

One might well ask whether there is a simple solution to this problem. For example, one might include *all* the available experimental data in the adjustment and increase the *a priori* assigned errors of each input datum before performing the least-squares adjustment so that all of the output values are consistent (i.e., ensure that the ratio of external error to internal error, or generalized Birge ratio R, is one—see Secs. III.A and III.B). However, we feel that this goes too far in the other direction and that by following such a procedure, one would, in effect, throw away information. A more reasonable approach might be one in which the *a priori* errors are not expanded, but all data are used which result in $R=1$. In most adjustments, including the present one, the input data are usually sufficiently censored that R turns out to be much less than unity. This can give the impression (misleadingly so) that all of the data are highly consistent and that the *a priori* errors have been overestimated. In any event, we feel that the adjusted values of the constants must always be viewed with caution. There is no guarantee that a situation similar to the present one involving e/h will not occur again.

D. Outline of Paper

The general plan of the paper is as follows: In Sec. II, we discuss the idea of auxiliary constants and stochastic input data, the quantities to be used as adjustable parameters, the general form of the equations to be used in the adjustments, and our philosophy and procedure for treating error. We next discuss all of the experimental input data, including auxiliary constants, *except* those data which *must* be derived from experiment by use of QED. In discussing the pertinent experiments we shall attempt to assess the results critically, but we shall leave detailed descriptions to the original papers. It is hoped that this will limit the length of the present paper without seriously reducing its readability. We also derive in Sec. II the individual equations used in the adjustments for each different type of input datum.

In Sec. III, we carry out a search for discrepancies among the data discussed in Sec. II by using equations like (1)–(3), and also by carrying out a large number of least-squares adjustments. A value for the fine-structure constant is then derived using what we find to be the most consistent or compatible set of input data.

In Sec. IV, the value of α derived in Sec. III is used to evaluate theoretical expressions for the Lamb shift and fine structure splitting in hydrogen, deuterium, and ionized helium, the hyperfine splitting in hydrogen, muonium, and positronium, and the magnetic moment anomaly of the electron and muon. The theoretical values for these quantities are then critically compared with the experimentally measured values after a reevaluation of the experiments themselves. This comparison provides the test of QED mentioned in Sec. I.B.

In Sec. V, we carry out a final least-squares adjustment using as input data the most reliable values of α discussed in Sec. IV together with all of the data used in Sec. III. The result is a set of best or recommended values for the fundamental physical constants.

We conclude with Sec. VI in which we summarize our findings and indicate where further experimental and theoretical work is needed, with particular attention to other potentially useful experiments which exploit quantum phase coherence effects in superfluids for the accurate measurement of fundamental constants.

II. Review of Experimental Data

A. Introduction

1. Auxiliary Constants, Stochastic Input Data, and Adjustable Constants

The input data used in a least-squares adjustment of the constants are generally classified into two groups (Cohen and DuMond, 1965). The first group, known as the auxiliary constants, contains quantities which have uncertainties sufficiently small that they can be considered as exactly known. The second group contains the more imprecise or stochastic input data. An example of an auxiliary constant is R_∞, the Rydberg constant for infinite mass, which has a one-standard-deviation uncertainty of about 0.1 ppm. (Throughout this paper, all uncertainties will be expressed in standard-deviation form unless otherwise noted—see Sec. II.A.4.) An example of a stochastic input datum is γ_p, the gyromagnetic ratio of the proton, which has an experimental uncertainty of about 4 ppm.

In order to carry out a least-squares adjustment, it is necessary to choose a subset of constants in terms of which all of the stochastic input data can be individually expressed (if necessary, with the aid of the auxiliary constants). It is the constants comprising this subset which are directly subject to adjustment and which we call the *adjustable constants.* In the present work the adjustable constants include the quantities α^{-1}, the inverse of the fine structure constant, e, the electron charge, and N, Avogadro's number. This choice is by no means mandatory but is computationally convenient. Any other complete set of constants in terms of which a series of observational equations could be formed, each containing only a single stochastic input datum but any number of auxiliary constants, would serve as well. (An adjustable constant can also be a stochastic input datum.) The choice of adjustable constants is therefore somewhat arbitrary. Note also that since the constants not chosen for direct adjustment are later obtained from appropriate combinations of the adjusted constants, optimum values of all of the constants are actually obtained.

2. *Conversion Factors as Adjustable Constants*

The national laboratories of the various countries maintain their own systems of electrical units by means of large groups of standard cells and precision resistors. For example, in the United States, the National Bureau of Standards (NBS) maintains a bank of 40 standard cells (known as the national reference group); its mean emf defines the U.S. legal volt, $\mathrm{V}_{\mathrm{NBS}}$ (Harris, 1964; private communication). Similarly, NBS maintains a bank of 10 1Ω precision resistors, the mean resistance of which defines the U.S. legal ohm, Ω_{NBS}. The ratio of the legal volt to the legal ohm is then the legal ampere:

$$\mathrm{A}_{\mathrm{NBS}} = \mathrm{V}_{\mathrm{NBS}}/\Omega_{\mathrm{NBS}}. \tag{4}$$

In practice, the NBS ohm is very nearly equal to the absolute ohm, while the NBS volt is about 9 ppm larger than the absolute volt (see Secs. II.B.3 and II.C.2). The ratio K of the NBS ampere to the absolute ampere is therefore

$$K \equiv \mathrm{A}_{\mathrm{NBS}}/\mathrm{A}_{\mathrm{ABS}} \approx 1.000009. \tag{5}$$

The one-standard-deviation uncertainty of K is several parts per million, so it cannot be used as an auxiliary constant. Thus, in addition to α^{-1}, e, and N, we also take K as one of our adjustable constants. Strictly speaking, K is not a fundamental constant but a conversion factor for relating an arbitrary electrical unit, i.e., the ampere as maintained by NBS in terms of standard cells and resistors, to the absolute SI ampere. However, this distinction is only conceptual, for K enters the adjustments in exactly the same manner as do the other adjustable constants.

There are two reasons why we choose to carry out our adjustments in terms of the electrical units maintained by NBS rather than, for example, the units maintained by the Bureau International des Poids et Mesures (International Bureau of Weights and Measures—BIPM). First, of the eight pieces of stochastic input data to be used in this work which involve electrical units, five were measured directly in terms of the units maintained by NBS, two were measured in terms of the British units, and one was measured in terms of the electrical units maintained by the USSR. Although international comparisons are carried out regularly to high accuracy (a few tenths of a part per million—see Sec. II.B.1), there is still a slight ambiguity in converting from one system of as-maintained electrical units to another. This is because the standard cells and resistors used to define the as-maintained units generally change with time, thus causing the relationships between the various as-maintained units to change with time. Consequently, the exact time period in which an experiment was performed becomes important (some of the data to be used date back 10 years). The use of units in terms of which the majority of the measurements have been made, i.e., NBS units, reduces to a minimum the uncertainties involved in converting from one set of units to another.

A second reason for working in NBS units is that NBS is the only standards laboratory in the world which has kept its as-maintained ampere under constant experimental surveillance for the last eight years. This has been done by measuring the gyromagnetic ratio of the proton as a function of time. In this method, a fixed current known in terms of the as-maintained electrical units is passed through a precision solenoid The proton precession frequency of a standard water sample inside the solenoid is measured and any change from previous measurements is attributed to a change in the as-maintained electrical units. As a result of this surveillance, it can be safely assumed that the NBS ampere has not changed by as much as 1 ppm since 1960 (see Sec. II.C.4) (Harris, 1964; private communication). Such a statement cannot be made about any of the other as-maintained amperes.

The discussion of the last two paragraphs emphasizes that K is not truly constant, but is expected to have a weak time dependence of as much as 0.5 to 1 ppm per decade. At present, this drift can be ignored since it is much less than the uncertainty in the direct measurements of K and the uncertainties in the other stochastic input data which are measured in terms of as-maintained units. This may not be the case in the future, and some type of "atomic" standard of voltage may become necessary. (The ohm is in comparatively good shape—see Sec. II.B.3.) The possible use of the ac Josephson effect as such a standard has recently been discussed (Taylor, Parker, Langenberg, and Denenstein, 1967).

In some exploratory adjustments involving x-ray data, we also include as an adjustable constant the x-unit-to-milliangstrom conversion factor Λ; but because of the over-all inconsistency and comparatively high uncertainty of the x-ray work, it is not used in the final adjustment which gives our best or recommended set of values for the constants. A recommended value for Λ is derived by combining the results of the final adjustment with the best of the x-ray measurements.

3. Form of the Operational Equations Used in the Adjustments

We shall attempt here to clarify some of the ideas presented in the two preceding sections with specific examples. Consider first the quantity $2e/h$. From the definition of the fine-structure constant, $\alpha = (\mu_0 c^2/4\pi)(e^2/\hbar c)$, it follows that

$$(\alpha^{-1})^{-1} e^{-1} = (\mu_0 c/4)(2e/h). \quad (6)$$

(We use SI units throughout because the as-maintained electrical units are expressed in that system; μ_0 is the permeability of free space and is by definition exactly equal to $4\pi \times 10^{-7}$ henry/m.) The units in terms of which $2e/h$ has been measured (Parker, Langenberg, Denenstein, and Taylor, 1969) are hertz per NBS volt, i.e., $\mathrm{Hz/V_{NBS}}$, while the units of $2e/h$ required in Eq. (6) are hertz per absolute volt, $\mathrm{Hz/V_{ABS}}$. In order to convert from $\mathrm{V_{NBS}}$ to $\mathrm{V_{ABS}}$, we introduce the ratio of the NBS ohm to the absolute ohm, $\Omega_{\mathrm{NBS}}/\Omega_{\mathrm{ABS}}$, and the constant $K \equiv \mathrm{A_{NBS}/A_{ABS}}$ to obtain

$$(\alpha^{-1})^{-1} e^{-1} K^1 N^0 = \frac{\mu_0}{4} \frac{c\Omega_{\mathrm{ABS}}}{\Omega_{\mathrm{NBS}}} \left(\frac{2e}{h}\right)_{\mathrm{NBS}}, \quad (7)$$

where the subscript NBS on $2e/h$ indicates that this quantity is to be expressed in terms of $\mathrm{V_{NBS}}$, i.e., as it has been experimentally measured. Note that we have succeeded in expressing a single stochastic input datum, in this case $2e/h$, in terms of auxiliary constants and our chosen set of adjustable constants. (The zero exponent of N simply indicates that this particular operational equation does not involve N; $c\Omega_{\mathrm{ABS}}/\Omega_{\mathrm{NBS}}$ can be taken as an auxiliary constant since its one-standard-deviation uncertainty is less than 0.4 ppm.) Similarly, it can be shown that (see Sec. II.C)

$$(\alpha^{-1})^0 e^1 K^{-1} N^1 = F_{\mathrm{NBS}},$$

$$(\alpha^{-1})^{-3} e^{-1} K^1 N^0 = \frac{\mu_0 R_\infty}{(\mu_p/\mu_B)} (\gamma_p)_{\mathrm{NBS}},$$

$$(\alpha^{-1})^{-3} e^{-2} K^0 N^{-1} = \frac{\mu_0 R_\infty}{(\mu_p/\mu_B)} \frac{1}{M_p{}^*} \left(\frac{\mu_p}{\mu_n}\right),$$

where $M_p{}^*$ is the mass of the proton in atomic mass units. [Throughout this paper we shall use the unified scale of atomic masses defined by taking the mass of ^{12}C to be exactly 12 atomic mass units (amu).] In each equation, an individual stochastic input datum (γ_p, F, or μ_p/μ_n) has been written in terms of auxiliary constants and the chosen adjustable constants.

The general form of these observational equations is clearly

$$\prod_{j=1}^{J} Z_j{}^{Y_{ji}} = A_i, \tag{8}$$

where i stands for the ith observational equation (total number N) or, equivalently, the ith stochastic input datum; Z_j is the jth adjustable constant (total number J), and Y_{ji} is the exponent of the jth adjustable constant in the ith observational equation. (Note that Y_{ji} is an integer and can be positive, negative, or zero.) A_i is the product of the ith stochastic input datum X_i and the combination of auxiliary constants a_i appropriate to the observational equation:

$$A_i = a_i X_i. \tag{9}$$

Also associated with each stochastic input datum X_i is its experimental uncertainty σ_i. In the usual least-squares procedure (Bearden and Thomsen, 1957; Cohen and DuMond, 1965), the weight factor for each observational equation is simply $1/\sigma_i{}^2$.

We are now in a position to make some general comments concerning auxiliary constants and stochastic input data. First, as has been emphasized by DuMond and Cohen (1953), the usual or classical least-squares method can only be applied to observational equations which are observationally independent of one another. If they are not, simple independent weights $1/\sigma_i{}^2$ cannot be assigned to the different equations, but rather the individual weights must be replaced by a weight matrix.* Thus, the uncertainties of the auxiliary constants which comprise the a_i must not be so large that two different observational equations which contain the same constants become correlated. Since errors add quadratically, a quantity having an uncertainty 5 to 10 times less than the typical uncertainties associated with the stochastic input data contributes only a few percent as much uncertainty and can therefore be safely used as an auxiliary constant. If the a_i are composed of such quantities, then the uncertainties of the A_i and X_i will be essentially the same, namely σ_i, and the weights of the observational equations will be given correctly by $1/\sigma_i^2$. (It is obviously pointless to form additional observational equations in order to treat auxiliary constants as adjustable constants, since their error is generally so small that the adjusted value would be identical to the experimental value.) It is now clear that K, which has an experimental uncertainty of several parts per million, must be classified as an adjustable rather than as an auxiliary constant. If it were not, it would give rise to large correlations between the various observational equations in which it appeared, and the relative weights of the equations would no longer be given by the inverse square of the uncertainty of the stochastic input datum contained in each equation. By taking K to be adjustable, we remove a serious deficiency of previous adjustments in which it was treated as an auxiliary constant, obtain a best value for K, and at the same time provide a means for unambiguously including the excellent high-field measurement of γ_p by Yagola, Zingerman, and Sepetyi (1966; 1962).

A second comment concerns the qualifications of a stochastic input datum [Bearden and Thomsen (1957)]. In general, such a quantity should have an uncertainty

---

* The generalized theory of least squares in which the individual equations are not independent has been discussed by Cohen (1953).

which is sufficiently small to allow it to carry some weight in the adjustment. For example, if the uncertainty in a particular stochastic input datum is 3 times that of another value of the same quantity, then it will only carry about 11% as much weight. Furthermore, on a purely statistical basis, its weight will probably be uncertain by more than this amount; if the uncertainty in estimating systematic errors is taken into account, the weight may be uncertain by as much as 20% or 30% (see Sec. II.A.4). Consequently, we take as a simple rule of thumb that it is meaningless to include any stochastic input datum if its error is more than 3 times the error of another similar datum obtained from either a direct measurement or from a combination of other data.

4. Treatment of Error

One of the most difficult tasks which faces any reviewer of the fundamental constants is that of ensuring that all error estimates are expressed on as equal a footing as possible. This is of the utmost importance since the weight any particular experiment carries in a least-squares adjustment depends on the inverse square of its uncertainty. In general, the correct estimation of systematic error is the major stumbling block for both experimenter and reviewer. To see why this is so, consider a typical experiment (assumed to follow the Gaussian error distribution) in which a certain quantity is measured a large number of times N, with each measurement X_i having a one-standard-deviation uncertainty σ_i due to random error. The weighted mean $\bar{X}$ and the uncertainty of the mean σ_m are given by (Young, 1962)

$$\bar{X} = \left(\sum_{i=1}^{N} \frac{X_i}{\sigma_i^2}\right) \Big/ \left(\sum_{i=1}^{N} \frac{1}{\sigma_i^2}\right); \qquad \frac{1}{\sigma_m^2} = \sum_{i=1}^{N} \frac{1}{\sigma_i^2}. \tag{10}$$

For identical σ_i, σ_m reduces to $\sigma_m = \sigma_i / N^{1/2}$. (For most experimental situations, all of the σ_i will be nearly the same and $N^{1/2}$ will be the approximate factor by which the uncertainty of the mean is reduced over the uncertainty of the individual measurements.) The total uncertainty σ_T conventionally assigned to the experimental result is the square root of the sum of the squares (root-sum-square, RSS) of the statistical uncertainty of the mean, σ_m, and estimates of the uncertainties σ_{sj} due to systematic effects which might influence the result:

$$\sigma_T = (\sigma_m^2 + \sigma_{s1}^2 + \sigma_{s2}^2 + \cdots)^{1/2}. \tag{11}$$

The σ_{sj} are usually called "systematic errors" or "possible systematic errors."* Now because of the factor $N^{1/2}$, σ_m is usually comparable to or less than the largest estimated systematic error component, and in the majority of cases, the total error of the experiment comes primarily from the systematic errors. Since estimates of these are somewhat subjective and are usually obtained from what can only be called educated guesses, the uncertainty in σ_T can be quite large. [From statistics, the one-standard-deviation uncertainty in a standard deviation, σ, derived from N measurements is $0.707\sigma/N^{1/2}$, which is about 10% for an N of 50 (Birge, 1932; Whittaker and Robinson, 1944).] If it is 15%, which would not appear to be unreasonable, then the uncertainty in the weight of the experiment in a least-squares adjustment will be 30%. This is much larger than the actual 11% weight the experiment would carry in an adjustment which contained a similar input datum with an uncertainty 3 times smaller. Thus we have the rule of thumb discussed in Sec. II.A.3.

---

* Equation (11) gives the standard deviation of the total error under the assumptions that the individual errors are mutually independent and that each is characterized by a distribution with zero mean and standard deviation σ_m or σ_{sj}. Although it is usually impossible to establish that the systematic errors actually meet these requirements, it is usually assumed that they do.

An equally important factor contributing to the uncertainty in the uncertainties of experimental results is the fact that different experimenters approach the estimation of systematic error with completely different philosophies. Some cautiously assign unreasonably large errors so that a later measurement will not prove their work to have been "incorrect." Others tend to underestimate the sources of systematic error in their experiment, perhaps because of an unconscious (or conscious) desire "to have done the best experiment." Such variation in attitude, although out of keeping with scientific objectivity, is nevertheless unavoidable so long as scientists are also human beings. It results, however, in quoted errors for different experiments which cannot be compared in a straightforward manner.

Because of this general error problem, we have recalculated wherever necessary and possible the experimental error to be assigned to a stochastic input datum. In some cases, use of an improved value of a particular parameter required to derive an experimental result from the original observational data has led to a change in both the result and its error. Such changes in the original result and/or error reported by an experimenter will be clearly noted in the text. All errors will be stated in terms of the standard deviation because of its more general applicability to the theory of least squares [a point emphasized by Cohen and DuMond (1965)]. In converting error expressed in terms of probable error (P.E., 50% confidence level) to standard deviation error, we have assumed $\sigma = 1.48\times(\text{P.E.})$. Although this is true only for the Gaussian error distribution (Whittaker and Robinson, 1944), it is the simplest and most plausible assumption to make. Occasionally, other terms have been used by experimenters to express error in addition to P.E., e.g., limit of error and average deviation,

$$\bar{\alpha} = \sum_{i=1}^{N} |X_i - \bar{X}|/N.$$

For these cases, we assume the limit of error is 3 times

the probable error or twice σ and that $\bar{\alpha}$ and σ are related by $\sigma = \bar{\alpha}(\pi/2)^{1/2} = 1.25\bar{\alpha}$ [Young (1962)].*

Before we conclude this section on the treatment of error, we note that numerieal values will be presented in several ways. Sometimes we give the results in the form XXX.XX±X.XX and at other times, in the form XXX.XX(XX). In the latter form, the uncertainty given in the parentheses corresponds to the uncertainty in the last digits of the main number. For convenience, we also give in many instances the uncertainty in parts per million. Any difference between the absolute error and the parts-per-million error is due to rounding. Frequently, the reader may find that the stated result of a calculation involving several quantities differs slightly from the result he would obtain using the numerical values given in the text for these quantities. The reason for this is that these quantities are presented with a number of significant figures appropriate to the assigned error, whereas we have performed the computations with numerical values having several additional digits in order to eliminate rounding error.

B. Auxiliary Constants

1. *Comparisons of As-Maintained Electrical Units*

In several instances, we will need to reexpress in terms of NBS electrical units experimental data originally obtained in terms of the as-maintained units of other national laboratories. To do this, we make use of the results of the international comparisons of electrical units carried out every three years or so at the BIPM in Sèvres, France. Since the uncertainty of these comparisons is only about 0.1 to 0.2 ppm, they can be taken as exact. In Tables I and II we give the results

---

* $\sigma = 1.25\,\bar{\alpha}$ only for a Gaussian distribution, but the numerical factor is not very sensitive to the form of the distribution; for a square distribution, for example, it is 1.15.

of the volt and ohm comparisons for the last 18 years.* The numbers in the tables are differences in μV and $\mu\Omega$, respectively, between the appropriate laboratory unit and the BIPM unit. Thus,

$$V_{\mathrm{LAB}} = V_{\mathrm{BIPM}} + \Delta\mu\mathrm{V}; \qquad \Omega_{\mathrm{LAB}} = \Omega_{\mathrm{BIPM}} + \Delta\mu\Omega. \quad (12)$$

(Because the Δ's are so small and laboratory units differ from absolute units by less than 20 ppm, the Δ's are for all practical purposes numerically the same whether expressed in absolute, BIPM, or laboratory units.) As an example of how these tables are used, suppose the relationships between the NBS and NPL (National Physical Laboratory) volts, ohms, and amperes are required for the year 1964. From Table I we have $V_{\mathrm{NPL}} = V_{\mathrm{BIPM}} + 3.1\ \mu\mathrm{V}$ and $V_{\mathrm{NBS}} = V_{\mathrm{BIPM}} - 2.2\ \mu\mathrm{V}$. Similarly from Table II, $\Omega_{\mathrm{NPL}} = \Omega_{\mathrm{BIPM}} - 3.50\ \mu\Omega$ and $\Omega_{\mathrm{NBS}} = \Omega_{\mathrm{BIPM}} - 0.25\ \mu\Omega$. Eliminating V_{BIPM} and Ω_{BIPM} yields

$$V_{\mathrm{NPL}} = V_{\mathrm{NBS}} + 5.3\ \mu\mathrm{V}; \qquad \Omega_{\mathrm{NPL}} = \Omega_{\mathrm{NBS}} - 3.25\ \mu\Omega. \quad (13a)$$

Since $A_{\mathrm{NBS}} = V_{\mathrm{NBS}}/\Omega_{\mathrm{NBS}}$ and $A_{\mathrm{NPL}} = V_{\mathrm{NPL}}/\Omega_{\mathrm{NPL}}$, we obtain

$$A_{\mathrm{NPL}} = A_{\mathrm{NBS}} + 8.55\ \mu\mathrm{A} \quad (13b)$$

or $A_{\mathrm{NPL}}/A_{\mathrm{NBS}} = 1.00000855$. In 1967 the corresponding number was 1.00000851, a decrease of 0.04 ppm. Such a small shift from one comparison to another is quite unusual.

The problem also arises as to how such relationships as are represented by Eqs. (13a) and (13b) can be deduced for years in which comparisons have not been

* The data through 1964 were kindly supplied by R. Driscoll and F. K. Harris, NBS, and were obtained from Comité International des Poids et Mesures, Procès-Verbaux. The 1967 data as well as Table III were kindly supplied by J. Terrien, Director, BIPM. See also Terrien (1969).

carried out, e.g., 1965. The simplest solution is to use the relationships resulting from the comparisons closest in time to the year in question. An alternate procedure, and one we shall follow whenever it appears necessary, is to interpolate between comparisons by assuming that only linear changes in the units have occurred. (This procedure is quite adequate since even in the worst cases the changes in the units between comparisons are only a few parts per million.) In order to interpolate correctly, knowledge is needed of the actual period of time during which the comparisons and experiments in question were carried out. Generally the comparisons require several months to complete, but they are made somewhat symmetrically with respect to some central date (the electrical units are first measured at the originating laboratory, then at BIPM, and finally remeasured at the originating laboratory.) In Table III we give the central dates of the comparisons listed in Tables I and II. See footnote on page 22.

We should emphasize here that implicit in all of our work is the assumption that the NBS units have remained unchanged over the last 10 years, i.e., that the constant K is time independent, even though it is clear from Tables I and II that the *relative* values of the various national as-maintained units are *not* time independent. The validity of this assumption has been established at the 1-ppm level for the NBS ampere as discussed in Sec. II.A.2. [It may be better than this (Harris, 1964; private communication).] Furthermore, measurements (indirect) of $\Omega_{NBS}/\Omega_{ABS}$ have uncertainties of only a few tenths part per million, and there is no indication of any appreciable drift in Ω_{NBS} (see Sec. II.B.3). It can therefore reasonably be assumed that the NBS as-maintained volt, ohm, and ampere have not changed by more than 1 ppm during the past decade. Since the fundamental-constant experiments requiring electrical units have uncertainties ranging from 2.4 to 10 ppm, even a drift in the NBS units as large as 1 ppm would be expected to have little effect on our least-squares adjustments (see Sec. II.A.3).

Table I. Relation between the units of emf as maintained by various countries and the Bureau International des Poids et Mesures (BIPM). The data are differences in microvolts between the lab and BIPM units (see text).

Lab	Country	1950	1953	1955	1957	1961	1964	1967
DAMW	E. Germany	−10.2	−2.8	+0.5	+1.1		−4.3	−6.80
PTB	W. Germany		−2.3	+0.6	+0.2	−0.1	−1.4	−0.64
NBS	USA	+0.8	−3.3	−0.7	−1.3	−1.9	−2.2	−2.58
NSL	Australia					+6.3	+5.5	+5.16
NRC	Canada		−3.1	−2.4	−0.8	−3.4	−1.7	−2.98
LCIE	France	−0.1	−1.8	−1.8	−2.1	−3.2	−3.1	−4.92
IEN	Italy					0	+0.8	−0.87
ETL	Japan	−3.5	−1.4	−2.0	−3.4	−2.9	−3.5	−2.66
NPL	Great Britain	+2.2	+3.2	+4.5	+5.2	+5.1	+3.1	+2.62
IMM	USSR (adjusted by 16 ppm in 1955)	+23.0	+22.3	+9.3	+8.4	+6.8	+7.8	+8.40

Table II. Relation between the units of resistance as maintained by various countries and the BIPM. The data are differences in micro-ohms between the lab and BIPM units (see text).

Lab	Country	1950	1953	1955	1957	1961	1964	1967
DAMW	E. Germany	+17.9	+22	+20.9	+16.9			
		(adjusted October 1957)			0.0	−2.9	−3.8	−0.39
PTB	W. Germany		+1.7	+2.0	+3.3	+3.75	+4.8	+5.06
NBS	USA	−0.4	−0.2	0.0	−1.0	−0.4	−0.25	−0.19
NSL	Australia					−3.5	−3.55	−3.63
NRC	Canada		−4.3	−5.2	−4.8	−4.15	−4.0	−2.94
LCIE	France	+1.0	−5.0	−6.2	−7.4	−8.5	−9.2	−12.22
IEN	Italy					0	+0.2	+0.89
ETL	Japan	−2.1	−1.0	−0.2	−0.4	−0.3	−0.3	+0.06
NPL	Great Britain	−2.6	−4.1	−3.4	−3.4	−3.4	−3.5	−3.50
IMM	USSR	+1.0	+0.8	+2.0	+0.4	−0.7	−0.9	−0.51

2. *Velocity of Light*, *c*

For the last 40 years, this important constant has had a rather checkered career characterized by measurements which appear widely divergent when compared with their assigned experimental errors. However, the situation has improved considerably over the last 15 years with the advent of the microwave-interferometer technique for determining c developed by Froome at NPL (1954; 1952). In his last and most accurate experiments, carried out at a frequency of $\sim$72 GHz (4-mm wavelength), Froome (1958) found

$$c = 299\ 792.50 \pm 0.10 \text{ km/sec } (0.33 \text{ ppm}). \quad (14a)$$

Just recently, Simkin, Lukin, Sikora, and Strelenskii (1967) in the Soviet Union completed a similar experiment at $\sim$36.8 GHz (8-mm wavelength), and obtained

$$c = 299\ 792.56 \pm 0.11 \text{ km/sec } (0.37 \text{ ppm}). \quad (14b)$$

(The quoted uncertainties for both experiments include random and systematic error.) The two microwave-interferometer values of c are in excellent agreement, thereby giving some hope (but of course not ensuring)

TABLE III. Central dates for the volt and ohm comparisons given in Tables I and II.

Comparison	Volt	Ohm
1950	19 February 1950	19 February 1950
1953	22 July 1953	26 October 1953
1955	24 July 1955	3 July 1955
1957	12 October 1957	12 October 1957
1961	6 January 1961	6 January 1961
1964	26 January 1964	26 January 1964
1967	18 February 1967	18 February 1967

that the various error-contributing factors in such experiments, e.g., diffraction, temperature, refractive index of air, scattered radiation, phase shift, etc., have been properly taken into account.

The second best method for obtaining c involves determining the time of flight over a measured base line. The modern version of this experiment (supposedly first attempted by Galileo, who stationed two men on distant hilltops with lanterns in hand) was developed by Bergstrand in Sweden (1957; 1950; 1949) and is known as the geodimeter method. It represents a significant improvement over the toothed wheel of Fizeau and the rotating mirror of Michelson (Bergstrand, 1956). A modulated Kerr cell is used as a light source and a photomultiplier tube modulated by the same oscillator is used as a detector. (The modulation frequency is ~10 MHz and the base lines are of order 10 km.) McNish of NBS (1962) has reviewed a large number of geodimeter measurements of c carried out by the United States Coast and Geodetic Survey (USCGS) as well as data gathered by Bergstrand and others. The measurements were made by many different observers under a variety of weather conditions, a wide range of geographical environments, at different times of the year, and over base lines measured with tapes calibrated in Australia, Great Britain, Sweden, and the United States. They therefore constitute a good sample for statistical analysis [but see McNish (1962)]. McNish concludes that the best value of c implied by the data is

$$c = 299\,792.6 \pm 0.25 \text{ km/sec (0.83 ppm)}, \qquad (14c)$$

in good agreement with the microwave-interferometer results.

Since the publication of McNish's survey, two other accurate determinations of c by the geodimeter method have appeared, one by Kolibayev (1965) in the USSR and the other by Grosse (1967) in Germany. Kolibayev reports measurements carried out from 1958 to 1963 over many different base lines using two different

TABLE IV. Summary of some velocity-of-light measurements made since 1948 (MWI, microwave interferometer; IRRS, infrared rotational spectrum; FLRC, fixed-length resonant cavity; VLRC, variable-length resonant cavity). (Probable errors have been converted to standard deviations by multiplying by 1.48.) The errors quoted for the Kolibayev and Grosse geodimeter measurements are statistical only.

Year of publication	Author	Method	c (km/sec)
1967	Simkin, Lukin, Sikora and Strelenskii	MWI	299 792.56±0.11
1967	Grosse	Geodimeter	299 792.5±0.05
1965	Kolibayev	Geodimeter	299 792.6±0.06
1950–1962	McNish (1962) summary of data of Bergstrand, USCGS, and others	Geodimeter	299 792.6±0.25
1958	Froome	MWI	299 792.50±0.10
1955	Florman[a]	RWI	299 795.1±1.5

1955	Plyler, Blaine, and Connor[b]	IRRS	299 792±6
1954	Froome [revised, Froome (1958)]	MWI	299 792.75±0.30
1952	Froome	MWI (first instrument)	299 792.6±0.7
1951	Aslakson[c]	Shoran	299 794.2±2.8
1950	Bol[d]	FLRC	299 789.3±1.0
1950	Essen[e]	VLRC	299 792.5±1.5
1949	Aslakson[c]	Shoran	299 792.4±3.6
1948	Essen and Gordon-Smith[f]	FLRC	299 792±4.5

[a] E. F. Florman, J. Res. Natl. Bur. Std. **54,** 335 (1955).

[b] E. K. Plyler, L. R. Blaine, and W. S. Connor, J. Opt. Soc. Am. **45,** 102 (1955).

[c] C. I. Aslakson, Trans. Am. Geophys. Union **32,** 813 (1951); **30,** 475 (1949); Nature **168,** 505 (1951); **164,** 711 (1949).

[d] K. Bol, Phys. Rev. **80,** 298 (1950).

[e] L. Essen, Proc. Roy. Soc. (London) **A204,** 260 (1950).

[f] L. Essen and A. C. Gordon-Smith, Proc. Roy. Soc. (London) **A194,** 348 (1948).

instruments. In this work, c was assumed to be 299 792.5 km/sec and then corrected so that the base-line lengths as measured with the geodimeter agreed with the results of measurements with invar wires. The average value of 23 separate determinations is given as

$$c = 299\ 792.6 \pm 0.06 \text{ km/sec},$$

where the quoted error is just the statistical or random error. Grosse also assumed the value $c = 299\ 792.5$ km/sec and found that base-line lengths measured with the geodimeter and with invar wires agreed to within approximately 0.15 ppm (0.05 km/sec in c). In view of the excellent agreement between these results and those surveyed by McNish, McNish's value may be taken as representative of the geodimeter work.

Other determinations of c, e.g., those made by the resonant-cavity method, radar measurement of base lines (Shoran technique), radio-frequency interferometer, infrared bands, ratio of electrical units, etc., tend to support the three values presented, but have significantly larger uncertainties and will not be mentioned further (see the summary in Table IV).* This is in keeping with the philosophy, adhered to throughout this paper, that it is incorrect to average data indiscriminantly (even if the average is a weighted one) which differ in value and uncertainty by large amounts. This is because two experimental results which differ by several times the RSS of their individual errors are contradictory; this implies that there may be systematic errors which have been overlooked in either or both of the experiments or the known systematic errors may have been grossly underestimated. Moreover [as emphasized by Cohen and DuMond (1965)], most experimenters rarely consider possible systematic errors which might shift their final result by as much as 10%

* For a detailed discussion, see, in particular, Bergstrand (1956) and also Cohen and DuMond (1965), McNish (1962), Froome (1956), and Bearden and Thomsen (1957; 1959).

of its random error. Thus, if two experimental results are combined, one with an error considerably greater than the other, the systematic errors not accounted for in the first may exceed the total error of the second. Including data with widely differing uncertainties may then do more harm than good because of the systematic errors the more imprecise data may introduce into the final average.

The weighted average of the two microwave-interferometer measurements, Eqs. (14a) and (14b), is

$$c = 299\,792.527 \pm 0.076 \text{ km/sec } (0.25 \text{ ppm}), \quad (15a)$$

while the weighted average of these two measurements and the value for c derived from the geodimeter work by McNish is

$$c = 299\,792.533 \pm 0.071 \text{ km/sec } (0.24 \text{ ppm}). \quad (15b)$$

[Throughout this paper, all weighted averages will be computed using Eq. (10).] However, we do not choose to use either of these values in our adjustments but instead adopt the Froome value, Eq. (14a), including its assigned error. There are two reasons for this. First, there is a slight possibility that a systematic difference of as much as 1 ppm exists in the calibration of base-line tapes (or wires) in the laboratories of the several countries involved in the geodimeter measurements (McNish, 1962). It therefore seems best to exclude the geodimeter results until this point is clarified. Second, the International Scientific Radio Union (U.R.S.I.) in its 12th General Assembly (1957) recommended for adoption a value of c identical to that of Froome but with a larger uncertainty (0.4 km/sec instead of 0.1 km/sec). The International Union for Geodesy and Geophysics (I.U.G.G.) did likewise, and it has become the accepted value used by all workers in the field. In view of the rather small difference (0.1 ppm) between the Froome value and the averaged values given in Eqs. (15a) and (15b), we do not think it wise to revise the adopted value of so important a constant for so small a change supported by so little evidence.

3. Ratio of the Absolute Ohm to the NBS Ohm, $\Omega_{ABS}/\Omega_{NBS}$

We will need a precise value for the ratio $\Omega_{ABS}/\Omega_{NBS}$ since it appears as an auxiliary constant in the observational equation derived for e/h, Eq. (7). Such a value can best be obtained from recent work by Thompson (1968; private communication) and coworkers at NSL (National Standards Laboratory, Australia) which utilized the calculable-capacitor technique to measure $\Omega_{NSL}/\Omega_{ABS}$. In this method, a capacitor is constructed in such a way that its capacitance can be accurately calculated from its mechanical dimensions. By a comparison of the impedance of the capacitor at a known frequency with that of the reference standard resistors, the values of the reference resistors can be directly established in terms of the prototype standards of length and time, i.e., in absolute units. However, this is strictly true only in the cgs-esu system of units. In SI units, there is the additional factor of ϵ_0, the permittivity of free space, in the equation which relates the capacitance to the capacitor dimensions. Since $\epsilon_0\mu_0 = 1/c^2$ and $\mu_0 = 4\pi \times 10^{-7}$ exactly, the capacitance varies as $1/c^2$. Hence, knowledge of c is required in the calculable-capacitor technique in order to determine a resistance in absolute (SI) ohms. (We note from the preceding section that the uncertainty in c^2 is 0.67 ppm.)

While the technique is obvious in principle, it has become practical only with the development by Thompson and Lampard (1956) (Thompson, 1959; Clothier, 1965) of a calculable capacitor (called a cross capacitor) which requires the measurement of but a single length. The form used by Thompson consists of four parallel cylindrical rods in a square array with a moveable center rod which defines the length. The length measurement is made using standard optical interferometry techniques. The capacitance of this structure is on the order of 0.25 pF and thus the actual comparison of the impedance of the capacitor with the resistance of a 1-ohm standard resistor requires several intermediate steps. First, a number of small fixed capacitors are measured by comparison with the calculable capacitor. These are then connected in parallel to form a 0.5-pF capacitor, and a 10:1 ratio bridge is used in four steps to build up from this capacitance to two 5-nF capacitors. The resistances of two 20-kΩ resistors are then derived from these capacitors using an ac bridge. The resistors are then connected in parallel and their ac–dc transfer characteristics determined. Finally, dc ratio techniques are used to step down the resistance to the 1-ohm level for comparison with the primary resistance standards.

In January 1964 and February 1967 Thompson found the following relationships to hold (Thompson, 1968; private communication):

$$1964: \quad \Omega_{\mathrm{NSL}} = \Omega_{\mathrm{ABS}} - 3.58 \pm 0.7\ \mu\Omega_{\mathrm{ABS}},$$

$$1967: \quad \Omega_{\mathrm{NSL}} = \Omega_{\mathrm{ABS}} - 3.80 \pm 0.7\ \mu\Omega_{\mathrm{ABS}}.$$

If one uses Table II, this implies

$$1964: \quad \Omega_{\mathrm{ABS}}/\Omega_{\mathrm{NBS}} = 1 + (0.28 \pm 0.7)\ \mathrm{ppm}, \tag{16a}$$

$$1967: \quad \Omega_{\mathrm{ABS}}/\Omega_{\mathrm{NBS}} = 1 + (0.36 \pm 0.7)\ \mathrm{ppm}. \tag{16b}$$

The 0.7-ppm uncertainty is the root sum square of two uncertainties. The first is the 0.2-ppm experimental error (both random and systematic) in the mechanical and electrical measurements, and the second is the 0.67-ppm error in c^2.

There are other experiments which agree with the NSL work, but their uncertainties are considerably larger. In 1960, Cutkosky (1961) at NBS used a calculable capacitor (but not in quite as refined a form as the more recent apparatus of Thompson) to obtain directly

$$\Omega_{\mathrm{ABS}}/\Omega_{\mathrm{NBS}} = 1 - (0.6 \pm 3.2)\ \mathrm{ppm}.$$

The value quoted by Cutkosky in his original paper was $\Omega_{\mathrm{ABS}}/\Omega_{\mathrm{NBS}} = 1 - (2.3 \pm 2.1)$ ppm where the 2.1-ppm uncertainty was a P.E. and did not include the uncer-

TABLE V. Summary of absolute-ohm determinations (TLC, Thompson–Lampard capacitor). The values given in the last column were obtained from the original experiments by using the comparisons of Table II closest in time to the date of the experiments. Probable errors and limits of error have been converted to standard deviations according to Sec. II.A.4. The error stated for the TLC experiments includes a 0.67-ppm uncertainty for c^2; c was taken to be 299 792.50±0.10 km/sec.

Year of publication	Laboratory and author	Method	$(\Omega_{ABS}/\Omega_{NBS})-1$ (ppm)
1968	NSL (1967), Thompson	TLC	0.36±0.7
1968	NSL (1964), Thompson	TLC	0.28±0.7
1967	NPL (1963, 1964) Rayner	Campbell	0.45±3.0
1964	ETL, K. Hara *et al.*[a]	Inductor	−1.05±8
1961	NBS, Cutkosky	TLC	−0.6±3.2
1957	NRC, Romanowski and Olson[b]	Modified Campbell	4.4±12
1954	NPL, Rayner (1967)[c]	Campbell	2.4±7.5
1938–1949	NBS, Thomas *et al.*[d] [see Cutkosky (1961)]	Wenner	3.0±4.5

[a] Comité Consultatif d'Electricité, Comité International des Poids et Mesures, May 1965, Document No. 9 (private communication).
[b] M. Romanowski and N. Olson, Can. J. Phys. **35**, 1312 (1957).
[c] G. H. Rayner, Proc. Inst. Elec. Engrs., (London) Pt. IV, **101**, 250 (1954).
[d] J. L. Thomas, C. Peterson, I. L. Cooter, and F. R. Kotter, J. Res. Natl. Bur. Std. **43**, 291 (1949).

tainty of c^2. Cutkosky later found (private communication) that he had added in with the wrong sign a 0.85-ppm correction for the frequency dependence of certain bridge resistors. Therefore, 2×0.85 ppm$=1.7$ ppm must be added to the original value. (We have converted to a σ by multiplying the original 2.1-ppm P.E. by 1.48, and the final error is the RSS of 3.1 ppm and 0.67 ppm.) In 1963 and 1964, Rayner (1967) at NPL, using Campbell's method (a calculable mutual inductor in a bridge arrangement), obtained $\Omega_{NPL}/\Omega_{ABS}=1-(3.7\pm3.0)$ ppm, where we have converted Rayner's original 2-ppm P.E. to a σ. Using the 1964 results from Table II, we find that this implies

$$\Omega_{ABS}/\Omega_{NBS}=1+(0.45\pm3.0)\text{ ppm}.$$

The values presented so far and four others are summarized in Table V.

We shall adopt the February 1967 Thompson value, Eq. (16b), because of its high accuracy and the fact that 1966–1967 was the time during which the e/h measurements were made (Parker, Langenberg, Denenstein, and Taylor, 1969). Again, we do not take a weighted average because of the large uncertainties of the other experiments compared with the experiments of Thompson.

It should be noted that the net auxiliary constant which enters Eq. (7) (the only important observational equation in which $\Omega_{ABS}/\Omega_{NBS}$ appears) is $c\Omega_{ABS}/\Omega_{NBS}$ rather than $\Omega_{ABS}/\Omega_{NBS}$. Since $\Omega_{ABS}/\Omega_{NBS}\sim1/c^2$, $c\Omega_{ABS}/\Omega_{NBS}\sim1/c$. The uncertainty in $c\Omega_{ABS}/\Omega_{NBS}$ is therefore only 0.4 ppm (RSS of 0.2 ppm and 0.33 ppm). This is 6 times smaller than the 2.4-ppm uncertainty of $2e/h$, and hence, $c\Omega_{ABS}/\Omega_{NBS}$ can safely be used as an auxiliary constant.

4. Acceleration Due to Gravity, g

The gravitational acceleration g is of course not really a constant since it varies widely from one location to another. Indeed, the gravitational gradient at the surface of the earth is $\approx$0.3 mgal/m or about 0.3

ppm/m (1 mgal$=10^{-5}$ m/sec^2). However, g can be determined to within a few tenths of a part per million for a well-defined site and is essentially a constant for that site. In the present work, we shall need the value of g at three locations: the National Bureau of Standards in Washington, D.C. where the current balance measurements of $K=A_{NBS}/A_{ABS}$ were performed, the National Physical Laboratory, Teddington, England where a similar measurement of A_{NPL}/A_{ABS} was carried out, and the Kharkov Institute of Measures and Measuring Instruments, USSR, where a high-field determination of the gyromagnetic ratio of the proton was performed (see Secs. II.C.2 and II.C.4). In each of these three experiments, g plays its usual role as a transfer constant or conversion factor for converting a known mass to a force.

Unfortunately, both the absolute measurement of g and the comparison of values obtained at different sites are very old and complex subjects which would require several volumes for adequate treatment. Cook (1965) has recently surveyed the field in a comprehensive review article to which we shall often refer. Here we explore the subject only far enough to establish the best probable values of g at the sites in question. Table VI summarizes the various pertinent measurements of g which we shall discuss.

Tate (1968, 1966) has measured g in Room 129 of the Engineering Mechanics Building at the new NBS near Gaithersburg, Maryland. The method used was that of direct free fall. The falling object was a 1-m fused quartz rod enclosed in a vacuum chamber which itself was falling. Tate found for a specific site in Room 129 to be known as NBS-2, $g(\text{NBS-2})=980\,101.8\pm0.45$ mgal. The difference between g at this site and g at the United States National Gravity base (top of east elevated pier) in the Department of Commerce Building, Washington, D.C. [to be denoted as $g(\text{CB})$] has been measured by the United States Coast and Geodetic Survey (USCGS) and, according to Tate (1968), is $g(\text{CB})-g(\text{NBS-2})=2.97$ mgal. (Such small differences in g are measured with spring-type gravity meters and are supposedly accurate to better than 0.1 mgal.) Thus, Tate's measurement implies

$$\text{Tate:}\quad g(\text{CB}) = 980\,104.77 \pm 0.45 \text{ mgal},$$

if one assumes that the transfer error is negligible.

Faller and Hammond (Hammond and Faller, 1967; Faller, 1967) of Wesleyan University, Scott Laboratory of Physics, Middletown, Connecticut, have built a portable instrument for the accurate measurement of g in which the falling object is one reflector of a laser interferometer of the Michelson type. For a site in Scott Laboratory to be known as Middletown A (MA), their preliminary result is* $g(\text{MA}) = 980\,305.31 \pm 0.10$ mgal (Faller and Hammond, private communication). The difference in g at this site and Tate's site, NBS-2, has been measured by the U.S. Air Force Cambridge Research Laboratories (AFCRL) and is given as $g(\text{MA}) - g(\text{NBS-2}) = 204.00$ mgal (Faller and Hammond, private communication). Thus, Faller and Hammond's measurement at Wesleyan implies

$$\text{Faller, Hammond:}\quad g(\text{NBS-2}) = 980\,101.31 \pm 0.10 \text{ mgal}.$$

Although this result is 0.49 mgal less than Tate's, it is consistent with his assigned error of ± 0.45 mgal. Faller and Hammond have also transported their apparatus to three other sites of interest and have obtained accurate measurements at those sites.* In May 1968, they measured g at a site to be known as NBS-3 in Room 01 of the Engineering Mechanics Building at NBS Gaithersburg. They obtained $g(\text{NBS-3}) = 980\;102.39 \pm 0.10$ mgal (Faller and

---

* While all of the Faller and Hammond values must be regarded as somewhat preliminary, it is unlikely that they will change by as much as 0.10 mgal or 0.1 ppm (Faller and Hammond, private communication). Repeatability is actually on the order of a few parts in 10^8. The uncertainty of 0.1 mgal has been temporarily assigned to take into account presently unestimated systematic errors. These workers hope to reduce the uncertainty by a factor of 2 or 3.

Table VI. Summary of measurements of the gravitational acceleration (TPFF, three-position free fall; SFM, symmetrical free motion; MPFF, multiple-position free fall). [Some of the data have been taken from Cook (1965).]

Year of publication	Location and author		Value implied at site indicated, (mgal)	Difference from Potsdam system[a] (mgal)	Method
1. 1968	NBS-3; Faller and Hammond[b]	CB:	980 104.23±0.10	13.77±0.10	TPFF, moving interferometer reflector
2. 1966	NBS-2; Tate (1968)	CB:	980 104.77±0.45	13.23±0.45	TPFF, photoelectric detection, falling enclosure
3. 1963	Princeton, N.J.; Faller	CB:	980 103.8±0.7	14.2±0.7	TPFF, moving interferometer reflector
4. 1936	NBS East Building; Heyl and Cook [revised, Cook (1965)]	CB:	980 101.8±1.3	16.2±1.3	Reversible pendulum
5. 1968	NPL; Faller and Hammond[b]	BFS:	981 181.86±0.10	13.61±0.12	TPFF, moving interferometer reflector
6. 1967	NPL; Cook [revised, Cook (private communication)]	BFS:	981 181.81±0.13	13.66±0.14	SFM, photoelectric detection

7. 1939	NPL; Clark [revised, Cook (1965)]	BFS:	981 183.2±0.7	12.3±0.7	Reversible pendulum
8. 1968	BIPM; Faller and Hammond[b]	SA:	980 925.95±0.10	13.81±0.12	TPFF, moving interferometer reflector
9. 1967 1968	BIPM; Sakuma	SA:	980 925.971±0.030	13.789±0.076	SFM, moving interferometer reflector
10. 1961	BIPM; Thulin	SA:	980 928.0±0.7	11.8±0.7	MPFF, photographs of ruled line standard
11. 1960	Ottawa; Preston-Thomas *et al.*	CFS:	980 605.9±1.0	13.2±1.1	TPFF, photographs of ruled scale
12. 1956	Leningrad; Agaletskii, Egorov (Woolard and Rose, 1963)	VNIIM: VNIIM:	981 918.7±1.0 981 922.4±2.1	12.1±1.2 8.4±2.2	Reversible pendulum MPFF, falling enclosure
13. 1956	Leningrad; Martsiniak (Woolard and Rose, 1963)	VNIIM:	981 921.5±1.7	9.3±1.8	MPFF, rod with photographic emulsion

[a] Based on g(Potsdam) =981 274 mgal. The Comité International des Poids et Mesures at its October 1968 session has recommended that as of 1 January 1969, g(Potsdam) be reduced to 981 260 mgal, thus defining a revised Potsdam system which is in better agreement with modern absolute measurements (Terrien, 1969),

[b] Faller and Hammond (private communication); see footnote at the bottom of page 37.

Hammond, private communication). Since $g(\text{NBS-2}) = g(\text{NBS-3}) - 1.13$ mgal as measured by Faller and Hammond using a gravity meter, their result gives for Tate's site NBS-2, $g(\text{NBS-2}) = 980\,101.26 \pm 0.10$ mgal. Since this value is only 0.05 mgal larger than the corresponding value obtained by transferring the Middletown A result to NBS-2, it may be concluded that both Faller and Hammond's measurements of g and the AFCRL transfers are well in hand. When transferred to the Commerce Building gravity base, Faller and Hammond's NBS-2 result becomes

$$\text{Faller, Hammond:} \quad g(\text{CB}) = 980\,104.23 \pm 0.10 \text{ mgal.}$$

We shall take this result as representing the Faller and Hammond value of $g(\text{CB})$ rather than that implied by the Middletown measurement because it does not require the relatively large transfer from Middletown A to NBS-2. In any event, the two values differ by only 0.05 ppm.

Faller (1963), using an apparatus similar to the one he and Hammond are presently using at Wesleyan, measured g at a site at Princeton University known as Princeton D(PD) (which is on top of a pier in the Palmer Physical Laboratory).† He found $g(\text{PD}) = 980\,160.4 \pm 0.7$ mgal (three low values rejected). A transfer between PD and NBS-2 by the AFCRL gives $g(\text{NBS-2}) - g(\text{PD}) = -59.57$ mgal (Faller and Hammond, private communication). Thus, Faller's value becomes at NBS-2 $g(\text{NBS-2}) = 980\,100.83 \pm 0.7$ mgal and at the Commerce Building gravity base,

$$\text{Faller:} \quad g(\text{CB}) = 980\,103.80 \pm 0.7 \text{ mgal.}$$

This is about 1 mgal less than the corresponding Tate value but only 0.43 mgal less than that of Faller and Hammond.

Heyl and Cook (1936) used reversible Kater pendulums in the subsubbasement of the old NBS

† The dropping distance in this experiment was only 5 cm as compared with the 1-m distance now being used by Faller and Hammond.

East Building in Washington, D.C., (NBS-EB) to obtain [after revision described by Cook (1965)] $g(\text{NBS-EB}) = 980\,082.3 \pm 1.3$ mgal. Assuming $g(\text{CB}) - g(\text{NBS-EB}) = 19.46$ mgal as measured by the USCGS and reported by Tate (1968), we obtain for the site CB

$$\text{Heyl, Cook:} \quad g(\text{CB}) = 980\,101.76 \pm 1.3 \text{ mgal.}$$

This result is between 2 and 3 mgal below the other three values of $g(\text{CB})$. However, it is not surprising in view of the large uncertainties inherent in the pendulum method [Cook (1965)].

At the National Research Council (NRC), Ottawa, Canada, Preston-Thomas, Turnbull, Green, Dauphinee, and Kalra (1960) measured g using a freely falling scale. The value obtained at NRC was $g(\text{NRC}) = 980\,613.2 \pm 1.0$ mgal, where the 1-mgal uncertainty was largely due to the fact that two different scales inexplicably gave results which differed by 1.5 mgal. The NRC measurement implies a value of g at the Canadian Fundamental Station (CFS) of

$$\text{Preston-Thomas:} \quad g(\text{CFS}) = 980\,605.9 \pm 1.0 \text{ mgal,}$$

where $g(\text{CFS}) - g(\text{NRC}) = -7.29$ mgal as given by Cook (1965) is used.

We now consider some European measurements. Cook (1967a; 1967b) has measured g at NPL in Teddington, England, at a site in Bushy House (BH), using the so-called symmetrical free motion or "upsy-daisy" method in which an object is projected upward, allowed to return to its original starting point, and is timed both going up and coming down. The value of g originally reported was $g(\text{BH}) = 981\,181.82 \pm 0.13$ mgal. However, Cook (private communication) recently revised this result to eliminate an error due to the use of an incorrect formula for the effect of the variation of g with height. His new result is $g(\text{BH}) = 981\,181.88 \pm 0.13$ mgal. At the nearby British Fundamental Station (BFS), this becomes

$$\text{Cook:} \quad g(\text{BFS}) = 981\,181.81 \pm 0.13 \text{ mgal.}$$

Faller and Hammond (1968; private communication) have also transported their apparatus to Cook's Bushy House site and obtained $g(\mathrm{BH})=981\,181.93\pm0.10$ mgal. This result exceeds Cook's revised result by 0.05 ± 0.16 mgal. [If two measurements of the same quantity are independent of each other, then from the law of propagation of errors, the standard deviation of the difference between them, σ_d, is simply $\sigma_d=(\sigma_1^2+\sigma_2^2)^{1/2}$.] The agreement is clearly quite good. When transferred to the BFS, the Faller and Hammond result becomes,

$$\text{Faller, Hammond:}\quad g(\mathrm{BFS})=981\,181.86\pm0.10\ \text{mgal}.$$

Clark (1939) determined g at NPL by the reversible pendulum method and obtained [revised value—see Cook (1965)]

$$\text{Clark:}\quad g(\mathrm{BFS})=981\,183.2\pm0.7\ \text{mgal}.$$

This exceeds the Cook and Faller and Hammond measurements by about 2 standard deviations. Again, because of the difficulties of the reversible pendulum method, the discrepancy cannot be taken too seriously.

Recently, Sakuma (1967) at BIPM, using a symmetrical-free-motion technique in which the projected object was the reflector of one arm of an interferometer, measured g at a site at BIPM called A_2. From 50 measurements carried out during August and September 1967, he obtained $g(A_2)=980\,925.675\pm0.014$ mgal, where the quoted error is just the statistical standard deviation of the mean. A series of 25 measurements carried out during the same months in 1968 gave $g(A_2)=980\,925.664\pm0.019$ mgal. The 1967 result exceeds that obtained in 1968 by only 0.011 ± 0.024 mgal, clearly demonstrating the great reproducibility of the measurements. Taking the weighted average of the two values yields $g(A_2)=980\,925.971\pm0.011$ mgal. The difference in g between the site A_2 and the standard reference site at BIPM, Sèvres Point A (SA), is reported by Sakuma (1967) to be $g(\mathrm{SA})-g(A_2)=0.30$ mgal. Thus

$$\text{Sakuma:} \quad g(\text{SA}) = 980\,925.971 \pm 0.030 \text{ mgal},$$

where the error has been increased to reflect the uncertainty of the transfer (Sakuma, 1967). Faller and Hammond (private communication) have also transported their apparatus to the BIPM and measured g on the pier designated Sèvres A in August 1968. Their result is

$$\text{Faller, Hammond:} \quad g(\text{SA}) = 980\,925.95 \pm 0.10 \text{ mgal}.$$

This value is only 0.021 mgal less than that obtained by Sakuma, and therefore the two measurements are in excellent agreement. Thulin (1961) [see Cook (1965)], also at BIPM, used a free-fall method in which the falling body was an engraved line standard to obtain g at a site A_1. He found $g(A_1) = 980\,927.25$ mgal± 0.7 mgal which becomes at SA

$$\text{Thulin:} \quad g(\text{SA}) = 980\,928.05 \pm 0.7 \text{ mgal}.$$

The difference between the Thulin and Sakuma (or Faller and Hammond) measurements is 2 mgal or 3 times Thulin's standard deviation. Cook (1965) has pointed out that this is quite surprising in view of the care with which Thulin worked.

There have been several Russian measurements of g, each carried out at the All Union Institute of Metrology (VNIIM) in Leningrad. Agaletskii and Egorov (1956) (Woollard and Rose, 1963) used a set of reversible pendulums and also a freely falling body within a falling chamber. Martsiniak (1956) (Woollard and Rose, 1963) used a freely falling rod. The results are given in the last three lines of Table VI. It is readily apparent that they are not very consistent among themselves. The pendulum measurement appears to be the best of the group (Cook, 1965).

In order to compare all of the different g measurements with each other, we use the so-called Potsdam gravity system. In this system, the reversible pendulum measurement of g made in Potsdam in 1906 by Kühnen and Furtwängler is used as the reference point for a world-wide gravity net. Measurements of the difference between g at Potsdam and at a great many other sites all over the world have been made using pendulums and spring-type gravity meters, and values of g have been assigned to these sites assuming g at Potsdam is 981 274 mgal (but see Footnote a, Table VI). However, the Potsdam measurement has been shown to be in error. This was first indicated by the measurements of Heyl and Cook at NBS (1936) and Clark at NPL (1939) which disagreed with their corresponding Potsdam-system values by significant amounts. Dryden (1942) reexamined the Potsdam measurement and pinpointed the problem as being due to the application by Kühnen and Furtwängler of several unjustified corrections. Dryden recommended a reduction of about 17 mgal in all the Potsdam-system values of g; this recommendation has become known as the Dryden reduction. The more accurate of the recent measurements seem to indicate that the Potsdam system is too high by between 13 and 14 mgal rather than 17 mgal. This is shown in Table VI, where we give the difference between the value of g at the site in question on the Potsdam system and the experimentally determined value at that site. In obtaining these differences, we have assumed that on the Potsdam system

$$g_P(\mathrm{CB}) = 980\ 118.0 \text{ mgal},$$

$$g_P(\mathrm{BFS}) = 981\ 195.47 \pm 0.06 \text{ mgal},$$

$$g_P(\mathrm{SA}) = 980\ 939.76 \pm 0.07 \text{ mgal},$$

$$g_P(\mathrm{CFS}) = 980\ 619.05 \pm 0.35 \text{ mgal},$$

$$g_P(\mathrm{VNIIM}) = 981\ 930.8 \pm 0.6 \text{ mgal},$$

where $g_P(\mathrm{CB})$ is the traditional Potsdam value at CB (Tate, 1968) and the others are those given by Cook (1965) based on the measurements of various workers. [Cook gives $g_P(\mathrm{CB}) = 980\ 117.59 \pm 0.31$ mgal.]

An examination of the next to last column of Table VI shows immediately the wide variability of the data. Discrepancies in the differences larger than the assigned errors are common. They are due in part to the experimental measurements themselves, but also to the inconsistencies in the Potsdam-system gravity net. The transfers between sites on either side of the Atlantic are particularly subject to great uncertainty because of the large differences in g between the sites, i.e., 500–1000 mgal. Although it is generally believed that differences in g on the order of several tens of mgal can be measured to an accuracy of a few tenths of a mgal, the uncertainty in the measurement of a large difference in g is thought to be 5 to 10 times greater (Cook, 1965). Indeed, Woollard and Rose (1963) have given g values on the Potsdam system for the various sites listed in Table VI which differ by several tenths of a mgal from those derived by Cook. One apparent inconsistency in the Potsdam system can be seen by comparing the difference $g_P(\mathrm{BFS}) - g_P(\mathrm{SA}) = 255.71 \pm 0.09$ mgal with the 255.91-mgal difference implied by the measurements of Faller and Hammond. The latter result should be quite reliable since it depends only on the difference between two absolute measurements reproducible to a few hundredths of a mgal and carried out by the same workers with the same apparatus. This discrepancy is responsible for the disagreement between the differences from the Potsdam system given on Lines 5 and 8 of Table VI. We also note that the Faller and Hammond values of g at SA and CB imply $g(\mathrm{SA}) - g(\mathrm{CB}) = 821.72 \pm 0.10$ mgal. When combined with $g_P(\mathrm{SA})$ as given by Cook it implies $g_P(\mathrm{CB}) =$ 980 118.04$\pm$0.12 mgal. Similarly, the Faller and Hammond measurements at BFS and CB imply $g(\mathrm{BFS}) - g(\mathrm{CB}) = 1077.63$ and $g_P(\mathrm{CB}) = 980\,117.84 \pm$ 0.12 mgal. Both of these values of $g_P(\mathrm{CB})$ are closer to the traditional value we have used, $g_P(\mathrm{CB}) = 980\,118.0$ mgal, than that given by Cook, $g_P(\mathrm{CB}) = 980\,117.59$. It appears that the only way to obtain a world-wide gravity net accurate to 0.1 mgal (0.1 ppm) is to perform high-accuracy absolute measurements, preferably with the same apparatus, at various sites around the world. Faller and Hammond's work is a commendable step in that direction.

In the present least-squares adjustments, we shall do the following: (1) Derive a value of g for the site of the NBS current balance on the assumption that the Faller and Hammond value of g at CB is correct (the transfer uncertainty from CB to the current balance site should be negligible). We use the Faller and Hammond measurement rather than that of Tate because of the former's higher accuracy and better consistency with other measurements such as Cook's, Faller's, and Sakuma's. (2) Derive a value of g for the site of the NPL current balance on the assumption that the Faller and Hammond value of g at BFS is correct. We use this value rather than Cook's because of its somewhat higher accuracy and excellent agreement with the very accurate measurement of Sakuma at BIPM. (3) Assume that the true value of g at the site of the high-field γ_p measurement at Kharkov is 13.80 mgal less than the Potsdam value of g at the same site. We adopt the 13.80-mgal difference because this is what is implied by a simple average of the Faller and Hammond and Sakuma measurements at Sèvres A. To estimate the error of this assumption, we note that the Potsdam value of g at VNIIM is given as 981 930.8 mgal with an uncertainty of 0.6 mgal (Cook, 1965). This value was derived from pendulum observations between Potsdam and Pulkova and from gravity-meter measurements between Pulkova and Leningrad. Thus, since the uncertainty in the transfer from Potsdam to Sèvres A is believed to be only a few tenths of a mgal (Cook, 1965), the value of g at Kharkov we have adopted should have an uncertainty of less than 1 mgal (1 ppm). This assumes of course that the transfer from Leningrad to Kharkov is known to a few tenths of a mgal, as would probably be the case if the connection were made by gravity meter. Note that an error in g as large as 1 ppm at the various sites in question will have only a negligible effect on our adjustments since the uncertainties in the stochastic input data which require values of g for their evaluation are between 6 and 10 ppm. Thus, g can safely be taken as an auxiliary constant. (For the NPL and NBS sites, the uncertainty in g is probably closer to 0.1 ppm than to 1 ppm.)

5. *Magnetic Moment of the Electron in Units of the Bohr Magneton*, μ_e/μ_B

The magnetic moment of the free electron in units of the Bohr magneton $\mu_B = e\hbar/2m_e$ or, equivalently, the free-electron g factor $g_s = 2\mu_e/\mu_B$ has been most accurately measured in a classic experiment by Wilkinson and Crane (1963). By measuring the difference between the electron spin precession frequency and the electron cyclotron frequency, these workers were able to determine directly the g factor anomaly a_e, where $g_s/2 = \mu_e/\mu_B = (1+a_e)$. Their result is

$$\mu_e/\mu_B = 1.001159622(27) \quad (0.027 \text{ ppm}).$$

Recently, Rich (1968a; 1968b; 1968c; see also Henry and Silver, 1969) has reexamined the data of Wilkinson and Crane and has applied corrections for relativistic effects not previously taken into account. Furthermore, he has carried out a new error analysis based on an error matrix formalism and has used an improved method involving the Lambe measurement of $g_s/g_p(H_2O)$ for converting a proton resonance frequency to a frequency f_0 required in the analysis (see the next section and also Sec. IV.A.1). His result is

$$\mu_e/\mu_B = 1.001159549(30) \quad (0.030 \text{ ppm}). \tag{17}$$

This represents a decrease of only 0.073 ppm in μ_e/μ_B but a 63 ppm decrease in a_e.

The quantity μ_e/μ_B can also be calculated from QED as a power series in α:

$$\frac{\mu_e}{\mu_B}=1+A_1\left(\frac{\alpha}{\pi}\right)+A_2\left(\frac{\alpha}{\pi}\right)^2+A_3\left(\frac{\alpha}{\pi}\right)^3+\cdots.$$

Schwinger (1949; 1948) was the first to calculate the coefficient of the first term. He obtained $A_1=\frac{1}{2}$. The second-order term (fourth order in perturbation theory) was first calculated by Karplus and Kroll (1950) and later corrected by Sommerfield (1957; 1958) and by Petermann (1957a; 1958a) [see also Smrz and Ulehla (1960) and Terent'ev (1962)]. They both obtained the same result, namely,

$$A_2=\tfrac{197}{144}+\tfrac{1}{12}\pi^2+\tfrac{3}{4}\zeta(3)-\tfrac{1}{2}\pi^2\ln 2\approx-0.32847897\cdots, \tag{18}$$

where ζ is the Riemann zeta function. The third-order term (sixth order in perturbation theory) has recently been estimated by Drell and Pagels (1965) using dispersion theory. From a formulation which gives the Schwinger term exactly and the fourth-order term approximately, they obtained $A_3\approx0.15$. Still more recently, Parsons (1968) has completed a calculation similar to that of Drell and Pagels but has included additional terms. He finds $A_3=0.13$. We may thus write

$$\mu_e/\mu_B=1+\tfrac{1}{2}(\alpha/\pi)-0.3285(\alpha/\pi)^2+0.13(\alpha/\pi)^3. \tag{19}$$

If Eq. (19) is evaluated using $\alpha^{-1}=137.0360$, a number we anticipate to be very close to both our WQED and final recommended or best values, we obtain

$$\mu_e/\mu_B=1.001159639(3)\ (0.003\text{ ppm}). \tag{20}$$

The quoted error is based on a 2-ppm uncertainty in α and a ±0.20 uncertainty in A_3.* (Note that $|A_4|$ would have to be about 30 in order to change μ_e/μ_B by

* This estimate of the uncertainty in A_3 is based on a suggestion of S. J. Brodsky (private communication), who has pointed out that a number of contributions, including all vacuum polarization terms, were omitted from Parson's estimate of A_3.

one digit in the last place.) Rich's revised experimental result for μ_e/μ_B, Eq. (17), is 0.090 ppm less than this theoretical value or three times his assigned experimental error. If instead we use the currently accepted value of α derived from the fine-structure splitting in deuterium, $\alpha^{-1} = 137.0388(6)$ (Cohen and DuMond, 1965), we obtain

$$\mu_e/\mu_B = 1.001159615(6) \quad (0.006 \text{ ppm}).$$

The revised experimental result for μ_e/μ_B is still 0.066 ppm less than this.

It is apparent that there is a significant discrepancy between the present experimental value of μ_e/μ_B and values calculated from the present theoretical formula using any plausible value of α. This discrepancy will be discussed in detail in Sec. IV.A.1. Here we face the problem of choosing a value of μ_e/μ_B to use as an auxiliary constant. Fortunately, although the discrepancy between the experimental and theoretical values ($\sim$0.1 ppm) is a serious matter when considered as a test of QED, it is completely negligible as far as the use of μ_e/μ_B as an auxiliary constant is concerned. We could equally well choose either value. For definiteness, we choose to use for μ_e/μ_B the theoretical value calculated with $\alpha^{-1} = 137.0360$, Eq. (20).* We do this because the theoretical expression for μ_e/μ_B as well as our chosen α value appear to be on somewhat firmer ground at present than the experimental result (see Sec. IV.A.1). Note that even if the value of α we have used to calculate μ_e/μ_B were found to be in error by 100 ppm, an extremely unlikely possibility, the resulting error in our adopted value of μ_e/μ_B would be less than 0.1 ppm. This is because the relative change in μ_e/μ_B is

---

* It should be noted that this choice does not invalidate our procedure for deriving a set of constants without essential use of QED theory. The only use we will make of μ_e/μ_B as an auxiliary constant is in the next section, where we use it to obtain a value for the auxiliary constant μ_p/μ_B. We will see there that we could equally well use a directly measured value of μ_p/μ_B without significantly altering our results.

only $\alpha/2\pi$ or 1.16×10^{-3} times as large as the relative change in α. (See also Notes Added in Proof.)

6. Magnetic Moment of the Proton in Units of the Bohr Magneton, μ_p/μ_B

The important auxiliary constant μ_p/μ_B has been measured in at least three different ways. The most unambiguous determination is that of Myint, Kleppner, Ramsey, and Robinson (1966). Using a hydrogen maser operating in an applied field of 0.35 T (1 tesla = 1 weber/m^2 = 10^4 G), these workers measured the ratio of the electron spin-flip frequency ν_a to the proton spin-flip frequency ν_b for the same value of applied field (see Fig. 2). The frequency ratio is of course identical with the g factor ratio $g_j(\mathrm{H})/g_p(\mathrm{H})$, where the proton g factor is referred to the *Bohr* magneton, i.e., $g_p = 2\mu_p/\mu_B$. Their result is

$$g_j(\mathrm{H})/g_p(\mathrm{H}) = 658.21049(20)\ (0.30\ \mathrm{ppm}). \quad (21)$$

To obtain μ_p/μ_B, both $g_j(\mathrm{H})$ and $g_p(\mathrm{H})$ must be corrected to their free-space values. The bound-state electron g factor $g_j(\mathrm{H})$ is related to g_s, the free-electron g factor, by the relativistic correction $g_j(\mathrm{H}) = g_s(1-\alpha^2/3)$ (Mott and Massey, 1965). The proton g factor in H, $g_p(\mathrm{H})$, can be corrected for diamagnetic shielding using the theory of Lamb (1941) with the result $g_p(\mathrm{H}) = g_p(1-\alpha^2/3)$. The correction factors cancel and it follows that

$$g_j(\mathrm{H})/g_p(\mathrm{H}) = g_s/g_p = 658.21049(20)\ (0.30\ \mathrm{ppm}). \quad (21a)$$

From the definitions of the g factors, g_s/g_p is also equal to μ_e/μ_p, so that

$$\mu_p/\mu_e = (g_s/g_p)^{-1} = 0.00151927083(46)\ (0.30\ \mathrm{ppm}).$$

Combining this result with our adopted value for $g_s/2 = \mu_e/\mu_B$ (preceding section), we obtain

$$\mu_p/\mu_B = 0.00152103264(46)\ (0.30\ \mathrm{ppm}), \quad (22)$$

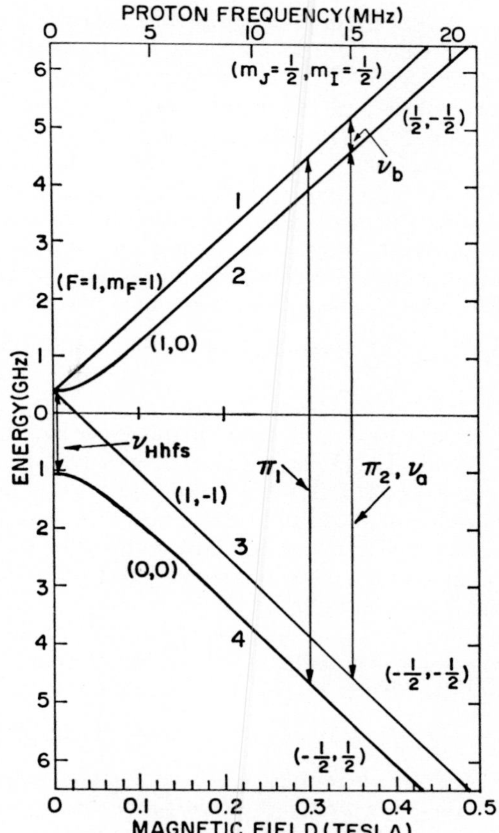

FIG. 2. Schematic diagram of the hyperfine Zeeman levels of the ground state of atomic hydrogen. At low fields the electron and nuclear (proton) spins are coupled and the quantum numbers are F and m_F. At high fields, the two spins are decoupled and the quantum numbers are m_J and m_I. For convenience, the states have been labeled 1, 2, 3, and 4. Transitions discussed in the text and which have been used to determine several important physical quantities are indicated. The magnetic field is also given in proton frequency units.

where we have assumed our value of μ_e/μ_B is exact.

The above measurement may be compared with that of Lambe (1959; 1969). Lambe's determination is the most accurate of the type in which one measures in the same magnetic field the ratio of the electron spin-flip frequency in H to the proton spin-flip frequency in either an H_2 gas, mineral oil, or water sample, thereby obtaining $g_j(\mathrm{H})/g_p(\mathrm{H_2, oil, H_2O})$. For the proton spin-flip transition, Lambe used a spherical H_2O sample, and for the electron spin-flip transition, a microwave absorption technique in a gas of atomic hydrogen with a molecular-hydrogen-gas buffer to reducé Doppler broadening. The transitions measured by Lambe are labeled π_1 and π_2 in Fig. 2. His final quoted result is

$$g_j(\mathrm{H})/g_p(\mathrm{H_2O}) = 658.2159088(436) \quad (0.066\ \mathrm{ppm}). \tag{23}$$

[The measurement error of the experiment, including both identifiable systematic errors and random errors, was 0.0000218, but Lambe quotes a value double this because of a possible dependence of $g_j/g_p(\mathrm{H_2O})$ on both discharge current and total pressure for data obtained from the π_1 transition.] The relativistic bound-state correction for $g_j(\mathrm{H})$ is, as noted above, $g_j(\mathrm{H}) = g_s(1-\alpha^2/3)$. Since $\alpha^2/3$ is only a 17.75-ppm correction, the particular value of α used is not critical; we take $\alpha^{-1} = 137.0360$. Thus Lambe's experiment gives

$$g_s/g_p(\mathrm{H_2O}) = 658.227593(44) \quad (0.066\ \mathrm{ppm}). \tag{24}$$

From the definitions of the g factors, $g_s/g_p(\mathrm{H_2O})$ is also equal to μ_e/μ_p', where the prime means "as obtained for protons in a spherical sample of water." Thus

$$\begin{aligned}\mu_p'/\mu_e &= [g_s/g_p(\mathrm{H_2O})]^{-1}\\ &= 0.00151923136(10) \quad (0.066\ \mathrm{ppm}).\end{aligned} \tag{25}$$

Using our adopted value for $g_s/2 = \mu_e/\mu_B$, we obtain

$$\mu_p'/\mu_B = 0.00152099312(10) \quad (0.066\ \mathrm{ppm}), \tag{26}$$

where we have again assumed our value for μ_e/μ_B is exact.

Equation (24) in combination with the value of g_s/g_p given by Myint *et al.*, Eq. (21a), implies

$g_p/g_p(H_2O) = 1 + (26.0 \pm 0.3)$ ppm or, equivalently, that the diamagnetic shielding constant $\sigma(H_2O)$ for protons in a spherical H_2O sample is (26.0 ± 0.3) ppm. Newell (1950)* gives $\sigma(H_2) = (26.6 \pm 0.3)$ ppm as a theoretical value for the shielding constant in H_2 gas, and Thomas (1950)† gives $\sigma(H_2O) - \sigma(H_2) = (-0.6 \pm 0.3)$ ppm as the experimentally measured difference between the shielding constants in H_2O and H_2 gas (all for spherical samples). Combining these two results yields a "theoretical" value for the shielding constant in water of $\sigma(H_2O) = (26.0 \pm 0.4)$ ppm which is identical with the purely experimental Lambe–Myint *et al.* result.‡ We shall therefore adopt, in all work requiring it,

$$\sigma(H_2O) = (26.0 \pm 0.3) \text{ ppm.}$$

We shall also adopt the Myint *et al.* value for μ_p/μ_B, Eq. (22), because it is so well supported by the Lambe measurement in combination with the theoretical estimates of $\sigma(H_2O)$.

In practice, we shall find that many experimental quantities are measured in terms of the precession frequency of protons in a spherical sample of water, e.g., γ_p and μ_p/μ_n. Such quantities may be included in our adjustments with greater accuracy if instead of using our adopted shielding correction and the Myint *et al.* value of μ_p/μ_B with their respective 0.3-ppm errors, we use the value of μ_p'/μ_B derived from the Lambe measurement, Eq. (26), with its 0.066-ppm error. We shall in fact follow this procedure wherever possible and take Eq. (26) as our adopted value for μ_p'/μ_B.

---

* This calculation, based in part on the work of Ramsey (1950), utilized some experimental data.

† Hardy has obtained $\sigma(H_2O) - \sigma(H_2) = (-0.6 \pm 0.15)$ ppm [see Liebes and Franken (1959)].

‡ Slightly different values for $\sigma(H_2)$ and $\sigma(H_2O) - \sigma(H_2)$ have also appeared in the literature. Harrick and coworkers (1953) give $\sigma(H_2) = (26.2 \pm 0.4)$ ppm and Gutowsky and McClure (1951) give $\sigma(H_2O) - \sigma(H_2) = (-0.3 \pm 0.3)$ ppm. However, these values agree with those of Newell and of Thomas to within the assigned uncertainties.

TABLE VII. Summary of experiments on μ_p/μ_B. The uncertainties quoted have in some instances been changed from P.E.'s to standard deviations. All data are for protons in spherical samples of H_2O. The uncertainties in μ_p/μ_B include the uncertainties in the shielding constants: $\sigma(H_2) = (26.6 \pm 0.3)$ ppm, $\sigma(H_2O) = (26.0 \pm 0.3)$ ppm, and $\sigma(\text{oil}) = (29.7 \pm 0.6)$ ppm. We have also used $g_j(H)/g_s = (1-\alpha^2/3) = 1-17.75$ ppm and $\mu_e/\mu_B = 1.001159639$ in obtaining the numbers given.

Year of publication and author	Quantity measured	Value	$(\mu_p/\mu_B)\times 10^3$	Error (ppm)
1966, Myint *et al.*	$g_j(H)/g_p(H)$	658.21049(20)	1.52103264(46)	0.3
1959, Lambe	$g_j(H)/g_p(H_2O)$	658.215909(44)	1.52103266(46)	0.3
1957, Geiger, Hughes, and Radford[a]	$g_j(D)/g_p(\text{oil})$	658.2169(9)	1.5210360(23)	1.5
1954, Beringer and Heald[b]	$g_j(H)/g_p(H_2)$	658.21600(20)	1.52103347(65)	0.4
1952, Koenig, Prodell, and Kusch[c]	$g_j(H)/g_p(\text{oil})$	658.2171(4)	1.5210355(13)	0.9
1968, Klein	$\omega_e/\omega_p(H_2O)$	657.46507(32)	1.52103290(87)	0.6
1963, Sanders, Tittel, and Ward[d]	$\omega_e/\omega_p(\text{oil})$	657.4620(36)	1.521046(84)	5.5
1959, Hardy and Purcell[e]	$\omega_e/\omega_p(H_2)$	657.4676(5)	1.5210280(12)	0.8
1959, Liebes and Franken	$\omega_e/\omega_p(\text{oil})$	657.4620(45)	1.521046(10)	6.9
1951, Gardner[f]	$\omega_e/\omega_p(\text{oil})$	657.475(8)	1.521016(19)	12

[a] J. S. Geiger, V. W. Hughes, and H. E. Radford, Phys. Rev. **105,** 183 (1957).
[b] R. Beringer and M. A. Heald, Phys. Rev. **95,** 1474 (1954).
[c] S. H. Koenig, A. G. Prodell, and P. Kusch, Phys. Rev. **88,** 191 (1952).
[d] J. H. Sanders, K. F. Tittel, and J. F. Ward, Proc. Roy. Soc. (London) **A272,** 103 (1963).
[e] W. A. Hardy and E. M. Purcell [as quoted in Cohen and DuMond (1965)].
[f] J. H. Gardner, Phys. Rev. **83,** 996 (1951).

Other experiments also tend to support our adopted values for μ_p/μ_B and μ_p'/μ_B but have larger uncertainties, and furthermore, some are complicated by uncertainties in the diamagnetic shielding correction for oil. In Table VII we summarize the pertinent measurements. The last group of experiments in the table determine μ_p/μ_B by measuring the ratio of the cyclotron frequency for the free electron, ω_e, to the resonance frequency of protons, ω_p, in either a water, mineral oil, or H_2 gas sample in the same magnetic field. By far the best measurement of this type is the recent one of Klein (1968) who determined ω_e by measuring the attenuation of X-band microwaves in a free-electron cloud. The half-width of the resonance was on the order of 5 ppm and the space-charge shift was reduced to about 1 ppm from the usual 50 ppm by using a large volume. Klein's result is

$$\frac{\omega_p(H_2O)}{\omega_e}=\frac{2\mu_p'B/\hbar}{2\mu_B B/\hbar}=\frac{\mu_p'}{\mu_B}$$

$$=0.0015209934(8)\ (0.5\ \text{ppm}). \qquad (27)$$

(For experiments such as Klein's in which a cylindrical H_2O sample doped with paramagnetic salt is actually used, the authors generally correct to a spherical pure H_2O sample.) In comparing this value of μ_p'/μ_B with that derived from the measurement of Lambe, Eq. (26), we see that they differ by only 0.2 ppm, well within the 0.5-ppm uncertainty of Klein's experiment.*

Using our adopted shielding constant $\sigma(H_2O)=26.0\pm0.3$ ppm, Klein's measurement also implies

$$\mu_p/\mu_B=0.00152103290(87)\ (0.6\ \text{ppm}). \qquad (28)$$

This value agrees with our adopted value, Eq. (22), to within 0.2 ppm. Although the uncertainties of the two results only differ by a factor of 2, we do not choose to average them because of the shielding-correction

---

* Klein's value of μ_p'/μ_B is the directly measured value referred to in footnote on page 49.

uncertainty inherent in Eq. (28). No such uncertainty is present in Eq. (22). (In any case, the weighted average would only differ from our adopted value by ~0.03 ppm.) Similar statements also apply to the measurement of Beringer and Heald, Table VII.

In converting the ratios ω_e/ω_p and g_j/g_p obtained using oil samples (Table VII) to free-space values, we have used as a shielding constant $\sigma(\text{oil}) = (29.7 \pm 0.6)$ ppm based on our adopted value of $\sigma(H_2O)$ and the value $\sigma(\text{oil}) - \sigma(H_2O) = (3.7 \pm 0.6)$ ppm as measured by Liebes and Franken (1959). However, the work of Thomas (1950) indicates that shielding constants can vary by 1–2 ppm for different oils. (See also Notes Added in Proof.)

7. *Atomic Masses and Mass Ratios*

We shall need several different atomic masses and mass ratios in the present work. For example, in deriving the Faraday constant F from the electrochemical equivalent of silver, we shall require the atomic masses of ^{107}Ag and ^{109}Ag. Similarly, in certain x-ray experiments the atomic masses of various isotopes of Si and other elements are required. In all such cases, we shall use the values given in the 1965 Mass Tables of Mattauch, Thiele, and Wapstra (MTW) (1965). These have been obtained from a least-squares analysis of both mass-spectroscopy and nuclear-reaction-energy data. The scale used is the unified scale, $^{12}C \equiv 12$.

We shall also need certain quantities involving M_p/m_e, the ratio of the free-proton mass to the free-electron mass. The ratio M_p/m_e can best be obtained from the ratio of the proton magnetic moment in units of the nuclear magneton to the proton moment in units of the Bohr magneton,

$$\frac{\mu_p'/\mu_n}{\mu_p'/\mu_B} = \frac{\mu_B}{\mu_n} = \frac{e\hbar/2m_e}{e\hbar/2M_p} = \frac{M_p}{m_e}.$$

For the proton moment in nuclear magnetons we take

$\mu_p'/\mu_n = 2.792707$, a value which we anticipate to be very nearly equal to our final least-squares adjusted value. It is also within 6 ppm of the relatively accurate measurement of Sommer, Thomas, and Hipple (see Sec. II.C.5). For μ_p'/μ_B, we use the value adopted in the preceding section. The result for M_p/m_e is therefore $M_p/m_e = 1836.1075$. It thus follows that the important quantity $1 + m_e/M_p$, which is required for the calculation of the atomic mass of the proton, the Rydberg constant for infinite mass, and various energy splittings in H, is

$$1 + m_e/M_p = 1.000544630.$$

It should be noted that even if the value of μ_p'/μ_n we have used were in error by 100 ppm, an amount which spans all known measurements of μ_p'/μ_n (see Table XIV), the quantity $1 + m_e/M_p$ would only be in error by the negligible amount of 0.05 ppm. A more realistic estimate of the uncertainty in $1 + m_e/M_p$ is 0.01 ppm (10 in the last two places).

The atomic mass of the proton M_p^* can be obtained from M_H^*, the atomic mass of the neutral H atom (we will use an asterisk to denote atomic mass) and the readily derived equation

$$M_p^* = \frac{M_H^*}{1 + m_e/M_p}\left[1 + \frac{\alpha^2}{2(2 + M_p/m_e)}\right].$$

For completeness, we have included an approximate correction for the binding energy of the hydrogen atom* even though it is small ($\sim$0.015 ppm) compared with the uncertainty in M_H^* [0.08 ppm, Mattauch, Thiele and Wapstra (1965)]. Using $M_H^* = 1.00782519(8)$ as given in the tables of Mattauch *et al.*† and $\alpha^{-1} = 137.0360$, we obtain

---

* The mass equivalent of the binding energy of the hydrogen atom, E_H, is $\approx -m_e\alpha^2/2(1 + m_e/M_p)$ and thus $E_H/M_H^* \approx -\alpha^2[2(2 + M_p/m_e)]^{-1}$.

† Two recent mass-spectrometer measurements give slightly different values. Matsuda and Matsuo (1968) find $M_H^* = 1.00782499(14)$ while Stevens and Moreland (1968) find $M_H^* = 1.00782501(3)$. However, these differ from the mass table value by only 0.2 ppm.

$$M_p^* = 1.00727661(8) \ (0.08 \text{ ppm}).$$

(Note that the atomic masses in the tables are for the neutral atoms.)

We shall also need the quantities $1+m_e/M_d$ and $1+m_e/M_\alpha$, where m_e/M_d is the ratio of the electron mass to the mass of the deuteron and m_e/M_α is the ratio of the electron mass to the mass of the α particle. It may be shown that‡

$$M_d/m_e = (M_D^*/M_H^*)(1+M_p/m_e) - 1,$$

$$M_\alpha/m_e = (M_{He}^*/M_H^*)(1+M_p/m_e) - 2,$$

where M_D^* and M_{He}^* are the atomic masses of neutral deuterium and helium, respectively. Using the masses given in the MTW tables and the value for M_p/m_e given above, we obtain

$$1+m_e/M_d = 1.000272450; \qquad 1+m_e/M_\alpha = 1.000137097.$$

The uncertainties in these quantities are probably less than 0.01 ppm or 10 in the last two places.

8. *Rydberg Constant for Infinite Mass, R_∞*

Cohen (1952) was the first to calculate this important auxiliary constant $\{R_\infty = [(\mu_0 c^2/4\pi)^2 m_e e^4](4\pi\hbar^3 c)^{-1}\}$, taking into account the effect of the Lamb shift. Using the spectroscopic data of Houston (1927) on H and ionized He, Chu (1939) on ionized He, and Drinkwater, Richardson, and Williams (DRW) (1940) on H and D, Cohen carried out a least-squares adjustment in order to obtain both R_∞ and the best value in absolute angstroms for the He line used as a wavelength standard by both Chu and Houston. The latter workers assumed

‡ The general expression for M_i/m_e, where M_i corresponds to the nucleus of a neutral atom M, is $M_i/m_e = [(M^*-E)/(M_H^*-E_H)](1+M_p/m_e) - Z$; E_H and E are the binding energies of H and the atom in question in atomic mass equivalents and Z is the atomic number. The binding energies are negative numbers. Neglecting E and E_H introduces negligible error compared with the uncertainty in M_p/m_e.

that their He reference line had a wavelength in air of $\lambda=5015.675$ Å. Cohen obtained $\lambda=5015.6778\pm0.0008$ Å and $R_\infty=109\,737.311\pm0.012$ cm^{-1}.

Martin (1959) recalculated R_∞ using $\lambda=5015.6779\pm 0.003$ Å as obtained from a weighted average of his own direct measurement of this wavelength and that of Series and Field. Martin also made somewhat different assumptions than did Cohen concerning the uncertainties in Chu's data and those of Drinkwater *et al.* However, the general procedure was the same; shifts in wave number from the "Dirac position" due to the Lamb shift were calculated, and then the differences between these new lines and a fictitious Balmer line defined by $W=R_i(n_0^{-2}-n_1^{-2})$ were computed. (Here, R_i is the appropriate Rydberg constant and all transitions involve either $n_1=3\rightarrow n_0=2$, $n_1=4\rightarrow n_0=2$, or $n_1=4\rightarrow n_0=3$.) Martin obtained

$$R_\infty=109\,737.312\pm0.008 \text{ cm}^{-1}.$$

In order to check these calculations, we have recalculated R_∞ using a somewhat different technique than that used by both Cohen and Martin. The method consists of directly comparing the measured lines with the lines calculated by Garcia and Mack (1965) from a complete theoretical equation for energy levels of hydrogenic atoms. This equation includes the fundamental Dirac term, the term due to the nonseparability of the Dirac equation in terms of reduced masses, a correction for both the finite size of the nuclear charge distribution and finite nuclear mass, and radiative corrections up to terms of order $\alpha(\alpha Z)^2$ and $\alpha(m_e/M_i)$ times the fine-structure splitting [see Erickson and Yennie (1965a; 1965b) for a complete discussion of these terms].* Garcia and Mack's calculation is based on

* The radiative and nuclear corrections do not play a significant role in the derivation of R_∞ and thus our procedure for deriving a set of constants without essential use of QED theory is not invalidated. If we were to ignore entirely the existence of these corrections and were simply to use the Dirac theory, the calculated value of R_∞ would differ by only several tenths of a part per million from the value we calculate here. This difference would have negligible effect on our results.

TABLE VIII. Theoretical line positions for various transitions in H, D, and $^4He^+$ as obtained from the calculation of Garcia and Mack (1965). The statistical line strengths are as defined in Condon and Shortley (1957).

Transition	Wave number of line (cm^{-1})		Relative line strength	Wave number of peak (cm^{-1})
Hα				
$3S_{1/2}$–$2P_{3/2}$	15 232.935678		25/128	15 232.935678
$3D_{3/2}$–$2P_{3/2}$	15 233.033406	α_2	1	15 233.065927
$3D_{5/2}$–$2P_{3/2}$	15 233.069540		9	
$3P_{1/2}$–$2S_{1/2}$	15 233.255771	α_3	25/24	15 233.259695
$3S_{1/2}$–$2P_{1/2}$	15 233.301551		25/256	
$3P_{3/2}$–$2S_{1/2}$	15 233.364177	α_1	25/12	15 233.388955
$3D_{3/2}$–$2P_{1/2}$	15 233.399279		5	
Hβ				
$4S_{1/2}$–$2P_{3/2}$	20 564.566605		1/4	
$4D_{3/2}$–$2P_{3/2}$	20 564.607821	β_2	1	20 564.620202
$4D_{5/2}$–$2P_{3/2}$	20 564.623066		9	
$4P_{1/2}$–$2S_{1/2}$	20 564.892761		45/32	
$4S_{1/2}$–$2P_{1/2}$	20 564.932478	β_1	1/8	20 564.950367
$4P_{3/2}$–$2S_{1/2}$	20 564.938495		45/16	
$4D_{3/2}$–$2P_{1/2}$	20 564.973694		5	
Dα				
$3S_{1/2}$–$2P_{3/2}$	15 237.080667		25/128	15 237.080667
$3D_{3/2}$–$2P_{3/2}$	15 237.178410	α_2	1	15 237.210941
$3D_{5/2}$–$2P_{3/2}$	15 237.214555		9	
$3P_{1/2}$–$2S_{1/2}$	15 237.400798	α_3	25/24	15 237.404727
$3S_{1/2}$–$2P_{1/2}$	15 237.446640		25/256	

$3P_{3/2}$–$2S_{1/2}$	15 237.509234	α_1	25/12	15 237.534045
$3D_{3/2}$–$2P_{1/2}$	15 237.544383		5	
He				
$4F_{5/2}$–$3D_{5/2}$	21 334.96004		1	21 334.96004
$4F_{7/2}$–$3D_{5/2}$	21 335.08205	2	20	21 335.08205
$4D_{5/2}$–$3P_{3/2}$	21 335.53599	1	245/32	21 335.53758
$4F_{5/2}$–$3D_{3/2}$	21 335.53845		14	

values for the fundamental constants given by DuMond and Cohen (1965). In particular, they took $\alpha^{-1}=137.0388$, $m_e=5.48597\times10^{-4}$ amu, and $a_0=5.29167\times10^{-9}$ cm (a_0 is the Bohr radius). While the present adjustment gives values for α^{-1}, m_e, and a_0 which differ from those used by Garcia and Mack, the differences are sufficiently small to have only a negligible effect (<0.001 ppm) on our derived value of R_∞.

Tables VIII–X summarize our calculations. (These tables are essentially identical in form to those given in the papers of Cohen and of Martin.) Table VIII gives the calculated theoretical wave numbers of the transitions of interest and their relative statistical line strengths.* The wave number corresponding to the intensity peak of a measured line is taken to be the center of gravity of the individual components which comprise the line (Cohen, 1952). However, our grouping of the lines differs from the grouping of Cohen and of Martin in that for the Hβ_2 line, we include the $4S_{1/2}$–$2P_{3/2}$ component, and for the Hβ_1 line, the $4P_{1/2}$–$2S_{1/2}$ component.† This choice is supported by the over-all

* The line strengths for the Hα and $D\alpha$ transitions are identical and were taken directly from Condon and Shortley (1957). The line strengths for the Hβ transitions are not the same as for $H\alpha$ as was assumed by Cohen, but can be calculated as outlined in Condon and Shortley. The values so obtained are given in Table VIII. The line strengths for the He transitions were similarly calculated.

† We wish to thank W. C. Martin for this suggestion.

TABLE IX. Comparison of experimental and theoretical line positions. The uncertainties are quoted as probable errors to facilitate comparison with the work of Cohen (1952) and of Martin (1959).

Component and observer	Corrected wavelength (air) (Å)	Vacuum wave number (cm^{-1})	Experiment minus theory (ppm)	R_i (cm^{-1})
$H\alpha_1$, Houston	6562.7147±0.0018	15 233.3888±0.0042	−0.010±0.275	109 677.575±0.030
$H\alpha_2$, Houston	6562.8510±0.0009	15 233.0724±0.0021	+0.427±0.138	109 677.623±0.015
$H\beta_1$, Houston	4861.2827±0.0013	20 564.9577±0.0055	+0.358±0.268	109 677.615±0.029
$H\beta_2$, Houston	4861.3605±0.0022	20 564.6286±0.0093	+0.408±0.453	109 677.621±0.050
He_1, Houston	4685.7056±0.0012	21 335.5315±0.0055	−0.283±0.257	109 722.236±0.028
He_2, Houston	4685.8056±0.0026	21 335.0762±0.0118	−0.274±0.555	109 722.237±0.061
He_1, Chu	4685.7043±0.0007	21 335.5375±0.0032	−0.006±0.151	109 722.267±0.017
He_2, Chu	4685.8038±0.0010	21 335.0844±0.0046	+0.111±0.214	109 722.279±0.024

Component and observer	Corrected wavelength (air) (Å)	Vacuum wave number (cm^{-1})	Experiment minus theory (ppm)	R_i (cm^{-1})
$H\alpha_1$, DRW[a]		15 233.3868±0.0032	−0.141±0.210	109 677.560±0.023
$H\alpha_2$, DRW		15 233.0670±0.0014	+0.070±0.092	109 677.584±0.010
$H\alpha_3$, DRW		15 233.2551±0.0063	−0.302±0.414	109 677.543±0.045
$D\alpha_1$, DRW		15 237.5317±0.0028	−0.154±0.184	109 707.403±0.020
$D\alpha_2$, DRW		15 237.2112±0.0013	+0.017±0.085	109 707.422±0.009
$D\alpha_3$, DRW		15 237.4127±0.0063	+0.523±0.413	109 707.477±0.045

[a] Drinkwater, Richardson, and Williams.

consistency of the recent measurements of Csillag (to be discussed shortly), who also used this same line grouping. The point is that these two unresolved components are sufficiently close to the center of the line to affect the position of the peak. It might also be argued that for the He_2 line (see Table VIII) the $4F_{5/2}$–$3D_{5/2}$ component should be included with the $4F_{7/2}$–$3D_{5/2}$ component. (Chu actually took this component into account in analyzing his experimental data, but Houston did not.) However, it appears to be far enough in the wing of the latter component to be excluded (Roesler and Mack, 1964, Fig. 6, components 9 and 10). This also seems to be the case for the $3S_{1/2} \rightarrow 2P_{3/2}$ component for both $H\alpha_2$ and $D\alpha_2$ (Drinkwater, Richardson, and Williams, 1940).

In Table IX we give the experimental data, compare them with the theoretical values, and for each measurement derive a value of the Rydberg constant R_i characteristic of the atom in question. The corrected wavelengths for both the Houston and Chu data are based on a value for the He wavelength standard of $\lambda(\text{air}) = 5015.6778 \pm 0.0001$ Å as measured by Terrien (1960) of BIPM. (This value is identical with that derived by Cohen and is very close to the weighted average used by Martin in his calculation.) We make the same assumptions regarding the errors to be assigned the data as does Martin for the reasons given in his paper. The third column of Table IX gives the vacuum wave numbers of the various lines. These were computed using the indices of refraction for standard air at the measured wavelengths as given by Coleman, Bozman, and Meggers (1960). (These tables were computed from the Edlén formula for the refractive index of standard air.) The data of Drinkwater, Richardson, and Williams do not need such conversion since these authors give their results in vacuum wave numbers directly. The fourth column of Table IX gives the difference in parts per million between the experimental and theoretical line positions. The last column gives the Rydberg constant for the atom in question,

TABLE X. Calculation of R_∞ from the data of Table IX.

Observer and pattern	R_∞ (cm^{-1})	R_∞ (cm^{-1})
Houston		
Hα	109 737.347±0.020	
Hβ	109 737.350±0.037	109 737.335±0.016
He	109 737.279±0.038	
Chu		
He	109 737.314±0.020	109 737.314±0.020
DRW[a]		
Hα	109 737.312±0.013	109 737.311±0.009
Dα	109 737.310±0.012	
Final weighted average	109 737.317±0.007	

[a] Drinkwater, Richardson, and Williams.

i.e., R_H, R_D, and R_{He}. These were obtained by correcting the values of R_H, R_D, and R_{He} used by Garcia and Mack by an amount equal to the differences between experiment and theory as given in the fourth column. This procedure follows from the fact that the line positions depend linearly on R_i.

In Table X we calculate R_∞ from the various R_i given in Table IX. We use the relation $R_\infty = R_i(1+m_e/M_i)$ and the values for the quantities $1+m_e/M_i$ derived in the preceding section. Note that the R_∞ corresponding to Houston's Hα data is the weighted average of the two R_∞ values corresponding to the $H\alpha_1$ and $H\alpha_2$ lines. This is also the case for the other data. Our final value is

$$R_\infty = 109\ 737.317 \pm 0.007\ \text{cm}^{-1}, \qquad (29)$$

which is not too different from that derived by Cohen and also Martin.

Recently, Csillag (1968; 1966; private communication) has derived a value of R_∞ from his measurements of six members of the Balmer series in deuterium (β, γ, δ, ϵ, ζ, η, corresponding to $n=4$ through $n=9$). He finds

$$R_{\rm D}=109\,707.417\pm0.003\ \mathrm{cm}^{-1},$$

where the quoted uncertainty is simply the random statistical error. Csillag observed two lines for each Balmer transition and derived values for $R_{\rm D}$ from the calculations of Garcia and Mack using a method equivalent to the one we have used here. In computing theoretical values for the centers of gravity of these lines from the tables of Garcia and Mack, Csillag used each of the three or four components which could possibly comprise the line, i.e., he included the $nS_{1/2}$–$2P_{3/2}$ and $nP_{1/2}$–$2S_{1/2}$ components as we have done (see Table VIII). His small statistical error gives much support to this choice. Combining Csillag's result for $R_{\rm D}$ with our adopted value of $1+m_e/M_d$ (preceding section), we obtain

$$R_\infty=109\,737.307\pm0.007\ \mathrm{cm}^{-1}, \qquad (30)$$

where the error has been expanded to include the following systematic error components: $0.003\ \mathrm{cm}^{-1}$ for possible pressure shifts of the reference lines of the $^{198}\mathrm{Hg}$ lamp used in the measurements and $0.001\ \mathrm{cm}^{-1}$ for the effects of nonstandard air (Csillag, 1968; 1966; private communication). Actually, we have increased the error somewhat more than is implied by these systematic errors because of the general complexity of this type of experiment. We believe Csillag's measurement should carry no more weight than all of the other experiments we have discussed combined. We shall thus adopt as the best value for R_∞ the weighted average of Eqs. (29) and (30),

$$R_\infty=109\,737.312\pm0.005\ \mathrm{cm}^{-1}.$$

The error quoted for this final value deserves some comment. We believe it to be rather optimistic, since the uncertainties assigned the various wavelength measurements (Table IX) do not include very critical estimates of the possible systematic errors. Furthermore, in all of the Rydberg-constant work, there is a possible uncertainty in the theoretical line positions due to the fact that the theoretical relative intensities of the components comprising the lines may not be realized in practice (Drinkwater, Richardson, and Williams, 1940; Roesler and Mack, 1964). Another factor which contributes to the uncertainty in our adopted value of R_∞ is the difficulty of assigning meaningful relative uncertainties to the various wavelength measurements. Reasonable arguments could probably be given for error assignments different from those of Martin and which we have used here. This would lead to a slightly different value of R_∞. Another problem of course is the slight ambiguity in deciding which components should comprise a given line. In view of these several difficulties, the value of R_∞ must be regarded as being uncertain by at least 0.01 cm^{-1} and perhaps even 0.02 cm^{-1} (0.1–0.2 ppm). Such a large uncertainty in this important constant would appear to be unnecessary in view of recent studies like that of Roesler and Mack on ionized He. These workers were easily able to resolve eight of the nine lines one might ever hope to resolve for the $n=4 \rightarrow n=3$ transition. (The full spectrum contains 13 lines.) A comparison of some of these lines with a wavelength standard is all that is required to obtain a new and highly accurate value for R_∞ which would be free of many of the uncertainties just discussed.

9. *Summary of the Auxiliary Constants*

Table XI gives our adopted values for all of the auxiliary constants discussed so far and which we shall use throughout the remainder of this paper. The uncertainties are given for the convenience of the reader only, since as far as our adjustments are concerned, the auxiliary constants are assumed to be exactly known (see Secs. II.A.1 and II.A.3). Note that the numerical values of μ_e/μ_B, $M_p{}^*$, and the three terms containing mass ratios may change slightly because our final adjusted values of α^{-1} and M_p/m_e may differ somewhat from the values we have used in our calculations. However, such changes will be entirely negligible, i.e., one or two digits in the last place ($\sim$0.001 ppm), and will have no effect on any of the adjusted values themselves.

C. Stochastic Input Data

1. Josephson-Effect Value of e/h

The determination of e/h by the present authors and Denenstein using the ac Josephson effect in superconductors is, of course, the primary motivation for this paper [Parker, Langenberg, Denenstein, and Taylor (1969)]. The measurement is based on a theoretical prediction by Josephson (1962; 1965) that if two weakly coupled superconductors are maintained at a dc potential difference V (strictly speaking, a chemical potential difference $\Delta\mu = eV$), there exists an alternating current between the superconductors with frequency

$$\nu = 2eV/h.$$

The two superconductors may be weakly coupled in a variety of ways, e.g., by the tunneling of superconducting electron pairs through an insulating barrier separating the superconductors. (Such structures are generally called Josephson junctions.) The Josephson frequency–voltage relation, $\nu = 2eV/h$, follows directly from the macroscopic phase-coherent nature of the superconducting state and is independent of both the method used to achieve weak coupling and the superconductors employed. In principle, e/h may be obtained by measuring the frequency of the radiation emitted by the ac Josephson current for a measured potential difference V, or by measuring the voltages at which microwave-induced steps appear in the dc junction current when the junction is irradiated with microwaves of known frequency ν. These steps are due to the non-

TABLE XI. Summary of auxiliary constants to be used in the present work. Note that some of these will be slightly changed as a result of our final adjustment (see text). (The prime on the magnetic moment of the proton means "for protons in a spherical sample of water.")

Quantity	Value	Units	Error (ppm)
c	299 792.50(10)	km/sec	0.33
$\Omega_{ABS}/\Omega_{NBS}$	1.00000036(70)		0.70
$c\Omega_{ABS}/\Omega_{NBS}$	299 792.61(12)	km/sec	0.39
g(CB)	980 104.23(10)	mgal	0.10
g(BFS)	981 181.86(10)	mgal	0.10
g(VNIIM)	981 917.00(1.00)	mgal	1.0
μ_e/μ_B	1.001159639(3)		0.003[a]
μ_p/μ_B	0.00152103264(46)		0.30
μ_p'/μ_B	0.00152099312(10)		0.066
μ_p/μ_e	0.00151927083(46)		0.30
μ_p'/μ_e	0.00151923136(10)		0.066
$\sigma(H_2O)$	26.0(3)	ppm	0.3
M_p^*	1.00727661(8)	amu	0.08
$1+m_e/M_p$	1.000544630(10)		0.01
$1+m_e/M_d$	1.000272450(10)		0.01
$1+m_e/M_\alpha$	1.000137097(10)		0.01
R_∞	10 973 731.2(1.1)	m^{-1}	0.10

[a] Calculated from theory; see text.

linear mixing of the applied radiation with the ac Josephson current and occur at voltages $V_n = nh\nu/2e$, where n is an integer. In practice, the microwave-induced-step technique is the more accurate of the two methods. The frequency measurement was straightforward since ν was $\sim$10 GHz, and hence the accuracy of the experiment was limited by the voltage measurement. Since V was usually of order 10^{-3} V ($n\sim 50$), a determination of e/h with an accuracy of several parts per million required measurements of V with an accuracy of

1 or 2 nV. This was made possible by a new potentiometer designed and manufactured by the Julie Research Laboratories. Parker and coworkers (Parker, Langenberg, Denenstein, and Taylor, 1969; Parker, Taylor, and Langenberg, 1967; Langenberg, Parker, and Taylor, 1968) showed that to within the 1–2-ppm precision of their experiments, the frequency–voltage ratio and therefore $2e/h$ is independent of a wide variety of experimental variables including microwave power and frequency, temperature, magnetic field, type of superconductors (Sn, Pb, Ta, Nb, and Nb_3Sn were used), and the method of achieving weak coupling. The voltage measurements were actually made in terms of the NBS volt as maintained during 1966–1967, and the final result is given as

$$2e/h = 4.835976(12) \times 10^{14}\ \mathrm{Hz/V_{NBS}}\ (2.4\ \mathrm{ppm}). \quad (31)$$

The quoted uncertainty includes both random error and estimates of systematic error. This value was calculated from data taken using both the radiation emission and microwave-induced-step techniques, but mostly the latter.

Recently, Clarke (1968) has carried out a high-precision differential experiment involving Josephson junctions of different materials. In these experiments, Clarke compared the differences in chemical potential across two different kinds of Josephson junctions biased on identical order steps. The junctions were irradiated with the same rf field and the chemical potential difference measured with a superconducting voltmeter based on the dc Josephson effect. Clarke found no difference in chemical potential between junctions of Sn, Pb, and In to within $1/10^8$ (0.01 ppm), the limit of resolution of the experiments. It may therefore be concluded that to this precision, junctions composed of different superconductors convert microwave radiation to chemical potential difference in exactly the same way. This result provides added experimental support for the validity of a determination of e/h using the Josephson effect.

Still more recently, Stephen (1968) and also Scully and Lee (1969) have considered possible frequency pulling effects in the radiation *emitted* by a Josephson junction. In all cases of interest, such effects appear to be entirely negligible, i.e., of order 0.001 ppm at most. Furthermore, Josephson (private communication) has noted that even the small shifts calculated by these workers may not be valid because the model Hamiltonian they assumed only allows departures from the Josephson frequency–voltage relation over the whole of the superconducting regions. Physically, one would expect departures only near regions where dissipation takes place, i.e., in the vicinity of the barrier. Since in most experiments, the electrical contacts to the junction are many coherence lengths away from the barrier, this distinction is of the utmost importance. In any event, the calculations of Stephen and of Scully and Lee are only applicable to radiation emission experiments, while the experiments of Parker *et al.* relied mainly on data taken with external radiation *incident* on Josephson junctions.

The observational equation for this input datum has already been derived in Sec. II.A.3 and is

$$(\alpha^{-1})^{-1}e^{-1}K^{1}N^{0}=\tfrac{1}{4}\mu_0(c\Omega_{\mathrm{ABS}}/\Omega_{\mathrm{NBS}})(2e/h)_{\mathrm{NBS}}. \quad (7)$$

Note that the observational equation for e/h in an adjustment in which Λ (the angstrom-to-kx-unit conversion factor) is included as an adjustable constant, would be identical to Eq. (7) except for an additional multiplicative factor of Λ^0 on the left side. Similarly, the equation appropriate to an adjustment in which data on N (Avogadro's number) are excluded would be the same as Eq. (7) but with N^0 deleted. Similar considerations also apply to the other observational equations. (See also Notes Added in Proof.)

2. *Ratio of the NBS Ampere to the Absolute Ampere,* $K\equiv A_{\mathrm{NBS}}/A_{\mathrm{ABS}}$

Accurate determinations of as-maintained amperes in absolute units date back over 60 years to the measurement in 1908 by Ayrton, Mather, and Smith at NPL. The general method used by these workers, that of measuring the force in absolute units between two coils of known dimensions carrying a current known in terms of the as-maintained ampere, is still the technique used today. The apparatus required for such an experiment is called a current balance and usually consists of two concentric and coaxial coils with the outer one fixed and the inner one suspended from one arm of a balance beam. The most recent determination of K is that carried out by Driscoll and Olsen at NBS (1968). They utilized a variation of the current balance known as the Pellat electrodynamometer. In this apparatus, the inner coil is at right angles to the outer coil and the torque on this coil is measured rather than the force. The result of Driscoll and Olsen's experiment is

$$K \equiv \mathrm{A}_{\mathrm{NBS}}/\mathrm{A}_{\mathrm{ABS}} = 1.0000093 \pm 6.6 \text{ ppm (P.E.)}, \quad (32)$$

where the quoted uncertainty is a probable error and is the RSS of the following *probable-error* estimates; (1) length of solenoid, 1 ppm, (2) pitch variations, 0.5 ppm, (3) diameter of rotatable coil, 5 ppm, (4) length of balance arm, 1.5 ppm, (5) determination of balancing mass, 2 ppm, (6) acceleration due to gravity, 1 ppm, (7) electrical standards, 0.5 ppm, (8) alignment of coils, 0.5 ppm, (9) coil temperature, 1 ppm, and (10) balance beam distortion, 3 ppm.

Driscoll and Olsen calculated K using a value for the gravitational acceleration at the site of the dynamometer of $g = 980\,081$ mgal. This was based on a value for g at the East Building subsubbasement gravity station of $g(\text{NBS-EB}) = 980\,083$ mgal. However, our adopted value for $g(\text{CB})$, that of Faller and Hammond, and the transfer $g(\text{NBS-EB}) - g(\text{CB}) = -19.46$ mgal as given by Tate (see Sec. II.B.4), indicate that $g(\text{NBS-EB}) = 980\,084.77$ mgal. This is an increase of 1.77 mgal or 1.81 ppm over the value used by Driscoll and Olsen. Since K depends on the square root of the measured torque and the torque depends linearly on g, we must increase the value of K given in Eq. (32) by 0.90 ppm. Thus,

$$K = 1.0000102 \pm 9.7 \text{ ppm}, \tag{33}$$

where the uncertainty is a standard deviation obtained by multiplying the 6.6-ppm uncertainty of Eq. (32) by 1.48 after first correcting for the now reduced uncertainty in g. (Driscoll and Olsen originally assigned a 2-ppm P.E. contribution to the uncertainty in the torque due to the uncertainty in g, whereas with Faller and Hammond's new g value, the uncertainty is reduced to $\sim$0.2 ppm.)

The present value of K obtained with the Pellat electrodynamometer may be compared with that obtained by Driscoll (1958) using essentially the same apparatus. He found

$$K = 1.0000138 \pm 12.2 \text{ ppm},$$

where we have again corrected both the original result and its uncertainty for the new value of g and have converted from a P.E. to a σ. (In all cases we have resorted to the original data to obtain an extra digit, thus reducing rounding errors.) The two dynamometer measurements agree well within their assigned uncertainties. However, following a recommendation by Driscoll (private communication) we shall not use the 1956 value as a stochastic input datum since the 1967 measurement supersedes it. This is a result of three major improvements in the dynamometer since 1956: (1) The rigidity of the beam arm was improved considerably, thereby reducing the distortion correction arising from a change in beam length and shift of the distributed mass of the beam from 37 to 16 ppm. (2) The two halves of the beam were better matched in length, thus reducing the uncertainty in determining the length of each half. (The total beam length could be measured directly, but the length of each half had to be obtained from the total length and the ratio of the lengths of each half. This ratio was determined by weighing and the more nearly matched the two halves, the greater the accuracy in the ratio.) (3) A new rotating inner coil was constructed of fused silica, resulting in a reduced temperature coefficient of expansion. All of these modifications greatly increase the confidence which can be placed in the 1967 result compared with that of 1956.

Driscoll and Cutkosky at NBS (1958) also measured K using a standard current balance with coaxial coils. They found

$$K = 1.0000092 \pm 7.7 \text{ ppm}, \tag{34}$$

where we have corrected both the original result and its uncertainty for the new value of g and have converted from a P.E. to a σ. (All three measurements of K at NBS were made at essentially the same location.) This result is in excellent agreement with the dynamometer measurements, a fact which is quite reassuring since the systematic errors are rather different in the two types of balances. For example, the dynamometer requires a precise measurement of the length of each half of the beam since the torque between the two coils is the product of these lengths and the measured forces. Furthermore, corrections for balance beam distortion which are necessary for the dynamometer are not required for the current balance.

The final measurement of K which we shall discuss in detail is that reported by Vigoureux (1965) at NPL. He used the NPL current balance to obtain $K_{\text{NPL}} \equiv \text{A}_{\text{NPL}}/\text{A}_{\text{ABS}}$, a quantity which can readily be converted to K using Tables I–III. The NPL current balance is of a somewhat different design than the NBS current balance in that there is a set of coaxial coils at both ends of the beam. [For a description of the NPL current balance, see Ayrton, Mather, and Smith (1908) and Vigoureux (1938; 1936).] This has certain advantages over the single set of coils used in the NBS balance. For example, checks on the accuracy of pitch and diametral corrections can readily be made by maintaining a current through the sets of coils such that the forces of the two coil systems exert torques in opposite directions. In addition, the sensitivity of the balance is doubled; this reduces the error in the force measurement. Vigoureux's original result was

$$K_{\text{NPL}} \equiv \text{A}_{\text{NPL}}/\text{A}_{\text{ABS}} = 1.0000147 \pm 3.6 \text{ ppm (P.E.)}, \tag{35}$$

where the 3.6-ppm uncertainty is a P.E. However, we will not use this value as it stands since it is based on Clark's old value of g(BFS) rather than Faller and Hammond's recent measurement which we have adopted (see Sec. II.B.4). Furthermore, Vigoureux did not take into account a correction to the effective diameter of the coils due to the fact that the wire is under strain. [The strain causes a variation in resistivity and hence current density over the cross section of each wire which differs from the natural "$1/r$" current distribution (Driscoll, 1958; Driscoll and Cutkosky, 1958).] Vigoureux also only took a simple average of the 70 separate measurements obtained from his two different series of runs. The first series of 40 measurements was made in October and November 1962, and the second series of 30 measurements was made from February to April 1963, *after* the coils had been taken down from the balance for dimensional measurements and replaced. Unfortunately, the mean of the two series differed by about 3 ppm and the scatter in the first series was much larger than that in the second. This implies the presence of an unknown systematic error. We therefore believe that since the two series are rather independent of one another, the final value should be the weighted average of their means. Thus:

$$\begin{aligned} &\text{Mean of first series:} && K_{\text{NPL}} = 1.0000135 \pm 6.8 \text{ ppm;} \\ &\text{Mean of second series:} && K_{\text{NPL}} = 1.0000166 \pm 4.3 \text{ ppm;} \\ &\text{Weighted mean:} && K_{\text{NPL}} = 1.0000157 \pm 3.6 \text{ ppm,} \end{aligned} \tag{36}$$

where the uncertainties are one standard deviation statistical errors.*

* Vigoureux (private communication) agrees that the value of K_{NPL} given in Eq. (36) is probably a better estimate of the current-balance results than the value given in Eq. (35).

The value of the acceleration due to gravity at the site of the balance used by Vigoureux was based on g(BFS) = 981 183.2 mgal as determined by Clark. The more accurate measurement by Faller and Hammond gives g(BFS) = 981 181.86 mgal, a decrease of 1.34 mgal or 1.37 ppm (see Sec. II.B.4). Thus, since $K_{\mathrm{NPL}} \propto g^{1/2}$, we must reduce Eq. (36) by 0.68 ppm. Moreover, Vigoureux (private communication) has estimated the strain-effect correction for his experiment from the data of Wells (1956) and concludes that an increase in K_{NPL} of +2 ppm is required. Combining this result with the correction due to the new g value implies that Eq. (36) must be increased by (2.0−0.68) ppm = 1.32 ppm. Thus

$$K_{\mathrm{NPL}} = 1.0000170 \pm 6.0 \text{ ppm}, \tag{37}$$

where the uncertainty quoted is a standard deviation and includes estimates of systematic error. It was obtained by replacing Vigoureux's original probable-error estimates of the uncertainty in the current (Vigoureux, 1965, Table 2) by standard-deviation errors and replacing his observational uncertainty and that due to lack of repetition by 3.6 ppm [one σ, Eq. (36)]. Furthermore, in addition to the systematic errors listed by Vigoureux, we include an uncertainty of 2.14 ppm (one σ) due to the uncertainty in the temperature of the coils and an uncertainty of 0.5 ppm (one σ) due to the uncertainty in the strain correction (Vigoureux, private communication).

In order to convert the NPL measurement to NBS units, we make use of Tables I–III. Vigoureux made his determinations in October and November 1962 and February to April 1963, in between the comparisons of 1961 and 1964. Thus, interpolating linearly (see Sec. II.B.1) using $A_{\mathrm{NPL}}/A_{\mathrm{NBS}} = 1+10.0$ ppm on 6 January 1961, and 1+8.55 ppm on 26 January 1964, we find that $A_{\mathrm{NPL}}/A_{\mathrm{NBS}} = 1+9.1$ ppm for the first series of measurements and $A_{\mathrm{NPL}}/A_{\mathrm{NBS}} = 1+8.9$ ppm for the second series. The mean is $A_{\mathrm{NPL}}/A_{\mathrm{NBS}} = 1+9.0$ ppm and hence Eq. (37) becomes

$$K = 1.0000080 \pm 6.0 \text{ ppm},$$

in good agreement with both NBS measurements, i.e., Eqs. (33) and (34).

In Table XII we summarize the various ampere measurements discussed as well as a measurement carried out by Curtis, Driscoll, and Critchfield (1942). The result of this measurement, which is included only for completeness, was revised by Driscoll and Cutkosky (1958) to conform to the new values assigned the volt and ohm standards in 1948. These workers also applied corrections for strain, weighting of individual turns, temperature gradient, and a new value of the gravitational acceleration. We have corrected Driscoll and Cutkosky's revised value for Faller and Hammond's new measurement of g.

The observational equation for K is simply

$$(\alpha^{-1})^0 e^0 K^1 N^0 = K. \tag{38}$$

3. *Faraday Constant, F*

In 1960, Craig, Hoffman, Law, and Hamer of NBS reported an extremely careful and painstaking determination of $E(\text{Ag})$, the electrochemical equivalent of silver, using a silver–perchloric acid coulometer. The method consisted simply of the electrolytic dissolution of metallic silver in aqueous solutions of perchloric acid containing initially a small amount of silver perchlorate. Metallic silver in sheet or rod form was used as the anode and was weighed before and after the passage of a current, known in terms of NBS electrical units, for a known time. This technique eliminated one of the main criticisms of the classic silver coulometer method, namely, the possibility that the deposited silver may contain occlusions of electrolyte, acid, or water. (In the classic method, silver was electrolytically deposited on platinum from an aqueous solution of silver nitrate.) The second major criticism usually leveled at the classic method is the possibility that a partial separation of the isotopes of silver may occur during the deposition. However, this has been shown not to be a pertinent

TABLE XII. Summary of absolute-ampere determinations.

Year of publication	Laboratory and author	Method	$K \equiv A_{NBS}/A_{ABS}$
1968	NBS, Driscoll and Olsen	Pellat electro-dynamometer	1.0000102±9.7 ppm
1958	NBS, Driscoll	Pellat electro-dynamometer	1.0000138±12.2 ppm
1958	NBS, Driscoll and Cutkosky	NBS current balance	1.0000092±7.7 ppm
1942	NBS, Curtiss, Driscoll, and Critchfield [revised, Driscoll and Cutkosky (1958)]	NBS current balance	1.000003±8 ppm
1965	NPL, Vigoureux	NPL current balance	1.0000080±6.0 ppm

criticism by the observed equality in the isotope ratio $^{107}Ag/^{109}Ag$ for electrolytically purified silver, natural silver from several sources, and for the silver in certified reagent-grade silver nitrate (Craig, Hoffman, Law, and Hamer, 1960). Craig *et al.* carried out 31 separate measurements from December 1956 to July 1958 on many silver samples characterized by widely differing metallurgical treatments. Eleven additional runs were made specifically to investigate the influence of oxygen dissolved in the silver and similarly, eleven runs were made to investigate the effect of hydrogen. Parameters studied in the various runs included impurity content, annealing procedure and atmospheres, duration of the runs, amount of silver dissolved, current density at the anode surface, potential of the silver anode during the electrolysis, and source of the silver used.

All of the 31 runs specifically made for determining $E(\text{Ag})$ were corrected by Craig *et al.* for the impurity content of the silver. These corrections were obtained using the results of spectrochemical analyses of the various silver samples and varied from a few tenths of a part per million for some of the electrolytically purified vacuum-annealed material to 21 ppm for some of the relatively impure mint silver. The size of these corrections depended critically on which of the following three assumptions were made concerning the state of the impurities in the silver: (1) All metallic impurities were in the form of oxides; (2) all metallic impurities were metallic; (3) only Cu and Fe were metallic and the rest were oxides. The actual correction applied to a particular run by Craig *et al.* was calculated on the assumption most appropriate to the metallurgical treatment given the sample. The applied correction differed from the mean of the three different possible corrections computed on the basis of the three different assumptions by a few tenths of a part per million to 15 ppm, depending on the size of the correction and sample treatment. [These differences are the impurity-correction uncertainties given in the last column of Table 10 of Craig *et al.* (1960). The signs of these differences or uncertainties were accidentally omitted in the original paper and are as follows (Hamer, private communication): Group 1, minus; group 2, plus; group 3 and 4, minus; remaining groups, plus.] The uncertainty in making an impurity correction once a particular assumption had been adopted was a few parts per million or less, depending on the magnitude of the correction.

Fifteen of the 31 runs were also corrected for dissolved hydrogen. It was observed that $E(\mathrm{Ag})$ depended on the length of the hydrogen treatment given a particular sample and that by extrapolating linearly to zero treatment time, it was possible to correct for this effect. (These corrections varied between 1 and 20 ppm.) The extrapolation was done separately with the data of group 1 and group 2 together [Table 10, Craig *et al.* (1960)], groups 6 and 7 together, and groups 11 and 12 together. The mean of group 1 and the mean of group 2 were used to define a linear $E(\mathrm{Ag})$-vs-treatment-time curve from which a value of $E(\mathrm{Ag})$ corresponding to zero treatment time was obtained (Hamer, private communication). The data of groups 11 and 12 were analysed in the same manner. For the runs in groups 6 and 7, the treatment times were so similar that the correction was assumed to be the same for both groups and was obtained using the slope of the $E(\mathrm{Ag})$-vs-treatment-time curve as given by the data of groups 1 and 2. (After correction, the means of groups 6 and 7 were averaged to give a single value.) The final result for the 15 "hydrogen runs" was obtained by a least-squares fitting procedure using the three values derived from the separate groups and is given by Craig *et al.* as

$$E(\mathrm{Ag}) = 1.117971(11) \times 10^{-6}\ \mathrm{kg/A_{NBS}\cdot sec}\ (9.6\ \mathrm{ppm}), \tag{39}$$

where the standard-deviation uncertainty is due to random error only. It is very important to realize that this result has been obtained by *ignoring* the differences in impurity-correction uncertainty among the data contained in the different groups. (The quoted uncertainty therefore includes no error estimate for impurity correction.) This procedure is open to some question since the impurity-correction uncertainties differ widely among the 15 runs; this implies that some type of weighted average would have been more appropriate. However, there is no obvious way to estimate the error to be assigned the impurity corrections and thus an unambiguous average cannot readily be obtained.

The hydrogen-corrected value may be compared with the mean of nine runs on vacuum-annealed (no hydrogen exposure) purified silver with an impurity correction of only 1 ppm and an impurity-correction uncertainty of 1 ppm. This result is

$$E(\mathrm{Ag}) = 1.1179722(70) \times 10^{-6}\ \mathrm{kg/A_{NBS} \cdot sec}\ (6.3\ \mathrm{ppm}). \tag{40}$$

Our previous criticisms notwithstanding, Eqs. (39) and (40) are in excellent agreement. [The statistical error quoted in Eq. (40) is that given by Craig *et al.* as calculated from $\sigma = \sigma_0/N^{1/2}$, where N is the number of runs entering into the mean (9 in this case), and σ_0 is the estimated standard deviation of a single run ($\sigma_0 = 21.146 \times 10^{-12}\ \mathrm{kg/A_{NBS} \cdot sec}$) as obtained by pooling the dispersions within the 12 groups of data listed in Table 10 of Craig *et al.* (1960). The error given in Eq. (39) is based in part on uncertainties calculated in this same way (Hamer, private communication).]

The remaining 7 runs were carried out with vacuum-annealed (no hydrogen exposure) impure silver requiring an impurity correction of between 5 and 8 ppm with an uncertainty of between 3 and 11 ppm. If the difference in impurity-correction uncertainty among these 7 runs is again ignored, their mean is

$$E(\mathrm{Ag}) = 1.1179741(80) \times 10^{-6}\ \mathrm{kg/A_{NBS} \cdot sec}\ (7.1\ \mathrm{ppm}), \tag{41}$$

where the quoted statistical error has been obtained from $\sigma = \sigma_0/N^{1/2}$ with $N = 7$ and does not include any uncertainty due to the impurity correction. This result is in excellent agreement with the two previous values, Eqs. (39) and (40), and implies that both the impurity and hydrogen corrections are probably well understood. However, we shall adopt the result of the 9 runs on the pure vacuum-annealed silver, Eq. (40), because it is not compromised by uncertainties due to large corrections for impurities and hydrogen treatment as are the other 22 runs.

It is an unfortunate circumstance that the impurity-correction uncertainties make it very difficult to derive unambiguously a weighted average from all the runs since the statistical uncertainty of such an average would surely be smaller than the uncertainty given in Eq. (40). [Equations (39) and (41) indicate that any average value for $E(\mathrm{Ag})$ would probably differ by a negligible amount from our chosen value, Eq. (40).] In essence, we are throwing away all of the information contained in $\frac{2}{3}$ of the measurements. On the other hand, we believe it is better to suffer a larger uncertainty than to combine data in an unjustifiable manner. We do point out that in calculating the error of Eq. (40) from $\sigma=\sigma_0/N^{1/2}$, advantage was taken of some information contained in the remaining 22 runs since σ_0 was estimated by Craig *et al.* from all of the data. Indeed, the standard deviation of the mean of the nine pure, no-hydrogen values calculated in the usual way from

$$\sigma_m{}^2=\left[\sum_{i=1}^{N}(X_i-\bar{X})^2\right]/N(N-1), \tag{42}$$

(Young, 1962) is 6.9 ppm as compared with the 6.3 ppm given in Eq. (40).

In order to obtain the total uncertainty to be assigned our adopted value of $E(\mathrm{Ag})$, we must combine its 6.3-ppm random error with the following individual systematic-error components (approximate 70% confidence-level estimates): (1) 0.01 sec or 0.5 ppm in timing the duration of a run, (2) 0.1 μV or 0.1 ppm for uncertainties in the emf of the standard cells, (3) 1 ppm for the temperature dependence of the standard resistors, and (4) 1 ppm for the impurity correction. In keeping with our usual procedure for combining random and systematic errors, the total error in $E(\text{Ag})$ is the RSS of these systematic errors and the random error. The final value for $E(\text{Ag})$ is therefore

$$E(\text{Ag}) = 1.1179722(72) \times 10^{-6}\ \text{kg}/\text{A}_{\text{NBS}}\cdot\text{sec}\ (6.5\ \text{ppm}). \tag{43}$$

In order to compute the Faraday constant from the measurement of the electrochemical equivalent of silver, we must have a value for the atomic mass of the silver used, since $F = M^*(\text{Ag})/E(\text{Ag})$. The main problem here is to determine the isotopic abundance ratio $r = {}^{107}\text{Ag}/{}^{109}\text{Ag}$. This ratio has been most accurately measured by Shields, Craig, and Dibeler (1960) and Shields, Garner, and Dibeler (1962).* Using isotopic standards prepared from nearly pure separated silver isotopes to calibrate their mass spectrometer, Shields, Craig, and Dibeler investigated the isotopic abundance of commercial silver nitrate and many different silver samples including the mint and electrolytically purified silver anodes used in the work of Craig *et al.* They found that the abundance ratio of the different samples was indistinguishable from the silver nitrate, and thus adopted r for this material as being characteristic of all silver. [This equivalence has been demonstrated to a factor of 5 better than the uncertainty in the r measurement for the silver nitrate which is in turn a factor of 2 more accurate than the r measurements for the various silver samples (Shields, private communication).] Shields, Craig, and Dibeler's final value is $r = 1.07547 \pm 0.00075$, where we have obtained the one-standard-deviation uncertainty by taking the RSS of one-half of their 0.00126 limit of error for the analytical measurements and one-half of their 0.00080 limit-of-error estimate for systematic errors. Shields, Garner, and

* We ignore the work of Crouch and Turnbull (1962) because of uncertainties involving an inexplicable mass discrimination effect.

Dibeler repeated and extended the work of Shields, Craig, and Dibeler with about one-half of the latter's analytical error, but the same systematic error. The final pooled value for all the silver nitrate work of both groups was reported by Shields, Garner, and Dibeler to be

$$r = 1.07597 \pm 0.00049 \text{ (450 ppm)}, \qquad (44)$$

where the error has again been obtained by taking the RSS of one-half the analytical and systematic limits of error.*

To compute the atomic mass of silver, $M^*(\text{Ag})$, we use the readily derived equation

$$M^*(\text{Ag}) = M^*(^{107}\text{Ag}) + [M^*(^{109}\text{Ag}) - M^*(^{107}\text{Ag})]/(1+r), \qquad (45a)$$

with

$$M^*(^{107}\text{Ag}) = 106.9050940(45) \text{ amu (0.042 ppm)},$$
$$M^*(^{109}\text{Ag}) = 108.9047560(50) \text{ amu (0.046 ppm)}, \qquad (45b)$$

as given by Mattauch *et al.* (1965). In combination, Eqs. (44), (45a), and (45b) yield

$$M^*(\text{Ag}) = 107.86834(23) \text{ amu (2.1 ppm)}. \qquad (45c)$$

Using Eqs. (43) and (45c) we find

$$F = 9.648570(66) \times 10^7 \text{ A}_{\text{NBS}} \cdot \text{sec/kmole (6.8 ppm)}.$$

To include the Faraday in our least-squares adjustment, we use the observational equation

$$(\alpha^{-1})^0 e^1 K^{-1} N^1 = F_{\text{NBS}}. \qquad (46)$$

---

* Dr. Shields has informed us that the value $r = 1.07597$ reported by Shields, Garner, and Dibeler (1962) is the result of all the data on silver nitrate including that obtained by Shields, Craig, and Dibeler (1960). Cohen and DuMond were apparently unaware of this and took a weighted average of both the Shields, Craig, and Dibeler and the Shields, Garner, and Dibeler values.

This follows from the definition of the Faraday, $F \equiv Ne$, and the introduction of $K \equiv A_{NBS}/A_{ABS}$ so that F can be expressed in NBS electrical units, i.e., as it was measured.

We shall see in detail later on that our knowledge of N is strongly influenced by the values of F and μ_p/μ_n used in the adjustments. Since there are several disparate values of μ_p/μ_n available, it is rather unfortunate that there is but one modern, high-accuracy measurement of F.* This lack is all the more regrettable because of the possible systematic errors inherent in experimental determinations of F. Questions relating to modification of isotopic abundances in the electrolytic process, the presence of inclusions in the deposited material, the effect of various impurities in the starting material, etc., must all be investigated carefully. While Craig *et al.* carried out what can only be described as a remarkably thorough experiment, additional measurements of F would make everyone feel more comfortable. Experiments presently underway at NBS to determine the Faraday with an iodine coulometer should help to alleviate this problem (Hamer, private communication).

4. *Gyromagnetic Ratio of the Proton, γ_p*

The gyromagnetic ratio of the proton is defined as the ratio of the angular precession frequency of a proton in a magnetic field B to the magnitude of the field. Thus,

$$\gamma_p = \omega_p/B.$$

(Note that since $\hbar\omega_p = 2\mu_p B$, $\gamma_p = 2\mu_p/\hbar$.) It is a most important stochastic input datum because, in combination with the Josephson-effect value of e/h, it essentially determines our WQED value of the fine structure constant. Two different techniques have been

---

* For a summary of the older measurements of the Faraday, see Hamer (1968). For an interesting history of the NBS as-maintained volt and its relationship to the Faraday, see Hamer (1967).

used for measuring γ_p. The more accurate method involves the free precession of protons, usually in a spherical water sample, in a small magnetic field of order 0.001 T. This field is obtained by passing a current known in terms of as-maintained electrical units through a precision solenoid; the field is then calculated from the current and the accurately known dimensions of the solenoid. Most of the earth's magnetic field is usually bucked out by large external Helmholtz coils, and any residual component of the earth's field along the axis of the solenoid is averaged out by reversing the current in the solenoid and remeasuring ω_p. The residual component normal to the solenoid field is usually so small that it can be neglected. (It typically contributes an uncertainty of about 0.1 ppm.) In order to raise the signal-to-noise level to the point where an accurate measurement of ω_p can be made, the water sample is usually prepolarized in a large magnetic field. The magnetization produced in the sample by this polarization is oriented normal to the solenoid field by the application of an rf pulse and then freely precesses around it, inducing a signal in a pickup coil at the frequency ω_p.

The second method uses resonant absorption and is usually carried out in a conventional electromagnet at a field of order 0.5 T. In this technique, the proton sample (usually cylindrical in shape) is located inside a small coil with axis normal to the field. An rf signal at a frequency ω_p is applied to the coil and resonant absorption of energy by the protons is detected by means of a change in the balance of an rf bridge containing the coil. In a more modern variant of the method, the coil is part of the tank circuit of a marginal oscillator, and the resonance is detected via a change of the oscillation amplitude. In the high-field method, the magnetic field is much more difficult to measure than in the low-field method, and its determination is the major source of error. The usual technique is to measure the force on a known length of wire carrying a current known in as-maintained electrical units. A long rectangular coil is suspended vertically from a balance with the lower end in the magnet gap. The vertical sides of the coil serve to bring the current to the force conductors formed by the lower horizontal portion of the coil. The fringing field of the electromagnet at the upper end of the coil is reduced to negligible levels by external coils. The force is, of course, obtained by weighing, and the acceleration due to gravity at the site of the balance must be known. (This apparatus is often called a Cotton balance.)

Before discussing the individual measurements of γ_p, we note a very important difference between the low- and high-field methods. In the low-field work, B varies *directly* with the current since it is simply equal to a solenoid constant times a current. If the current is known in terms of as-maintained electrical units, those of NBS for example, then the magnetic field will also be known in NBS units. However, the field may be corrected to absolute units by introducing the constant $K \equiv \mathrm{A}_{\mathrm{NBS}}/\mathrm{A}_{\mathrm{ABS}}$. Recalling that this means the unit of current maintained by NBS is K times larger than the absolute ampere, it is evident that the magnetic field expressed in absolute units is K times larger than the field expressed in terms of NBS units. Since γ_p varies inversely with field, it follows that γ_p as measured in NBS units must be reduced by the factor K to obtain γ_p in absolute units. Hence

$$\gamma_p{}^{\mathrm{ABS}} = \gamma_p{}^{\mathrm{NBS}}(\mathrm{low})/K, \tag{47a}$$

where $\gamma_p{}^{\mathrm{NBS}}(\mathrm{low})$ means "measured in a calculable low-field solenoid in terms of NBS units." The situation is *opposite* for the high-field method. Here, B is obtained from the measured force on a current-carrying conductor and varies *inversely* with the current in the conductor, since the force F on a conductor carrying a current i in a field B normal to the conductor's length L is $F = BiL$. We therefore find, in a manner similar to that just given,

$$\gamma_p{}^{\mathrm{ABS}} = K\gamma_p{}^{\mathrm{NBS}}(\mathrm{high}). \tag{47b}$$

There are two important points to note about Eqs. (47a) and (47b). First, since $K \simeq 1.000009$ (as indicated by the various measurements summarized in Table XII), we expect that $\gamma_p^{\mathrm{NBS}}(\mathrm{low})$ should be about 18 ppm greater than $\gamma_p^{\mathrm{NBS}}(\mathrm{high})$. Second, by combining high- and low-field measurements, it is possible to obtain an independent measurement of K. That is,

$$K = [\gamma_p^{\mathrm{NBS}}(\mathrm{low})/\gamma_p^{\mathrm{NBS}}(\mathrm{high})]^{1/2} \qquad (47c)$$

In principle, this method of determining K can provide an accuracy significantly greater than a current balance. Also note that a value of γ_p in absolute units can be obtained from the geometric mean of a high- and a low-field measurement.

Table XIII summarizes the most important measurements of γ_p made over the last 20 years. The first high-accuracy free-precession determination was reported by Driscoll and Bender (1958). The experiment was carried out at the Fredericksburg (Virginia) Magnetic Observatory of the U.S. Coast and Geodetic Survey. The spherical water sample was polarized in the 0.5-T field of an electromagnet and then shot down a pneumatic tube into the precision solenoid some distance away. The precession frequency was obtained by measuring the time required for a given number of cycles of the signal induced in the pickup coil. The Fredericksburg work consisted of two separate series of measurements using two different precision solenoids. One series was carried out in July 1958 with a solenoid wound on a fused silica form,* and the other in August 1958 with a solenoid wound on a Pyrex form (Driscoll, private communication). The results are (in units of 10^8 rad/sec$\cdot$T$_{\mathrm{NBS}}$)

* This solenoid was also used in the Pellat electrodynamometer measurements of 1956 and 1967 (Driscoll and Olsen, 1968; Driscoll, private communication).

γ_p', Fredericksburg, 1958:

silica form: 2.675148,

Pyrex form: 2.675145,

average: 2.6751465. (48)

(As before, the prime means "for protons in a spherical sample of H_2O.") The two measurements agree to within 1.1 ppm. Since the repeatability or statistical error of an individual γ_p measurement is a few parts in 10^7 ($\sim$0.3 ppm) (Driscoll, private communication; 1964; Driscoll and Olsen, 1968), this agreement implies that the techniques for measuring the dimensions of a solenoid are well under control.

The γ_p measurements were continued at NBS Washington using the same silica solenoid and the following values obtained for the period January 1960–1967 (Driscoll, private communication; 1964; Driscoll and Olsen, 1968)† (in units of 10^8 rad/sec$\cdot$T$_{NBS}$):

γ_p', Washington, 1960–1967:

1960:	2.6751560,	1964:	2.6751546,
1961:	2.6751566,	1965:	2.6751554,
1962:	2.6751559,	1966:	2.6751551,
1963:	2.6751545,	1967:	2.6751557,

average, 1960–1967: 2.6751555. (49)

These measurements were carried out in a program to monitor the stability of the NBS ampere, and their agreement demonstrates our previous statement that the NBS ampere has probably not changed by as much as 1 ppm over the last decade; the difference between the largest and smallest values (1961 and 1963) is less

† γ_p' is equal to 2π times ν_p'/H_0 as given in the next to last column of Table I in the 1964 paper of Driscoll.

TABLE XIII. Summary of measurements of the gyromagnetic ratio of the proton. (The prime means "for protons in a spherical sample of water.")

Year of publication	Laboratory and author	Approximate magnetic field (tesla)	γ_p' (10^8 rad/sec·T_{NBS})	Error (ppm)
	Low Field			
1957	Univ. of Cologne, Wilhelmy	0.010	2.67550(12)	45
1958–1968	NBS, Fredericksburg, 1958, Driscoll and Bender	0.0012		
	Silica form		2.675148	
	Pyrex form		2.675145	
	Average		2.6751465	
	Washington, 1960–1967, Driscoll and Olsen			
	Silica form		2.6751555	
	Gaithersburg, 1968, Driscoll and Olsen			
	Silica form		2.6751526	
	NBS final average (weights of 1:2:2—see text)		2.6751525(99)	3.7

1962	NPL, Vigoureux	0.0010–0.0020	2.6751440(70) (error to be expanded to 5.8 ppm—see text)	2.6
1965–1968	ETL, 1968, Hara, Nakamura, Sakai, and Koizumi, (1965 value of 2.6751654 shown to be incorrect—see text)	0.00096	2.6751384(86) (error probably 2 or 3 times 3.2 ppm—see text)	3.2
1959–1968	VNIIM, 1968, Studentsov, Malyarevskaya, Shifrin [includes corrected data reported by Yanovskii and Studentsov (1962) and Yanovskii, Studentsov and Tikhomirova (1959)—see text]	0.00005–0.0001	2.6751349 (error unknown but probably ~10 ppm—see text)	
	High Field			
1950	NBS, Thomas, Driscoll and Hipple	0.47	2.675231(26)	9.7
1961	PTB, Capptuller	0.28	2.67525(10)	37
1962–1966	KhGIMIP, 1966, Yagola, Zingerman and Sepetyi	0.24–0.47	2.675105(11) (error to be expanded to 7.4 ppm—see text)	4.0

than 0.8 ppm. In comparing the average Fredericksburg value with the average Washington value, Eqs. (48) and (49), we see that the latter exceeds the former by 3.4 ppm. This is quite surprising in view of the $\sim$0.3-ppm repeatability of the measurements and implies the presence of a systematic error. Although the magnetic environment at Fredericksburg was considerably better than that at Washington, the electrical standards were less well known. Drifts in standard cell emf due to large changes in room temperature caused some difficulty (Driscoll, private communication). Within the last year Driscoll and Olsen (1968) (Driscoll, private communication) have moved the γ_p apparatus to the nonmagnetic facility at the new NBS in Gaithersburg (Harris, 1966) and have completed a new series of measurements using the same silica solenoid.* (The excellent magnetic environment of this facility is demonstrated by the 3-sec free-precession decay time of the 3.8-cm-diam proton sample as compared with the $\sim$1.5-sec decay time observed in Washington.) Although Driscoll and Olsen did not prepolarize the proton sample, their signal-to-noise ratio of 3:1 was large enough so that they could measure the proton precession frequency to an accuracy of a few tenths of a part per million with a reasonable number of measurements. (The fluctuations in the individual measurements of ω_p were $\sim$1 ppm.) Driscoll and Olsen also measured γ_p using the so called nuclear-induction method in which the proton sample is contained in a coil with axis perpendicular to the axis of the precision solenoid. The coil is driven by a weak ac electric field of frequency $\omega(\nu \sim 52$ kHz), and when $\omega \simeq \omega_p$, proton coupling of the

* Very few of the measurements of γ_p at NBS have been accompanied by simultaneous measurements of the dimensions of the precision solenoid. In fact, the dimensions of the silica solenoid have only been measured about three times since it was constructed in the early 1950's. However, the dimensions apparently remain constant to within a few tenths of a part per million. Changes in dimensions due to changes in temperature are taken into account by using an experimentally determined temperature coefficient of expansion.

drive coil and an orthogonal pickup coil (previously adjusted for zero coupling) results in a continuous nuclear-induction signal in the pickup coil. Driscoll and Olsen obtained the same average value using both methods, but the statistical error for the free-precession value was ~ 0.1 ppm while for the nuclear-induction value, it was ~ 0.4 ppm. Their final result is

Gaithersburg, 1968:

$$\gamma_p' = 2.6751526 \times 10^8 \text{ rad/sec} \cdot \text{T}_{\text{NBS}}. \tag{50}$$

This value exceeds that obtained at Fredericksburg by 2.3 ppm but is only 1.1 ppm less than the Washington value.

Because all of the NBS measurements of γ_p' were made with essentially the same apparatus, they are not independent determinations. We must therefore combine the three separate NBS values of γ_p', Eqs. (48)–(50), in order to obtain a single representative value. After careful consideration, we have decided to average them together with the Gaithersburg and Washington values carrying equal weight, but the Fredericksburg value carrying only half as much weight (i.e., in the ratio 2:2:1). The reason for giving so little weight to the Fredericksburg work is the relatively poor control the experimenters had over the electrical standards. The reason for weighting equally the result obtained at Washington, where the magnetic environment was rather poor, and the result obtained at Gaithersburg is the comparatively large number of measurements carried out at Washington over an eight year period. Our average value is

NBS average:

$$\gamma_p' = 2.6751525(99) \times 10^8 \text{ rad/sec} \cdot \text{T}_{\text{NBS}} \ (3.7 \text{ ppm}). \tag{51}$$

A case could perhaps be made for other weightings, e.g., 3:2:1 or 1:1:1 for Gaithersburg, Washington, and Fredericksburg, respectively. However, the means for these weightings differ from Eq. (51) by less than 0.4 ppm.

Some comments concerning the quoted standard-deviation error are in order. Driscoll gives the following *probable-error* (P.E.) estimates of the uncertainties in a γ_p measurement: (1) Mean pitch of solenoid (i.e., solenoid length), 2 ppm, (2) pitch variations, 1 ppm, (3) paramagnetic materials near the sample, 1 ppm, (4) electrical standards, 1 ppm, (5) magnetic-field contribution of the compensating coils used to increase the homogeneity of the field, temperature of coils, maladjustment of coils, precession frequency, and solenoid diameter, less than 1 ppm each. An examination of these uncertainties shows that only (1)–(3) are purely systematic; the remainder are seen to be essentially random when considered in the light of the large number of measurements made, the three different locations, and the time period involved. Thus, we adopt as a one-standard-deviation systematic error, 1.48 times the RSS of the first three errors, i.e., 3.6 ppm. To this we must add a statistical or random error. A reasonable estimate of this uncertainty may be obtained by computing the standard deviation of the mean, σ_m, of the three values of γ_p' using Eq. (42), but taking into account our previously assigned weights. We find $\sigma_m = 0.6$ ppm, and thus the total RSS uncertainty in γ_p' is 3.7 ppm.

Vigoureux (1962) at NPL reported a free-precession measurement of γ_p similar to that carried out at NBS. However, instead of polarizing the sample in a magnet far removed from the precision solenoid and shooting it down a tube, Vigoureux used an additional polarization coil inside the solenoid. Actually, a long precision solenoid capable of providing the required magnetic field intensity and homogeneity was not available. In its place, Vigoureux used two comparatively short solenoids separated by a gap adjusted to give maximum uniformity at the center of the gap. One troublesome problem was that the distance between the midplane of an individual solenoid and reference lines engraved on plugs in the flanges of the solenoid was not accurately known. As a result, the distance between the midplanes of the two solenoids which was required for the calculation of the field could not be accurately determined. This difficulty was circumvented by turning the solenoids around half way through the measurements so that the original outer ends faced each other; the average of the two means obtained with the solenoids in the two orientations is independent of the location of the plugs. The mean of Vigoureux's 20 measurements of γ_p' with the plain faces of the coil outward is 2.6750771×10^8 rad/sec$\cdot$T$_{NPL} \pm 0.6$ ppm, while the mean for the plain faces inward is 2.6752642×10^8 rad/sec$\cdot$T$_{NPL} \pm 0.8$ ppm. Here, the errors are one-standard-deviation statistical uncertainties. The mean of these two measurements gives the required value of γ_p' and is

$$\gamma_p' = 2.6751707 \times 10^8 \text{ rad/sec} \cdot \text{T}_{NPL} \qquad (52)$$

with a total random error of 1.0 ppm.

In order to use Vigoureux's measurement in our adjustments, it must be converted to NBS electrical units. The NPL work was carried out in January and February 1961, very close to the 6 January central date of the 1961 comparisons (see Table III). Using the 1961 values as given in Tables I and II, we find that $A_{NPL}/A_{NBS} = 1.000010$. This implies that γ_p' as expressed in NPL units, Eq. (52), must be reduced by 10 ppm. The result is

$$\gamma_p' = 2.6751440 \times 10^8 \text{ rad/sec} \cdot \text{T}_{NBS}, \qquad (53)$$

3.2 ppm less than the average NBS value, Eq. (51). In view of the 3.7-ppm uncertainty in the NBS result, this difference is quite reasonable.

In principle, the total error of the NPL experiment should be obtained in the usual way, i.e., by taking the RSS of the 1-ppm random error and the individual systematic errors. The RSS of the limit-of-error estimates of the systematic errors given by Vigoureux (1962) is 4.8 ppm. The one-standard-deviation systematic error is just half of this or 2.4 ppm. Thus, the final error of the experiment including the 1-ppm random error would be 2.6 ppm. However, the NBS work clearly demonstrates the difficulty of obtaining reproducible results in different locations. Since the NPL measurement was carried out at just one location over a time span of only two months, and because of the nature of the precision solenoid used and the relatively short free-precession decay time (Vigoureux, 1962), we believe the NPL result should not carry any more weight in our adjustments than the average NBS value, Eq. (51). In fact, we believe it should carry the same weight as either the NBS Gaithersburg or Washington result.* The appropriate adjustment of the relative weights can be achieved either by increasing the error assigned the NPL result to $(5/2)^{1/2}$ times the error assigned the NBS average value, i.e., to 5.8 ppm, or by decreasing the error assigned the NBS average value to $(5/2)^{-1/2}$ times the error assigned the NPL value, i.e., to 1.6 ppm. (This follows from the fact that the Gaithersburg, Washington, and Fredericksburg values were weighted in the ratio 2:2:1 in order to obtain our average NBS value and that in a least-squares adjustment, the weight of an input datum is simply $1/\sigma_i^2$.) In view of the differences between the NBS values obtained at Gaithersburg, Washington, and Fredericksburg and the systematic errors given by Driscoll, we do not believe a reduction of the NBS error is justified. We therefore choose to increase the error assigned the NPL result and shall adopt

NPL:

$$\gamma_p' = 2.6751440(156) \times 10^8 \text{ rad/sec}\cdot\text{T}_{\text{NBS}} \ (5.8 \text{ ppm}). \tag{54}$$

Another low-field measurement of γ_p requiring detailed discussion was carried out from 1965 to 1968 by Hara, Koizumi, Nakamura, and Imaizumi (private

* Dr. Vigoureux (private communication) is in general agreement with this point of view.

communication; 1968) at the Electrotechnical Laboratory (ETL), the national standards laboratory of Japan. These workers used a prepolarizing coil inside the precision solenoid as did Vigoureux, but did not compensate for the earth's field. Instead, they measured it using a second proton resonance apparatus. One of their main problems was adjusting the position of a correction coil inside the precision solenoid. The coil was required to increase the field homogeneity which was initially rather low because the precision solenoid had a length-to-diameter ratio of only one. Hara, Koizumi, *et al.* carried out three series of runs from February to April 1965. Their results for γ_p', in units of 10^8 rad/sec·T_{ETL}, are

$$\begin{array}{lll} \text{I} & \text{10 February 1965:} & 2.6751582, \\ \text{II} & \text{18 March 1965:} & 2.6751614, \\ \text{III} & \text{21 April 1965:} & 2.6751702. \end{array} \tag{55}$$

Using the total number of measurements for each series as weights (185, 148, and 154, respectively), Hara, Koizumi, *et al.* give as an average

$$\gamma_p' = 2.6751630(85) \times 10^8 \text{ rad/sec} \cdot T_{ETL} \text{ (3.2 ppm)}. \tag{56}$$

The quoted standard-deviation error is the RSS of the following error estimates as given by Hara, Koizumi, *et al.*: (1) solenoid diameter and length, 0.7 ppm each, (2) wire diameter, 0.4 ppm, (3) solenoid temperature, 0.3 ppm, (4) current distribution, 1.4 ppm, (5) correction coil alignment, 2.0 ppm, (6) calibration of standard cells and standard resistors, 0.3 ppm each, (7) current measurement, 1.0 ppm, (8) temperature of cells, 0.8 ppm, (9) fluctuation of earth's field and measurement of earth's field, 1.0 and 0.1 ppm, respectively, and (10) determination of precession frequency, 0.3 ppm. In obtaining their values of γ_p', Hara, Koizumi, *et al.* assumed that the current was uniformly distributed over the cross section of the solenoid wire. If the natural or "$1/r$" current distribution had been assumed

(Driscoll, 1958; Driscoll and Cutkosky, 1958), then the field at the center of the solenoid would be 0.68 ppm larger, and γ_p' would have to be reduced by this amount. If the effects of wire strain were to be included, this reduction might increase to 1.3 ppm. The 1.4-ppm current-distribution error assigned by Hara, Koizumi, *et al.* is meant to take into account these various possibilities; they do not believe they have sufficient knowledge of the state of their wire to apply a correction for nonuniform current density. (Note that for the γ_p measurements at NPL and NBS, this correction is entirely negligible because the length-to-diameter ratios of the solenoids used were much larger than one.)

The ETL result may be converted to NBS units by the use of Tables I–III. Extrapolating linearly between the 1964 and 1967 comparisons with $A_{ETL}/A_{NBS} = 1-1.25$ ppm for 26 January 1964 and $A_{ETL}/A_{NBS} = 1-0.33$ ppm for 18 February 1967, we find $A_{ETL}/A_{NBS} = 1-0.90$ ppm for the midperiod of the measurements. This implies Eq. (56) must be increased by 0.90 ppm and the final result is

ETL, 1965:

$$\gamma_p' = 2.6751654(85) \times 10^8 \text{ rad/sec} \cdot T_{NBS} \text{ (3.2 ppm)}, \tag{57}$$

in only fair agreement with the NBS and NPL values. However, recently Hara, Nakamura, Sakai, and Koizumi (1968) repeated the measurement of γ_p at ETL and discovered that the 1965 results were probably incorrect. They found that if the air flow through their solenoid was insufficient, warm air due to heat dissipated in the prepolarizing coil reached the sensor used to monitor the temperature of the solenoid. In effect, the sensor was measuring the air temperature, not the solenoid temperature. Hara, Nakamura, *et al.* corrected this problem by enclosing the solenoid in a thermally insulated box and blowing temperature regulated air through the prepolarizing coil inside the solenoid. They claim that if the amount of air flow is sufficient, no change in γ_p can be observed for different prepolarizing coil currents. Other improvements in the new γ_p measurements include a new determination of the coil dimensions and reconstructed prepolarizing, pickup, and 90° pulse coils. Their γ_p' results for three series of measurements carried out in April 1968 are (in units of 10^8 rad/sec$\cdot$T$_{\text{ETL}}$)

$$\begin{array}{lll} \text{I} & \text{3 April 1968:} & 2.6751373, \\ \text{II} & \text{4 April 1968:} & 2.6751383, \\ \text{III} & \text{16 April 1968:} & 2.6751400. \end{array} \tag{58}$$

These values are about 9 ppm less than those obtained in 1965, Eq. (55). Using the number of measurements for each series as weights, (224, 280, and 168, respectively), Hara, Nakamura, *et al.* give as an average

$$\gamma_p' = 2.6751384(86) \times 10^8 \text{ rad/sec}\cdot\text{T}_{\text{ETL}} \text{ (3.2 ppm)}. \tag{59}$$

The quoted error is the RSS of the error estimates given by Hara, Nakamura, *et al.* which are identical to those given for the 1965 measurements, except for the following changes: (1) A 2.0-ppm uncertainty has been added to take into account fluctuations in the diameter of the solenoid, (2) the uncertainty in the current measurement has been reduced from 1.0 to 0.5 ppm, and (3) the uncertainty due to field coil alignment has been reduced from 2.0 to 1.0 ppm.

Because these new measurements of γ_p were made in 1968, we use the results of the 1967 comparisons to convert to NBS units. From Tables I and II we find $A_{\text{ETL}}/A_{\text{NBS}} = 1 - 0.33$ ppm and therefore

ETL, 1968:

$$\gamma_p' = 2.6751392(86) \times 10^8 \text{ rad/sec}\cdot\text{T}_{\text{NBS}} \text{ (3.2 ppm)}. \tag{60}$$

This result is only 1.8 ppm less than that of Vigoureux,

Eq. (54), well within the error assigned by Hara, Nakamura, *et al.* to the ETL experiment. It is less than the average NBS value, Eq. (51), by 5.0 ppm, which is about equal to the 4.9-ppm RSS of the errors assigned the two experiments. The probability for this difference to occur by chance is about 30%. The agreement between the ETL, NPL, and NBS results is therefore reasonable. However, we do not choose to use the ETL result in our adjustments. The reasons for this decision are purely experimental and are as follows: (1) Uncertainties in the temperature of the solenoid are still, in our opinion, a possible source of significant error. While no change in γ_p was observed for different prepolarizing coil currents "if the amount of air flow is sufficient," blowing air through the precision solenoid leads one to suspect that the temperature of the solenoid and hence its dimensions are not really well known. (2) Hara, Nakamura, *et al.* observed a disturbing thermal instability in the diameter of the solenoid. Measurements indicated that the diameter might vary as much as 10 ppm depending on the solenoid's thermal history. This would lead to a possible 5-ppm variation in γ_p since for the dimensions of their solenoid (length and diameter equal), the field varies as the square root of the diameter. (3) In their experiments, Hara, Nakamura, *et al.* observed that γ_p varied by 3–4 ppm depending on the direction of the current in the prepolarizing coil. The reason for this is unknown, and the authors simply averaged the data for the two directions. Until all of these problems are clarified, (including the uncertainty in the current distribution), the ETL result must be considered suspect. Its true error may well be several times the quoted 3.2 ppm.

Another low-field determination of γ_p requiring discussion was carried out between 1958 and 1968 at the All Union (Mendeleev) Institute of Metrology in Leningrad (VNIIM), the national standards laboratory of the Soviet Union. In this work, pairs of Helmholtz rings were used instead of a precision solenoid. The spherical H_2O sample was prepolarized in a 0.26-T field produced by a prepolarization coil positioned between the Helmholtz rings. The magnetic fields produced by the rings ranged from 0.5×10^{-4} to 10^{-4} T. The first report on this work was made by Yanovskii, Studentsov, and Tikhomirova in 1959. They gave the result $\gamma_p' = 2.67520(15) \times 10^8$ rad/sec·T_{USSR} (56 ppm). The measurements were continued by Yanovskii and Studentsov, and in 1962 they reported the value $\gamma_p' = 2.67506(5) \times 10^8$ rad/sec·T_{USSR} (19 ppm). The quoted errors and the 52-ppm difference between the two values indicate that at this stage these experiments were of relatively low precision.

The results of further measurements have been reported recently by Studentsov, Malyarevskaya, and Shifrin (1968a; 1968b). These workers made several improvements in the earlier apparatus and studied possible sources of systematic error. In order to investigate possible shielding effects due to the prepolarizing coil, γ_p measurements were carried out in the absence of the coil by prepolarizing the proton sample in an electromagnet and shooting it into the Helmholtz rings as was originally done in the NBS experiments. No shielding effects due to the prepolarization coil were detected. Scatter of as much as 100 ppm had been observed in the heterodyne frequency measurements of the early experiments of Yanovskii and coworkers. This was traced by Studentsov, Malyarevskaya, and Shifrin to amplitude-dependent phase shifts in the heterodyne system which caused errors in the measurement of the frequency of the exponentially decaying free-precession signals. The problem was aggravated by the small precession frequencies (2–4 kHz) arising from the small magnetic field of the Helmholtz rings. Studentsov *et al.* replaced the heterodyne system by a set of two or three simultaneously operating frequency counters, and the remaining frequency-measurement error was estimated from the discrepancies between the readouts of these counters to be about 5 ppm. During the period 1960–1968, a total of 12 different pairs of Helmholtz rings were employed in the γ_p measurements. The dimensions of all the rings were remeasured with improved techniques by Studentsov *et al.* between 1966 and 1968, and the revised values were used to correct the earlier γ_p results made by Yanovskii *et al.*, who used nine of the present 12 sets of Helmholtz rings. The result of all measurements during the period 1960–1968 is reported by Studentsov *et al.* as

$$\gamma_p' = 2.6751625 \times 10^8 \text{ rad/sec}\cdot\text{T}_{\text{USSR}}, \tag{61}$$

with a statistical uncertainty of 1.7 ppm.

To convert this value to NBS units we use Tables I and II. We note from these tables that the ratio $\text{A}_{\text{USSR}}/\text{A}_{\text{NBS}}$ increased by 3.0 ppm during the period 1957–1967. This presumably reflects a 3-ppm drift in the USSR as-maintained ampere, since the NBS ampere appears to have remained constant to well within 1 ppm over this period [see Eq. (49) and preceding data]. Although the details of the temporal distribution of the measurements are not given by Studentsov *et al.*, most of the measurements appear to be more or less uniformly distributed over the interval between the 1961 and 1967 comparisons. We therefore take a simple average of the 1961, 1964, and 1967 comparisons with the result $\text{A}_{\text{USSR}}/\text{A}_{\text{NBS}} = 1.0000103$. Applying this to Eq. (61), we find

$$\text{VNIIM:}\quad \gamma_p' = 2.6751349 \times 10^8 \text{ rad/sec}\cdot\text{T}_{\text{NBS}}. \tag{62}$$

This result is 1.6 ppm less than the ETL value, 3.4 ppm less than the NPL value, and 6.6 ppm less than the NBS value.

Despite the considerable amount of experimental effort represented by the VNIIM result, we have decided to omit it from the present adjustments because there appears to be no way to assess the actual uncertainty and therefore the weight to be assigned the result, at least with the information available at the time of writing. The value given in Eq. (61) is a simple unweighted average of the means of a series of measurements for each pair of Helmholtz rings, each series extending over a number of years. Studentsov *et al.* (1968) give no indication whether the larger uncertainties of the early measurements (e.g., those due to errors in the early heterodyne frequency measurements) have somehow been taken into account. No mention was made of any estimate of possible systematic error from any source. (It was noted that the 5-ppm estimated frequency-measurement error "has been included with the random error," which is given as 1.7 ppm.) We might guess the total systematic error to be of the same order as that in the experiments of Driscoll or Vigoureux, but there are indications that this could be a rather dangerous assumption. For example, Studentsov *et al.* (1968) quote a series of measurements of γ_p made in a particular Helmholtz pair for each of the years 1960, 1961, 1962, 1966, and 1967. The spread between the maximum and minimum values is 16 ppm. The comparable yearly measurements of Driscoll at NBS over the same period [Eq. (49) and preceding data] have a spread of less than 0.8 ppm. The total spread of the means of the series of measurements from which Eq. (61) was obtained is 25 ppm. In the presence of this kind of scatter, it is highly unlikely that a systematic error of, say, 5 ppm would even have been noticed, much less identified and eliminated. (In contrast, the few tenths of a part per million random error in the NBS measurements permitted a critical search for systematic errors of order 1 ppm and less.) Clearly, unless it becomes possible to obtain much more detailed information about the sources of error in the experiments of Studentsov *et al.*, it would be rash to include them in an adjustment.

We conclude our discussion of the low-field measurements with a brief mention of the results of Wilhelmy (1957) and Kirchner and Wilhelmy (1955). Using a low-quality solenoid of 1 m length and 6 cm diameter, these latter workers obtained $\gamma_p' = 2.67549(12) \times 10^8$ rad/sec$\cdot$$T_{\mathrm{PTB}}$ (45 ppm). (The Physical-Technical Laboratory or PTB is the German national standards laboratory.) An earlier, less accurate experiment by

Wilhelmy in 1955 gave $\gamma_p' = 2.67562(24) \times 10^8$ rad/sec·T_{PTB} (90 ppm). The value given in Table XIII is the 1957 result converted to NBS units using the results of the 1957 comparisons. Because of the large uncertainty of Wilhelmy's result and its obvious inconsistency with the other low-field measurements, it will not be used in the present adjustments.

The first high-field measurement of γ_p was reported in 1950 by Thomas, Driscoll, and Hipple of NBS (1950a; 1950b). [Note that the numbers given in Tables 4 and 5 of the first paper are in NBS units (R. Driscoll, private communication).] The rectangular coil of their Cotton balance contained nine turns of wire and was wound on a glass form. The measured force on the coil was related to the field at the proton resonance sample by carefully plotting out the field distribution in the gap as well as the region occupied by the coil and numerically integrating the field over the coil. Clearly, constancy of the field distribution with time as well as reasonable homogeneity is required if the experiment is to be successful. The value of γ_p obtained by Thomas, Driscoll, and Hipple for their "standard oil sample" was 2.675231×10^8 rad/sec·T_{NBS}. This was calculated using a value for acceleration due to gravity at the site of the balance of $g = 980\,081$ mgal which is based on a value of g at the East Building gravity station of $g(\text{NBS-EB}) = 980\,083$ mgal. Faller and Hammond's measurement gives $g(\text{NBS-EB}) = 980\,084.77$ mgal, an increase of 1.81 ppm. Since γ_p varies inversely with the force and therefore with g, the original result of Thomas *et al.* must be *decreased* by this amount. However, the proton precession frequency of the "standard oil sample" used by Thomas, Driscoll, and Hipple was later found by Thomas (1950) to be 1.9 ppm less than the precession frequency of a spherical water sample.* This implies that the result of

* Thomas gives 28.1 ppm as the diamagnetic correction for the "standard oil sample" based on $\sigma(H_2) = 26.8$ ppm. The work discussed in Sec. II.B.6 shows that $\sigma(H_2) = 26.6$ ppm and $\sigma(H_2O) = 26.0$ ppm. The shielding constant for the standard sample is therefore 1.9 ppm larger than for a spherical H_2O sample.

Thomas *et al.* must be *increased* by this amount to bring it to the desired form, that is, as measured for protons in a spherical sample of water. The two corrections are seen essentially to cancel one another, leaving the original value unchanged. The final result of the experiment is then

$$\gamma_p' = 2.675231(26) \times 10^8 \text{ rad/sec} \cdot \text{T}_{\text{NBS}} \text{ (9.7 ppm)}. \tag{63}$$

The original uncertainty quoted by Thomas *et al.* was 22 ppm including both random and systematic errors and was believed to be "several times the probable error." If, as implied by the description of the experiment, it is 3 times the probable error or a limit of error, then the one-standard-deviation uncertainty is half this or 11 ppm. The 9.7 ppm of Eq. (63) has been obtained by subtracting out from the original error assignments of Thomas *et al.* the uncertainties arising from the gravitational acceleration, the ratio of the NBS ampere to the absolute ampere, and ferric ions in the proton sample.

In comparing the result of Thomas *et al.* with the low-field measurements, particularly those carried out at NBS and NPL, it is immediately evident that there is a gross discrepancy. Instead of being on the order of 18 ppm *less* than these low-field measurements as implied by Eqs. (47) and Table XII, it is 30 ppm *larger*. The Thomas, Driscoll, and Hipple measurement is therefore inconsistent by about 50 ppm, 5 times the standard deviation of the difference. This is quite surprising in view of the great care with which Thomas *et al.* worked. The exact origin of the discrepancy remains unknown, but in light of a similar high-field measurement carried out in the Soviet Union (to be discussed below), it may perhaps be attributed to uncertainties in the positioning of the measuring coil and to inhomogeneities of the magnetic field. Because of this large and obvious inconsistency, the measurement of Thomas *et al.* will not be used in our adjustments (see Sec. II.C.7).

In 1961, a high-field measurement of γ_p was reported by Capptuller at PTB (1961a; 1961b; 1964). Capptuller attempted to reduce the errors resulting from the measurement of the dimensions of the balance coil by using a coil which could be accurately varied in width with the aid of quartz length standards. He obtained the value $\gamma_p'=2.67522\times 10^8$ rad/sec$\cdot$T$_{PTB}$. This result was calculated using a value of the gravitational acceleration at the site of the balance based on the Potsdam system (Capptuller, private communication) [i.e., g(balance) = 981 266 mgal]. However, the Potsdam system is too high by about 13.80 mgal or 14.1 ppm (see Sec. II.B.4) and therefore the value of g used by Capptuller should be reduced by this amount.* Since γ_p as obtained by the high-field method varies inversely with g, the original γ_p' value quoted by Capptuller must be increased by 14.1 ppm, giving $\gamma_p'=2.67526\times 10^8$ rad/sec$\cdot$T$_{PTB}$. Capptuller carried out his measurements in 1959 and 1960, between the 1957 and 1961 comparisons of electrical standards. From Tables I and II we find that for 1957, $A_{PTB}/A_{NBS}=1-2.8$ ppm, while for 1961, $A_{PTB}/A_{NBS}=1-2.35$ ppm. The mean is $A_{PTB}/A_{NBS}=1-2.6$ ppm, and we therefore reduce the PTB value of γ_p' by this amount in order to express it in NBS units. Hence

$$\text{PTB:}\quad \gamma_p'=2.67525(10)\times 10^8\ \text{rad/sec}\cdot\text{T}_{NBS}\ (37\ \text{ppm}),$$

where the quoted standard-deviation error is that given by Capptuller. We have not concerned ourselves with the exact period of time during which the experiments were carried out or with the fact that Capptuller used a cylindrical proton sample containing paramagnetic ions since such corrections would be less than 1 ppm. This is completely negligible compared with the assigned uncertainty, which comes mainly from the

* The value of g so obtained is 981 252.2 mgal. This is in excellent agreement with a recent absolute measurement of g at PTB which gave $g=981\ 252\pm 2$ mgal (Capptuller, private communication).

correction to the effective width of the coil, as obtained by integrating the field over the coil. Because of its large uncertainty, we shall ignore Capptuller's measurement in our adjustments. It should be noted, however, that it is inconsistent with the low-field determinations in the same manner and by about the same amount as the measurement of Thomas, Driscoll, and Hipple.

The most recent and also the best high-field measurement of γ_p is that carried out in the Soviet Union at the Kharkov State Institute of Measures and Measuring Instruments (KhGIMIP) by Yagola, Zingerman, and Sepetyi (YZS) (1966; 1962) from 1960 to 1966. These workers used balance coils which differed from those used in previous investigations in that they consisted of only one or two turns. This permitted a more accurate determination of the effective width of the turns. The initial work reported in 1962 was carried out on two coils, No. 1 and No. 2, with one and two turns, respectively (turn A and turn B for coil No. 2). Coil No. 1 gave for γ_p' 2.674998, while the mean value obtained from coil No. 2, turn A, turn B, and turns A and B together, was 2.675072 (all in units of 10^8 rad/sec·T_{USSR}). The mean of all four measurements was 2.675054, but it was apparent that there was a systematic error present, probably related to coil No. 1. As a result, the measurements were continued, and in 1966 new results using a third coil, also with two turns, were reported. The mean value for γ'_p obtained with coil No. 3, turn A, turn B, and turns A and B together, was 2.675070, in excellent agreement with the results from coil No. 2. Yagola *et al.* concluded that the result from coil No. 1 contained an inadmissable systematic error and could be rejected on the basis of its unexplicable 28-ppm difference from the mean of the results of the other two coils and the fact that the measurements with coil No. 1 were carried out when the technique and equipment were first being investigated. Thus, based on coils No. 2 and 3 only,

$$\gamma_p' = 2.675071 \times 10^8 \text{ rad/sec} \cdot T_{USSR}. \tag{64}$$

Several corrections to Eq. (64) must now be considered. First, the proton resonance sample used by Yagola *et al.* was not a sphere of water but was a cylinder containing a $0.1M$ solution of $NiSO_4 \cdot 7H_2O$. They established experimentally that there were no shifts in ω_p exceeding 1 ppm in this solution "as compared with the resonance in mineral oil which has been thoroughly studied by many investigators." We have estimated the correction to be applied to Eq. (64) due to its shape and the presence of paramagnetic ions by assuming (1) a cylinder of infinite length and hence a demagnetization factor of $\alpha = 2\pi$, (2) a susceptibility for the ions $\chi_{\text{ion}} = 0.45 \times 10^{-6}$ cgs units as calculated from the data of Dickinson (1951) (his Table VI), and (3) a susceptibility for H_2O of $\chi_{H_2O} = -0.72 \times 10^{-6}$ cgs units (*Handbook of Physics and Chemistry*, 1968). Using the relation

$$(B_i - B_0)/B_0 = [(4\pi/3) - \alpha]\chi,$$

where B_0 is the applied field and B_i the actual field, we find that Eq. (64) must be decreased by 0.56 ppm. Because of our imperfect knowledge of the exact sample shape, we will apply a correction of -0.28 ppm and include an uncertainty of 0.28 ppm in the final error assigned the experiment. A second correction to Eq. (64) which must be considered is that due to a change in the value of the gravitational acceleration used by Yagola *et al.* These workers calculated γ_p using a value for g at the site of their balance 11 mgal below the corresponding Potsdam system value. According to the discussion of Sec. II.B.4, this value of g is too large by 2.80 mgal or 2.85 ppm since the Potsdam system is too high by 13.80 mgal rather than 11 mgal. Since γ_p varies inversely with g, this implies that Eq. (64) must be increased by 2.85 ppm. The total correction to be applied to Eq. (64) is therefore $(2.85 - 0.28)$ ppm = 2.57 ppm and the result is

$$\gamma_p' = 2.675078 \times 10^8 \text{ rad/sec} \cdot \text{T}_{\text{USSR}}. \qquad (65)$$

In order to use the Kharkov measurement in our adjustments, it must first be expressed in NBS units. The main experiments were carried out from about

1961 to 1964 (Zingerman, private communication), a period which spans both the 1961 and 1964 comparisons. From Tables I and II we find for 1961, $A_{USSR}/A_{NBS}=1.0000090$, while for 1964, $A_{USSR}/A_{NBS}=1.00001065$. Because the experiments were carried out more or less uniformly over this period, we take the mean of these two values and obtain $A_{USSR}/A_{NBS}=1.000098$. Since a high-field measurement of γ_p varies directly with the current, Eq. (65) must be increased by 9.8 ppm. The final result is

KhGIMIP:

$$\gamma_p'=2.675105(20)\times 10^8 \text{ rad/sec}\cdot T_{NBS} \text{ (7.4 ppm)}. \tag{66}$$

A discussion of the stated standard-deviation error is in order. Yagola *et al.* give 4 ppm as the one-standard-deviation uncertainty of their experiment, including both random and systematic errors. Taken at face value, it means that the value of K implied by this measurement and the NPL and NBS low-field results [see Eq. (47c)] would carry significantly more weight in the present adjustments than all three of the direct current-balance measurements we shall use, combined (see Table XVI). Since the high-field measurement of γ_p' is itself a form of current-balance experiment, a certain amount of skepticism seems warranted. Indeed, the difficulties inherent in this type of experiment have been clearly demonstrated by the experiences of Thomas *et al.* at NBS, Capptuller at PTB, and the large discrepancy between the results obtained by Yagola *et al.* using coil No. 1 and those obtained using coils No. 2 and No. 3. Although Yagola *et al.* have taken extraordinary pains to track down and eliminate systematic errors in their Cotton balance measurements of magnetic field, we are not convinced that their data clearly justifies the claimed 4-ppm total uncertainty. Inspection of their results (1966; 1962) indicates that the agreement between values of γ_p' obtained using different turns on a particular coil is significantly poorer than might be expected, implying a systematic error may well be present. For example, the mean value of γ_p' obtained from 10 measurements using coil No. 3, turn A, is (in units of 10^8 rad/sec·T_{USSR}) 2.675083±2.0 ppm, where the error is the statistical standard deviation of the mean computed in the usual way [Eq. (42)]. For turn B, $\gamma_p'=2.675054\pm1.8$ ppm. The turn-A result exceeds that obtained from turn B by 11 ppm, which exceeds the standard deviation of the difference σ_d by a factor of 4. The probability for this to occur by chance is ~0.006%. A similar situation obtains for coil No. 2. The result for turn A is given as $\gamma_p'=$ 2.6750777, and for turn B, $\gamma_p'=2.675063$. Again the turn-A result exceeds that obtained from turn B, the difference being 5.2 ppm. In addition, for both coils there is a disturbing similarity in the relationship between results obtained with turns A and B together and those obtained with turns A and B separately. For turns A and B together, coil No. 3 gave $\gamma_p'=$ 2.675073, 2.8 ppm less than the turn-A result but 7.1 ppm larger than the turn-B result. Similarly, for coil No. 2, turns A and B together gave $\gamma_p'=2.675077$, identical to the result obtained from turn A but 5.2 ppm larger than the result obtained from turn B.

In view of these facts, we believe that a more reasonable estimate of the true error of the experiment can be obtained by the following procedure: We interpret as standard-deviation errors rather than the claimed limits of error the "noneliminated systematic-error" components listed by Yagola *et al.* for the determination of the effective width of a turn [Yagola *et al.* (1962) Table 1]. This is justified, we believe, since determining the effective width is probably the main factor contributing to the apparent discrepancies just discussed. The RSS of these errors is 6.6 ppm. To this we add RSS the remaining "noneliminated systematic errors" listed by Yagola *et al.* but interpreted as limits of error as claimed, i.e., two standard deviations. This decision is based on the fact that these errors mainly include contributions due to the gravitational acceleration, comparisons of standard cells and resistors, frequency errors, etc., all of which are sufficiently familiar that the interpretation of the errors given by Yagola *et al.* as standard deviations rather than limits of error would be obviously incorrect. Thus, the total systematic error given by Yagola *et al.* is 7.1 ppm. To this we add RSS our own error estimates as follows: (1) 0.5 ppm for possible drifts in the as-maintained electrical units, (2) 0.3 ppm for paramagnetic ions and sample shape, and (3) 2.0 ppm as representative of the random error in the experiments. (This is typically what is observed for a series of measurements using one turn of a particular coil.) The total error is therefore 7.4 ppm.

In comparing the high-field result of Yagola *et al.* with the average NBS low-field value, Eq. (51), we see that it is 17.9 ppm less, implying via Eq. (47c) that $K=1.0000089$. This value of K is in excellent agreement with the direct current-balance measurements listed in Table XII. The Kharkov measurement is therefore quite consistent with the NBS low-field work. If we compute K via Eq. (47c) using both the low- and high-field measurements carried out in the Soviet Union, Eq. (62) and (66), we find $K=1.0000056$.

The observational equation for γ_p' can be readily derived using the following relationships:

$$\gamma_p'=\omega_p'/B; \qquad \hbar\omega_p'=2\mu_p'B; \qquad \mu_B=e\hbar/2m_e;$$
$$R_\infty=(\mu_0 c^2/4\pi)^2(m_e e^4/4\pi\hbar^3 c); \qquad \alpha=(\mu_0 c^2/4\pi)(e^2/\hbar c). \tag{67}$$

The result is

$$\frac{\alpha^3}{e}=\frac{\mu_0 R_\infty}{(\mu_p'/\mu_B)}\gamma_p'. \tag{68}$$

(We use values obtained for protons in spherical samples of water because doing so permits the use of the more accurate Lambe result for μ_p'/μ_B—see Sec. II.B.6.) As it stands, Eq. (68) is valid only if γ_p' is expressed in absolute units. Introducing the conversion constant K, the observational equation for γ_p' measured in NBS units by the *low-field* method is

$$(\alpha^{-1})^{-3}e^{-1}K^{1}N^{0} = \frac{\mu_0 R_\infty}{(\mu_p'/\mu_B)} (\gamma_p')_{\mathrm{NBS}}, \tag{69}$$

and similarly, the correct observational equation for a *high-field* determination of γ_p' in NBS units is

$$(\alpha^{-1})^{-3}e^{-1}K^{-1}N^{0} = \frac{\mu_0 R_\infty}{(\mu_p'/\mu_B)} (\gamma_p')_{\mathrm{NBS}}. \tag{70}$$

5. *Magnetic Moment of the Proton in Units of the Nuclear Magneton*, μ_p/μ_n

The important quantity μ_p/μ_n ($\mu_n \equiv e\hbar/2M_p$) can be obtained by measuring the ratio of the proton spin-flip or precession frequency $\omega_p = 2\mu_p B/\hbar$ to the cyclotron frequency of the proton, $\omega_c = eB/M_p$, in the same magnetic field B. Thus,

$$\frac{\omega_p}{\omega_c} = \frac{2\mu_p B/\hbar}{eB/M_p} = \frac{\mu_p}{e\hbar/2M_p} = \frac{\mu_p}{\mu_n}.$$

There have been seven measurements of μ_p/μ_n over the last 20 years which must be considered. These are summarized in Table XIV and will now be discussed in some detail.

Sommer, Thomas, and Hipple (STH) (1951) [see also Hipple, Thomas, and Sommer (1949)] reported the final result of a measurement of μ_p/μ_n which utilized a device called an omegatron. While the measurement of ω_p is comparatively straightforward and can be done to 1 or 2 ppm, the determination of ω_c poses real experimental difficulties. The omegatron was developed to overcome some of these difficulties, in particular, the inherent low resolution of the cyclotron. In the omegatron of Sommer *et al.*, a variable-frequency rf electric field is applied at right angles to the magnetic field. At resonance, ions of a selected charge-to-mass ratio are accelerated by the rf field and spiral outward until they attain a radius of 1 cm where they strike a collector. An ion of a different charge-to-mass ratio cannot reach this radius unless the rf frequency is changed accordingly. (Conversely, the frequency of the rf field may be held constant and the magnetic field varied.) The width of the resonance curve depends on the time of flight of the resonant ions and may have a total width at half-maximum of a few parts in 10^5 for protons.

Unfortunately, the omegatron has its own peculiarities which contribute to the uncertainty in ω_c. One of the main problems is that the observed cyclotron resonance frequency deviates from the simple relation $\omega_c = eB/M_p$ because of radial electrostatic fields. These fields come from the applied trapping voltage required to prevent drift of the protons or other ions along the magnetic field, as well as from space charge within the omegatron. In order to correct for this effect, Sommer *et al.* took advantage of the fact that the shift is proportional to the mass of the ions. By observing resonance for protons at a certain frequency, and then in quick succession, resonance for another ion (e.g., H_2^+) at a second frequency under the same operating conditions, the shift could be experimentally determined by linear extrapolation to zero mass. Difficulties may arise, of course, because the space charge may not be the same for the second ion and the magnetic field may drift. Nevertheless, Sommer *et al.* were able to obtain consistent results over a wide range of operating conditions for the ion pairs H^+ with H_2^+, D_2^+, and H_2O^+. The good agreement of the correction factors determined from measurements on the triad of masses H^+, H_2^+, and D_2^+ also gave much added confidence in the method. The value reported by Sommer *et al.* is

$$\omega_p^*/\omega_c = \mu_p^*/\mu_n = 2.792685 \pm 0.00006, \qquad (71)$$

where ω_p^* is the observed resonance frequency in the "standard oil sample" and the uncertainty is "several times the estimated probable error" and includes both random and systematic errors.

The proton precession frequency in the "standard oil sample" has been found by Thomas (1950) to be 1.9 ppm less than the precession frequency in a spherical water sample.* Equation (71) must therefore be increased by 1.9 ppm so that it is in the desired form, i.e., as determined for protons in a spherical sample of water. This will be denoted as before by a prime. Hence, the Sommer *et al.* result becomes $\mu_p'/\mu_n = 2.792690$. The standard-deviation uncertainty to be assigned this value is open to some question. If we assume the ± 0.00006 given by Sommer *et al.* is 3 times the probable error or 2σ (i.e., a limit of error), then the one-standard-deviation uncertainty would be ± 0.00003 or 10.7 ppm. On the other hand, the average deviation for all 45 measurements is given as ± 0.000025, which implies a one-standard-deviation statistical or random error in the mean of $(1.25)(0.000025)/(45)^{1/2} = \pm 0.0000047$ or 1.7 ppm (a Gaussian distribution has been assumed; see Sec. II.A.4). Although Sommer *et al.* give no detailed list of systematic errors, we may make the following estimates from their paper and a discussion by Thomas, Driscoll, and Hipple (1950a) of the same magnet and resonance probe used by Sommer *et al.*: (1) shielding of the magnetic field by the copper–nickel electrodes of the omegatron, 1.5 ppm, (2) effect of the magnetic field associated with the filament current, 3.5 ppm, (3) measurement of ω_p, 3 ppm, and (4) field drifts, 1.5 ppm. (These are meant to be conservative 70% confidence-level estimates.) The RSS of the systematic errors and the 1.7-ppm random error is 5.3 ppm, about half the error obtained by assuming that the uncertainty quoted by Sommer *et al.* is a limit of error. However, because of the electric-field and space-charge problems inherent in the omegatron, the several parts-per-million asymmetry observed by Sommer *et al.* in their cyclotron resonance curves, and other characteristic limitations of the omegatron (to be discussed below when we review the similar measurement of μ_p'/μ_n by Petley and Morris), we shall assume the uncertainty given by Sommer *et al.* is in fact a limit of error. The final result of the experiment is therefore

---

* See footnote on page 104.

TABLE XIV. Summary of measurements of the magnetic moment of the proton in units of the nuclear magneton. (The prime means "for protons in a spherical sample of water.")

Publication date	Authors	Method and approximate magnetic field (tesla)	μ_p'/μ_n	Error (ppm)
1949–1951	Sommer, Thomas, and Hipple	Omegatron, 0.47	2.792690(30)	11
1950–1956	Trigger, Jeffries, and Bloch	Inverse cyclotron, 0.53, 0.7, 0.96	2.79267(10)	36
1955–1963	Sanders, Turberfield, Collington, and Dellis	Inverse cyclotron, 0.24	2.792701(73)	26
1961	Boyne and Franken	Cyclotron, 0.8–1.3	2.792832(55)	20
1965	Mamyrin and Frantsuzov	Mass spectrometer, 0.13	2.792794(17)	6.2
1967	Petley and Morris	Omegatron, 0.47	2.792746(52)	19
1967	Marion and Winkler	Nuclear reaction energies	2.79260(13)	45

$$\mu_p'/\mu_n = 2.792690(30) \text{ (11 ppm)}. \qquad (72)$$

A preliminary measurement of μ_p/μ_n was reported in 1950 by Bloch and Jeffries, and the final result in 1951 by Jeffries. These workers used a small cyclotron (8.5-cm diam) in a field of about 0.5 T. The cyclotron was operated in an inverse or decelerating mode in which protons were formed in an external source and injected into the outer portion of one of the dees. The protons traversed a slowly shrinking spiral path identical to the usual cyclotron trajectory, but in the backward direction, and were finally detected by a probe near the center of the dees. A resonance curve was obtained by measuring probe current as a function of the frequency of the rf dee voltage V_d. The inherently low resolution (large linewidth) characteristic of a cyclotron was considerably improved in the experiments of Jeffries and Bloch by operating V_d at a frequency which was n times the resonance frequency where n, an odd integer, was about 9. The width of the detected resonance decreases with increasing n mainly because the timing or phase relation between the protons and V_d becomes more critical by a factor of n. Fractional linewidths of 10^{-4} were typically observed.

The main problem with the method is determining exactly what portion of the observed resonance curve corresponds to ω_c. Jeffries and Bloch developed a simple first-order linear theory which related ω_c to the frequency at which the probe current went to zero on the high-frequency side of ω_c for a given value of dee voltage V_d, and also to the point of discontinuity in the derivative of the probe current on the high-frequency side. The two methods gave results which were in general agreement, but the first could be applied more accurately and was the main technique used. In practice, the point at which the probe current went to zero was obtained as a function of V_d for fixed n and probe position. Theoretically, this point approaches ω_c as V_d approaches a certain cutoff value of the voltage, V_C, determined by the thickness of the injection plates.

Unfortunately, the signal amplitude also decreases as V_d approaches V_C, and the data must be extrapolated to V_C. The extrapolation is on a very steeply rising curve of ω_c versus V_d and is rather uncertain. More important, the basic theory is so uncertain that Jeffries includes a 71-ppm systematic error for lack of knowledge of the exact position of resonance. This corresponds to $\frac{1}{3}$ the total width of the observed resonance curves for large n and is by far the major error-contributing factor; the random or statistical uncertainty in the mean value of μ_p/μ_n obtained from 17 runs was only 4.3 ppm and the remaining systematic errors were estimated to be less than 15 ppm. The final value quoted is $\mu_p'/\mu_n = 2.79242(2)$ (72 ppm). We will not attempt to correct for the cylindrical shape of the proton resonance sample and the paramagnetic ions contained therein because of this large uncertainty. (The net correction is only of order 1 ppm.)

The work of Jeffries was later repeated by Trigger (1955) with a completely redesigned apparatus in which measurements could be made at different values of magnetic field. Using the same theory as did Jeffries, Trigger obtained essentially the same numerical result, namely $\mu_p'/\mu_n = 2.79244$. However, Trigger (1956; private communication) later reanalyzed his data with the help of a new theory in which a set of three coupled nonlinear equations was used to describe the cyclotron orbits. The theory predicted a frequency shift dependent upon the voltage V_d with the result that Trigger's original value was too low by about 0.00023 (82 ppm). The final value is

$$\mu_p'/\mu_n = 2.79267(10) \ (36 \text{ ppm}), \qquad (73)$$

where most of the uncertainty still comes from determining the point in the resonance curve which corresponds to ω_c. Equation (73) has been obtained by subtracting from the value of μ_p/μ_n given by Trigger (1956) the 27-ppm correction included by Trigger (1955) for the diamagnetic shielding of protons in H_2O. We have also applied a -1.4-ppm correction calculated from the data of Dickinson (1951) because the proton resonance sample used was a cylinder containing a $0.01M$ solution of $FeCl_3$ rather than a spherical sample of pure H_2O. In comparing Trigger's revised result with the measurement of Sommer *et al.*, Eq. (72), we see that the two are in good agreement. However, we will not utilize Trigger's result in our adjustments because of its relatively large error and the uncertainty of the error; the theory is sufficiently suspect that it is difficult to say whether the quoted 36-ppm error should be regarded as a probable error or a standard deviation.

Another measurement obtained with the decelerating or inverse cyclotron was reported by Sanders and Turberfield (ST) in 1963.* In their version of the instrument, the dee geometry of Bloch and Jeffries was modified to an arrangement in which the two dees (about 7 cm in radius) were separated by a center electrode about 3 cm wide. Thus, an orbiting proton crossed four gaps per revolution rather than the usual two. The two dees were grounded and an rf voltage V_d was applied between them and the center electrode. The width of the central section was chosen so that the transit time of a proton (or other ion) across it was initially slightly less than one cycle of V_d for eighth-harmonic operation (less than 2 cycles for 16th harmonic operation). Ions crossing the electrode are decelerated, provided their phase is in a certain range relative to the phase of V_d, and spiral into orbits of smaller radii. As the orbit radius decreases, the transit time for crossing the central electrode increases, gradually approaching one cycle of V_d (eighth harmonic) or two cycles (16th harmonic). When this occurs, the net deceleration is zero and an asymptotic or stable orbit is reached. Such a situation is in marked contrast to the Bloch and Jeffries inverse cyclotron, in which the cyclotron orbits continually decrease. The reduction of the orbit radius to a value close to the asymptotic orbit corresponds to

* Preliminary measurements were reported by Collington, Dellis, Sanders, and Turberfield (1955) and Sanders (1957) but will be ignored in favor of the results reported in the final paper.

resonance and is observed by placing a probe outside the position of the asymptotic orbit and varying the frequency of V_d in the region of the eighth or 16th harmonic.

The problem still remains as to what point of the resonance curve corresponds to ω_c. Sanders and Turberfield employed a technique similar to that of Jeffries and Bloch in that plots were obtained of the high-frequency limit of the resonance curve versus V_d. However, because of the presence of a leakage field from the injector, the value of V_d at which the high-frequency limit was equal to ω_c was shifted to higher values of V_d. Thus the theoretical curves which were fitted to the experimental data in order to obtain ω_c had a comparatively small slope, which made it somewhat easier to determine ω_c. The final result of several series of measurements on protons using eighth- and 16th-harmonic operation, several different probe positions, and two injector plates of different thickness for each series is given as $\omega_c/2\pi = 3.580795$ MHz for a magnetic field in which the proton spin precession frequency was 10 MHz. These numbers imply the value 2.792676. However, this is not the final result of the experiment since a +10-ppm field inhomogeneity and relativistic mass correction must be applied to it, as well as a net −1-ppm correction because the proton resonance sample was a cylinder containing a 0.02M solution of $MnSO_4$ rather than a spherical sample of pure water. [As before, this correction may be calculated using the data of Dickinson (1951).] The final value of μ_p'/μ_n is therefore

$$\mu_p'/\mu_n = 2.792701(73) \ (26 \text{ ppm}), \tag{74}$$

where the quoted uncertainty has been converted from the probable error given by Sanders and Turberfield and contains both random error and estimates of possible systematic error. The main error-contributing factor is the 25-ppm uncertainty in the determination of ω_c from the resonance curves and is primarily systematic, since the random uncertainty in ω_c is only

7 ppm. (A similar but less accurate measurement on H_2^+ ions, after mass correction, gave the same result to within 2 ppm.) Although the final error assigned by Sanders and Turberfield may seem rather large in view of the small statistical uncertainty, we believe it is quite reasonable when considered in the light of the experiences of Trigger, Jeffries, and Bloch, and the 13-ppm difference between the final value reported by Sanders and Turberfield (1963) and their earlier measurements (Collington *et al.*, 1955; Sanders, 1957). Sanders and Turberfield's result, Eq. (74), agrees quite well with the two experiments previously discussed.

In 1961, Boyne and Franken (BF) reported a measurement of μ_p/μ_n in which ω_c was determined by measuring cyclotron resonance absorption in a dilute, centimeter-sized cloud of thermal energy ions. A weak rf electric field produced by a marginal oscillator was applied to the cloud normal to the magnetic-field direction. When the frequency of the oscillator was set at ω_c, the ions absorbed energy and the output of the oscillator decreased. The main difficulty of the method is that the observed cyclotron resonance differs from the relation $\omega_c = eB/M_p$ due to electrostatic shifts which are on the order of several hundred parts per million. This situation is similar to that which obtains in the omegatron but is several times worse. Boyne and Franken corrected for the shifts by measuring ω_c for several values of magnetic field in the range 0.8–1.25 T, and then extrapolating to infinite field. This procedure follows from the fact that if the electrostatic field is independent of the magnetic field, then

$$\omega_c^* = \omega_c(1 - \Gamma/B^2), \tag{75}$$

where ω_c^* is the experimentally measured cyclotron frequency, ω_c is the actual frequency, B is the field, and Γ is a function of the electrostatic fields only.

In practice, Boyne and Franken used H_2^+ ions rather than protons for the cyclotron resonance because of their higher cross section for production. (They were generated by electron-impact ionization.) It was also experimentally more convenient to measure the nuclear resonance frequency of deuterons, ω_d, instead of ω_p. Thus, the experimentally determined quantity was $\omega_c(H_2^+)/\omega_d$. This can be converted to the required $\omega_p(H_2O)/\omega_c(H^+)$ using the accurately known mass ratio $M(H^+)/M(H_2^+)$ and the frequency ratio ω_p/ω_d via the equation

$$\frac{\mu_p'}{\mu_n}=\frac{\omega_p(H_2O)}{\omega_c(H^+)}=\left(\frac{\omega_d}{\omega_c(H_2^+)}\right)\left(\frac{\omega_p(H_2O)}{\omega_d}\right)\left(\frac{M^*(H^+)}{M^*(H_2^+)}\right).$$

The final result of 24 separate runs is given by Boyne and Franken as $\omega_c(H_2^+)/\omega_d=1.165956$, where ω_d was obtained for deuterons in a cylindrical D_2O sample doped with $CuCl_2\cdot 2H_2O$ (0.33M solution). The ratio of ω_d in this sample to ω_p in a 1:1 cylindrical sample containing a 0.01M $CuCl_2$ solution was measured in a separate experiment and found to be $\omega_p/\omega_d=6.514411\pm$ 0.5 ppm. (The difference between this proton sample and an ideal spherical water sample is only a few tenths of a part per million and can be ignored.) Combining these measurements with the value $M(H_2^+)/M(H^+)=$ 2.00054463, which follows from the value of M_p/m_e used in Sec. II.B.7, yields

$$\mu_p'/\mu_n=2.792832(55)\ (20\text{ ppm}). \qquad (76)$$

The quoted error is that given by Boyne and Franken and arises from the determination of ω_c. It has three sources: (1) line-shape symmetry, 3 ppm, (2) magnetic-field inhomogeneity, 5 ppm, (3) possible deviations from linearity and the slope–intercept correlation effect, 18 ppm. Clearly, the uncertainty is mainly due to item (3) (to be discussed below).

In comparing the result of Boyne and Franken with that of Sommer, Thomas, and Hipple, Eq. (72), it is immediately evident that the two measurements are in serious disagreement. Indeed, they differ by 51 ppm as compared with the 22-ppm standard deviation of their difference. The probability for this to occur by chance is about 2%. In view of the large electrostatic shifts in the experiment of Boyne and Franken and the required extrapolation from the comparatively low magnetic fields used to infinite field, this work is suspect. Indeed, a disturbing correlation was observed between the slope of the linearly extrapolated straight line obtained for each run and the intercept of the line which determines ω_c. In order to understand this correlation, Boyne and Franken considered several physically realistic models which might cause the electrostatic-field term Γ in Eq. (75) to vary like $\Gamma = \Gamma_0 + \Gamma_1 B$ or $\Gamma_0 + \Gamma_1 B^{-1}$. They concluded that shifts in the extrapolated intercepts of ω_c/ω_d and hence μ_p'/μ_n as large as 15 ppm could be produced without introducing experimentally observable curvatures in the linear portions of the data. On this basis, they estimated that such effects could not produce an error greater than 25 ppm without conflicting with the large majority of their data and that the 18-ppm uncertainty assigned for the correlation effect was a realistic 70% confidence-level estimate. However, Petley (private communication) claims that if Γ is assumed to vary like $\Gamma_0 + \Gamma_1 B$, the data of Boyne and Franken, including that from rejected runs, can be fitted well within the experimental scatter with an intercept which yields the Sommer *et al.* value of μ_p'/μ_n (or any intermediate value) with little difficulty. In any case, this problem significantly reduces the confidence which can be placed in the experiment of Boyne and Franken.

A measurement of μ_p/μ_n employing a so-called magnetic resonance mass spectrometer was reported in 1965 by Mamyrin and Frantsuzov (1968; 1965; 1964) of the USSR. In this device, ions are accelerated while traversing a three-element modulator consisting of three screen grids. The two outer grids are grounded and an rf voltage applied to the central grid. In their first traversal of the modulator, ions having a correct phase relationship with the applied voltage receive a velocity increment such that they can pass through a slit after traversing half an orbit. These ions then traverse the second half of their orbit, reenter the modulator, receive a second velocity increment, and exit to a collector through a second slit laterally displaced from the first. If the frequency of the oscillator, ω_{osc}, equals $n\omega_c$, where n is an integer (about 200 in these experiments), then the ions make their second pass through the modulator in the same phase with the rf voltage as for their first pass, thereby receiving the identical velocity increment. As a result, only if the second slit is correctly placed with respect to the first will the ions be able to enter the collector. The cyclotron frequency is measured for only a single ion revolution (i.e., from the first to the second pass through the modulator) as compared with several thousand revolutions in the omegatron and several hundred in the cyclotron. Nevertheless, the resonance curve is narrower than in either of these instruments. The advantage of the single orbit is that the entire trajectory of an ion can be traced and an accurate theory constructed to correct for the effects of perturbing electric and magnetic fields.

In the resonance spectrometer of Mamyrin and Frantsuzov, it was found that the relationship between ω_{osc} and ω_c was actually $\omega_{osc}=(n+k)\omega_c$, where $k \ll 1$ and is independent of n. It arises because the second velocity increment is not strictly identical to the first and because of the finite length of time spent in the modulator. The quantity k was theoretically calculated from six transcendental equations which are the exact equations of motion of the ions. It was found that $k=-0.0028$, assuming the ions traversed the modulator in $\frac{3}{4}$ of an rf half-cycle. For $n=196$, this meant that the two outer grids had to be separated by 1.35 mm, and for $n=98$, by 2.70 mm. Since k is so small and n so large, the relationship between ω_{osc} and ω_c may be written as $n\omega_c=\omega_{osc}(1-k/n)$. This equation provides a means for checking the calculation of k, since for different n, the oscillator frequency and grid spacing must be varied accordingly in order to get resonance at ω_c.

Shifts in ω_c were also observed due to electric fields arising from space-charge and contact-potential differences. They varied between −3 and +20 ppm, considerably less than those observed in the omegatron of Sommer *et al.* The shifts were corrected by exactly the same method used in the omegatron experiments, i.e., by measuring ω_c for ions of different mass and extrapolating to zero mass. (Mamyrin and Frantsuzov used ions of He and Ne.) Again, this is only valid if the electric fields are the same for the different ions. The final result of 13 series of measurements made during four runs is reported to be (Mamyrin and Frantsuzov, 1968; 1965)

$$\mu_{p}'/\mu_n = 2.792794(17) \ (6.2 \text{ ppm}), \qquad (77)$$

where the quoted uncertainty is that given by Mamyrin and Frantsuzov and is the RSS of a 3.7-ppm random error and estimates of individual systematic-error components totaling 4.9 ppm RSS (Mamyrin and Frantsuzov, 1968; 1965). Although the proton resonance sample was a weak aqueous solution of $CuCl_2$, it was spherical in shape and thus no correction is required. [The "q" term of Dickinson (1951) contributes less than 0.1 ppm and can be ignored.]

In comparing this result with the others already discussed, we see that it agrees with the measurement of Boyne and Franken (which is thought to be suspect for the reasons previously given), is in mild disagreement with the measurement of Trigger and of Sanders and Turberfield, and is in significant disagreement with the omegatron measurement of Sommer *et al.*; the result of Mamyrin and Frantsuzov exceeds that obtained with the omegatron by 37 ppm as compared with the 12-ppm standard deviation of the difference. The probability for this to occur by chance is about 0.25%. Mamyrin and Frantsuzov were quite aware of these discrepancies and carefully searched for systematic errors by performing various geometrical, electrical, and magnetic checks. No such errors were uncovered and the discrepancy remains. We shall investigate its implications in Sec. III.

In 1967, a measurement of μ_p/μ_n was reported by Petley and Morris (PM) (1968a; 1967) at NPL.

These workers used an omegatron very similar to that of Sommer *et al.* but of the quadrupole type with hyperbolic electrodes (Petley and Morris, 1968b; 1965). This arrangement increases the resolution of the omegatron by increasing the time of flight of the resonant ions. The time of flight is usually limited by the drift of the center of the ion orbit in the plane normal to the magnetic field. The drift causes the ions to hit the wall rather than the collector and is due to the trapping voltage which is required to prevent ion loss arising from motion in the direction of the magnetic field. The hyperbolic omegatron gives the ions slower drift velocities for the same trapping voltage, thus increasing resolution. However, departures from the hyperbolic form can lead to resonance asymmetry and a resonance width which is slightly dependent on the direction of frequency scan, and to a lesser extent, line-splitting and harmonic effects. These last two effects can arise from the nonuniform-rf-field characteristic of this omegatron. (Splitting the hyperbolic electrodes helps to improve the uniformity of the rf field.)

In their work, Petley and Morris measured ω_c for hydrogen ions (H_2^+), deuterated hydrogen ions (HD^+), and deuterium ions (D_2^+) and corrected for shifts in ω_c due to electrostatic fields by extrapolating to zero mass in the usual way. [Again, conversion to ω_c for protons introduces no additional uncertainty because of the high accuracy ($\sim$0.1 ppm) of the required mass ratios.] One of the main problems in this work was the $\sim$40-ppm skewness observed in the resonance curves due to variations of the electrostatic and magnetic fields over the ion orbits. Petley and Morris corrected for this effect by calculating values of μ_p/μ_n from the 10%, 50%, and 90% amplitude points of each resonance curve for each of the three ion species and extrapolating to zero mass for each of the three amplitude points separately. A linear equation giving the zero mass value of μ_p/μ_n as a function of per cent amplitude was then fitted to these three points. Another problem was that the position of the ion resonances shifted by about 18 ppm when the current in the omegatron filament was reversed. Petley and Morris corrected for this effect by fitting a linear equation giving the zero mass value of μ_p/μ_n vs percent amplitude to data obtained for each direction of filament current and taking the mean of the two equations. The value given by this mean equation evaluated at the 100% amplitude point yielded (after further correction amounting to 16.7 ppm) the final value of μ_p/μ_n,

$$\mu_p'/\mu_n = 2.792746(52) \ (19 \text{ ppm}). \tag{78}$$

Petley and Morris actually used a cylindrical proton resonance sample containing oil. However, they corrected to a spherical water sample using their measured value of 3.5 ppm for the difference in shielding between oil and water, and the usual -1.5 ppm correction for sample shape as obtained from $(B_i - B_0)/B_0 = [(4\pi/3) - 2\pi]\chi_{H_2O}$ (Dickinson, 1951).

The quoted 19-ppm standard-deviation error is that given by Petley and Morris and includes both random and systematic error. The three major contributions to the error were uncertainties in (1) the corrections for magnetic-field inhomogeneity over the ion orbits, 12 ppm, (2) the zero-mass extrapolation used to eliminate resonance shifts due to electrostatic fields, 10 ppm, and (3) screening of the applied magnetic field by the omegatron electrodes, 7 ppm. The RSS of these three errors alone is 17 ppm, very nearly equal the final 19 ppm quoted for the experiment. In the omegatron experiment of Sommer, Thomas, and Hipple, the magnetic-field inhomogeneity was on the order of 10 ppm, somewhat better than the 30 ppm in the Petley and Morris experiment. The electrostatic shifts were also about 25% less in the Sommer *et al.* experiment, and the total electrode shielding effect was measured to be less than 2 ppm as compared with the 8 ppm observed by Petley and Morris. Sommer *et al.* also observed only about half the shift in the resonance curve with filament current direction observed by Petley and Morris. In view of these differences, we believe the uncertainties assigned to the two experiments reasonably reflect their relative accuracies.

The result of Petley and Morris lies midway between that of Sommer, Thomas, and Hipple and that of Mamyrin and Frantsuzov, being 20 ppm larger than the first and 17 ppm smaller than the second. Since the assigned error is 19 ppm, it is not in disagreement with either of these experiments and thus does not give much information concerning which of the two values of μ_p/μ_n may be more nearly correct.

The last value of μ_p/μ_n to be considered is that reported by Marion and Winkler (1967; Marion, 1968). These workers noticed that there were slight differences in the values of certain proton–nuclear reaction energies obtained with their time-of-flight beam analyzer (or velocity gauge) at the University of Maryland and values obtained at the University of Zurich with an absolute magnetic analyzer. These differences could be traced to the effective value of μ_p/μ_n used in the analysis of the magnetic-analyzer data, and this led to the realization that the two experiments combined provided a new means of measuring μ_p/μ_n. The velocity gauge measures the velocity v_p of the incident protons via the relation $v_p=D/T$, where D is the distance traveled and T is the flight time. The magnetic analyzer determines the magnetic rigidity $B\rho$ of the protons, which is closely related to v_p since

$$B\rho=\frac{M_p v_p}{e[1-(v_p/c)^2]^{1/2}}. \tag{79}$$

Here, ρ is the orbit radius of curvature.

In practice, B is measured in terms of the spin precession frequency of protons, ω_p'. When B and v_p are eliminated from Eq. (79) via the relations $\hbar\omega_p'=2\mu_p'B$, $\mu_n=e\hbar/2M_p$, and $v_p=D/T$, it becomes

$$\frac{\mu_p'}{\mu_n}=\frac{\rho\omega_p'[1-(D/cT)^2]^{1/2}}{(D/T)}. \tag{80}$$

With the exception of c, all of the quantities contained

on the right side of Eq. (80) are measured in the two experiments.

Using the original data obtained from magnetic-analyzer and velocity-gauge measurements of v_p for the $^{27}Al(p, \gamma)^{28}Si$ resonance (992 keV), and the $^{7}Li(p, n)^{7}Be$ neutron threshold (1881 keV), Marion and Winkler (1967) deduced the value $\mu_p'/\mu_n = 2.79259(12)$ (44 ppm). However, Marion (private communication) at the University of Maryland has recently obtained new velocity-gauge data and gives as a final result

$$\mu_p'/\mu_n = 2.79260(13) \text{ (45 ppm)}. \tag{81}$$

The error quoted is that given by Marion and is intended to be one standard deviation. The difference between the value of μ_p'/μ_n given in Eq. (81) and the effective value used in obtaining reaction energies from the magnetic analyzer explains why the two methods originally gave different results.

While the relatively large uncertainty assigned the Marion and Winkler value of μ_p'/μ_n [Eq. (81)] precludes its use in our least-squares adjustments, it is in agreement with (and therefore gives added support to) the lower values of μ_p'/μ_n as obtained by Sommer *et al.*, Trigger, Sanders and Turberfield, and Petley and Morris. Future reaction-energy measurements are planned which may result in a value of μ_p'/μ_n having an uncertainty of only 20 ppm. Such a measurement would be of some importance in view of the differences between the lower values of μ_p'/μ_n and the higher values obtained by Boyne and Franken and by Mamyrin and Frantsuzov. Figure 3, in which we graphically compare all of the measurements of μ_p'/μ_n we have discussed (as well as others to be discussed shortly), clearly shows these differences.

The observational equation for μ_p/μ_n is very similar to that for γ_p and follows from Eqs. (67) and the fact that $\mu_n \equiv e\hbar/2M_p$ and that $NM_p = M_p^*$. The final result is

$$(\alpha^{-1})^{-3}e^{-2}K^0N^{-1} = \frac{\mu_0 R_\infty}{(\mu_p'/\mu_B)} \frac{1}{M_p{}^*} \frac{\mu_p'}{\mu_n}. \qquad (82)$$

(See also Notes Added in Proof.)

6. X-Ray Experiments

There is a large body of data from the field of x rays which bears on the fundamental constants. Unfortunately, much of it is inconsistent and of relatively large uncertainty. As a result, no such data will be included in our final adjustment to obtain the best WQED value of α nor in our final adjustment to obtain the best values of all of the constants. However, we would like to extract from the various x-ray measurements any information which might be contained therein concerning Avogadro's number. This could be quite important, since such information would shed some light on the discrepancies among the μ_p/μ_n data discussed in the previous section. We would also like to obtain a best or recommended value for Λ, the angstrom-to-kx-unit conversion factor. We shall therefore briefly review what is considered to be the best of the available x-ray data. However, our reevaluation of these experiments has not been as critical or thorough as for the other stochastic input quantities because of the limited usefulness of the x-ray work. We have leaned heavily on the reviews of Bearden and coworkers which were made in conjunction with their compilation of new tables of x-ray wavelengths and energy levels (Bearden, 1967; Bearden and Burr, 1967). [See also Cohen and DuMond (1965).]

High-accuracy x-ray measurements have long been plagued by ambiguities in the definition of the fundamental x-ray unit of length.* Such a unit is necessary because x-ray wavelengths can be compared with great accuracy, but measurements of wavelengths in absolute

---

* For a recent review of the history of the x-unit, see Thomsen and Burr (1968).

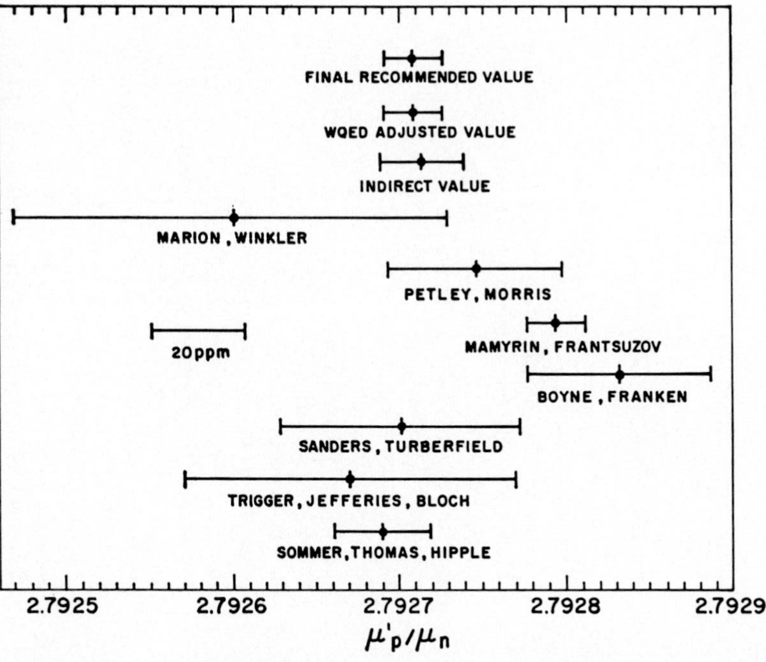

FIG. 3. Graphical comparison of the several direct measurements of μ_p'/μ_n listed in Table XIV and the indirect value derived in Sec. III from the Faraday. The error bars indicate one standard deviation. The measurements are ordered chronologically with the oldest at the bottom. Also shown are the WQED and final recommended adjusted values (see Secs. III.C and V.B).

length units is very difficult. An arbitrarily defined unit, called the x-unit (xu) and intended to be very nearly 10^{-11} cm, was therefore introduced into the field by Siegbahn nearly half a century ago [see, for example, M. Siegbahn (1925)]. It plays the same role as the as-maintained electrical units, and a conversion constant is required to relate the x-unit to the absolute length unit just as K is required to relate the as-maintained NBS ampere to the absolute ampere. The conversion constant is called Λ and is defined as the ratio of a wavelength expressed in angstroms (10^{-10} m) to the same wavelength expressed in kilo-x-units (kxu),

$$\Lambda \equiv \lambda(\text{Å})/\lambda(\text{kxu}).$$

We shall see that $\Lambda \approx 1.002$, so this definition implies that the kxu is about 1.002 times larger than the angstrom.

The choice of an operational definition of the x-unit has been subject to considerable controversy and confusion. For many years, x-ray wavelengths were measured in terms of the lattice spacing of calcite, i.e., the x-unit was defined by assigning the value $d_1 = 3.02904$ kxu for the first-order lattice constant of calcite at 18°C (Sandström, 1957). The flaw here of course is that every "good" calcite crystal does not have the same lattice constant. Variations in lattice constant as large as 20 ppm occur even among highly selected crystals, and similar variations may occur from point to point within a single crystal (Bearden, 1965b). It therefore became the practice to use specific x-ray emission lines as wavelength standards. Two supposedly equivalent working definitions of the x-unit emerged, one based on $\lambda(\text{Mo } K\alpha_1) \equiv 0.707831$ kxu and the other on $\lambda(\text{Cu } K\alpha_1) \equiv 1.537400$ kxu. Recent precision measurements of Bearden and coworkers (Bearden, 1965a; 1965b; Bearden, Henins, Marzolf, Sauder, and Thomsen, 1964; Henins and Bearden, 1964) and by Cooper (1965) have shown that these two definitions are incompatible. They found that if $\lambda(\text{Cu}K\alpha_1)$ is assigned the value 1.537400 kxu, then the experimental value of $\lambda(\text{Mo}K\alpha_1)$ is 0.7078448 kxu, a change of 20 ppm. In an attempt to clear the air, Bearden proposed that the x-unit be replaced by a completely new unit, the "angstrom star" (Å*), based on assigning $\lambda(\mathrm{W}K\alpha_1) \equiv 0.2090100$ Å* (Bearden, 1965a). The tungsten $K\alpha_1$ line was chosen for narrowness, symmetry, and reproducibility. The assigned numerical value was based on the wavelength measurements of Bearden, Henins *et al.* (1964), and Bearden's (1965b) evaluation of Λ based primarily on Cohen and DuMond's 1963 adjusted values of the constants and was chosen to make Å* equal to Å. The conversion factor Λ^* relating the two is defined like Λ, i.e.,

$$\Lambda^* = \lambda(\text{Å})/\lambda(\text{Å}^*).$$

According to Bearden (1965b), $\Lambda^* = 1.000000 \pm 5$ ppm (P.E.). However, we shall see that the results of the present adjustment imply that Λ^* may actually exceed unity by as much as 20 ppm. It would thus appear that there is no particular advantage is using Å* and Λ^* in the present discussion, especially since much of the experimental data in the literature refers to the x-unit defined by $\lambda(\mathrm{Cu}K\alpha_1) \equiv 1.537400$ kxu.* Thus, although the older definition of the x-unit based on taking the calcite grating constant to be 3.02904 kxu lies somewhere between the $\mathrm{Cu}K\alpha_1$ and $\mathrm{Mo}K\alpha_1$ definitions of the x-unit (Thomsen, private communication), we shall adopt the $\mathrm{Cu}K\alpha_1$ scale for use here.

In general, the various x-ray experiments fall into one of four distinct categories which can be characterized as measurements of either hc/e, N, Λ, or λ_C, the Compton wavelength of the electron. The most important of these measurements are summarized in Table XV and will now be briefly discussed.

(*a*) *Measurements of hc/e.* In principle, hc/e can be

---

* This is not to say that the new scale is without merit. On the contrary, it represents the first x-ray scale actually defined in terms of a particular x-ray wavelength and as such is reproducible to within a few parts per million. Furthermore, Bearden's new table of x-ray wavelengths based on the Å* unit probably represents the most consistent table of its kind ever produced.

determined by measuring the shortest-wavelength x-ray emitted by an x-ray tube with an accelerating voltage V across it. This follows from the simple relation $eV=h\nu=hc/\lambda$ or $hc/e=V\lambda$. The quantity $V\lambda$ is often referred to as the voltage–wavelength conversion product and λ the short-wavelength limit or SWL of the continuous x-ray spectrum. In practice, λ is measured in kx-units and V in terms of an as-maintained volt, for example, that of NBS. There are two experimental difficulties which tend to limit the accuracy of hc/e as determined from SWL experiments. First, the electron energy is not simply given by the voltage measured across the tube because contact potentials and the initial thermal energy of the electrons also contribute. Secondly, in a solid-target x-ray tube, fine structure associated with the electron energy band structure of the target material is observed at the short-wavelength cutoff, making it very difficult to define the cutoff accurately. Spijkerman and Bearden (1964) attempted to circumvent this second difficulty by using a Hg vapor jet as a target (Bearden, Huffman, and Spijkerman, 1964). The technique proved quite successful and a simple empirical relation was developed to describe the shape of the observed isochromat (a plot of the variation of x-ray intensity for a given wavelength near cutoff as a function of tube voltage). However, the use of a gas target did require an additional correction for the work function of the anode. The final result of the Spijkerman and Bearden experiment is

$$hc/e=12373.15\pm0.41\ \mathrm{V_{NBS}\cdot kxu}\ (33\ \mathrm{ppm}), \qquad (83)$$

where we have converted the original probable error to a standard deviation. This value is the weighted average of four separate measurements made with two different tubes, and has been reexpressed in terms of the NBS volt using the original conversion factor adopted by these authors to convert from the NBS volt to the absolute volt.

A measurement of the same quantity but by a rather different method has been carried out by Hagström, Hörnfeldt, Nordling, and Siegbahn (1962). Their technique consists of photoejecting electrons from the same shell in a converter material using the characteristic $K\alpha$ radiation of two different elements. The photoelectrons are brought to a common focus in a beta spectrometer by accelerating or retarding them with a voltage V. Thus, the experiment measures the difference in energy between the two x-ray lines, that is, $eV = h(\nu_1 - \nu_2) = hc(\lambda_1^{-1} - \lambda_2^{-1})$ or $hc/e = V\lambda_1\lambda_2/(\lambda_2 - \lambda_1)$. Hagström *et al.* carried out their experiment by ejecting electrons from the K shell of Mn using Cu$K\alpha_1$ and Mo$K\alpha_1$ radiation. Taking $\lambda(\text{Mo}K\alpha_1) = 0.7078448$ kxu and $\lambda(\text{Cu}K\alpha_1) = 1.537400$ kxu, we find that the result of their experiment (average of four runs) is

$$hc/e = 12373.35 \pm 1.5 \text{ V}\cdot\text{kxu (120 ppm)}. \qquad (84)$$

The error quoted is that given by Hagström *et al.* but may be closer to twice the probable error rather than a standard deviation. The experiment was of a preliminary nature and consequently no attention was paid to the difference between absolute and as-maintained electrical units. We shall therefore not use this result here. However, it may be noted that the measurement is in surprisingly good agreement with the work of Spijkerman and Bearden, thereby providing some added support for that experiment.

The observational equation for hc/e is very similar to that for the quantity $2e/h$ as measured by the ac Josephson effect. From the definition of the fine-structure constant, $\alpha = (\mu_0 c^2/4\pi)(e^2/\hbar c)$, it follows that

$$(\alpha^{-1})^1 e^1 = (2/\mu_0 c^2)(hc/e).$$

This expression is valid only if hc/e is expressed in terms of absolute units. To convert it to a form which is valid for hc/e measured in terms of the NBS volt and kx-unit, we introduce the conversion constants K, Λ, and also the quantity $\Omega_{\text{ABS}}/\Omega_{\text{NBS}}$. The final result is

$$(\alpha^{-1})^1 e^1 K^{-1} N^0 \Lambda^{-1} = \frac{2}{\mu_0 c^2} \frac{10^{-10}}{\Omega_{\mathrm{ABS}}/\Omega_{\mathrm{NBS}}} \left(\frac{hc}{e}\right)_{\mathrm{NBS,kxu}}, \quad (85)$$

where $(hc/e)_{\mathrm{NBS,kxu}}$ is expressed as measured, i.e., in NBS volts and kx-units.

(*b*) *Measurements of N*. In principle, Avogadro's number N can be obtained from the simple relation $N = fM/\rho d^3\Phi$, where M is the molecular weight, ρ is the density, f is the number of atoms per unit cell, d is the lattice constant, and Φ is a dimensionless geometrical factor, all characteristic of a particular crystal species with an assumed ideally perfect lattice. (This technique is often referred to as the x-ray crystal-density method or XCRD.) The quantity Φ depends on the shape of the unit cell and is unity for cubic crystals. For noncubic crystals, it can be obtained from x-ray measurements of the angles between various atomic planes, but does not require knowledge of the wavelength of the x-ray used. However, the lattice constant d in the above equation is in absolute units, while in practice, d is measured in kx-units. The relationship becomes therefore

$$N\Lambda^3 = fM/\rho d_{\mathrm{kxu}}{}^3\,\Phi,$$

and the experiment actually measures $N\Lambda^3$ rather than N.

The most precise measurement of $N\Lambda^3$ has been carried out by Henins and Bearden using Si (Henins and Bearden, 1964; Henins, 1964). These workers made 46 measurements on a wide variety of Si samples using Cu$K\alpha_1$ radiation and 34 similar measurements using Cu$K\alpha_2$ radiation. The two series gave the same result, and the final value is*

$$N\Lambda^3 = 6.059768(95) \times 10^{26}\ \mathrm{kmole}^{-1}\ (16\ \mathrm{ppm}),$$

* These authors actually give a value for Λ as calculated using $N = 6.02252 \times 10^{26}$ kmole$^{-1} \pm 11$ ppm (P.E.), the 1963 adjusted value (Cohen and DuMond, 1965). The result we give for $N\Lambda^3$ was obtained by working backwards from their Λ and this N.

TABLE XV. Summary of various x-ray measurements (SWL, short-wavelength limit; ΔE, energy difference between two x-ray lines; XCRD, x-ray crystal density; PRG, plane ruled grating; CRG, concave ruled grating; EPA, electron–positron annihilation).

Publication date and authors	Type of experiment	Quantity measured	Value	Error (ppm)
1964, Spijkerman and Bearden	SWL	hc/e	12373.15(41) $V_{NBS}\cdot$kxu	33
1962, Hagström, Hörnfeldt, Nordling, and Siegbahn	ΔE	hc/e	12373.35(1.5) V·kxu	120
1965, Bearden	XCRD (calcite)	$N\Lambda^3$	6.05972(23) $\times 10^{26}$ kmole^{-1}	37
1964, Henins and Bearden	XCRD (Si)	$N\Lambda^3$	6.059768(95) $\times 10^{26}$ kmole^{-1}	16
1955–1957, Smakula, Kalnajs, and Sils [revised, Henins and Bearden (1964)]	XCRD (various)	$N\Lambda^3$	6.06002(39) $\times 10^{26}$ kmole^{-1}	65
Various [revised, Bearden (1965a); Henins and Bearden (1964)]	XCRD (Mo$K\alpha_1$–various)	$N\Lambda^3$	6.05966(53) $\times 10^{26}$ kmole^{-1}	87
1931, Bearden [see Bearden (1965b)]	PRG	Λ	1.002030(50)	50
1938–1940, Tyrén [revised, Edlén and Svensson (1965)]	CRG	Λ	1.002030(38)	38
1962–1964, Knowles	EPA(H_2O)	λ_C	24.21263(92) $\times 10^{-3}$ kxu	38
	EPA(Ta)	λ_C	24.21421(36) $\times 10^{-3}$ kxu	15

where we have converted from a P.E. to a standard deviation. The main error-contributing factor is the 15-ppm uncertainty in the molecular weight of Si which is due to the uncertainty in the relative abundances of its isotopes. The uncertainty in the lattice constant in kx-units is less than 1 ppm while the error in the crystal density is only 4.5 ppm (Henins, 1964). Since Si is cubic, $\Phi \equiv 1$. [Henins and Bearden's result was calculated using atomic masses as given in the 1961 mass tables (Konig, Mattauch, and Wapstra, 1962). For Si, the differences between the 1961 values and those given in the 1964 tables (Mattauch, Thiele, and Wapstra, 1965) are about 0.1 ppm, which is negligible.]

A similar experiment was carried out by Bearden (1965a) on calcite with the result

$$N\Lambda^3 = 6.05972(23) \times 10^{26} \text{ kmole}^{-1} \text{ (37 ppm)}.$$

Since calcite is rhombohedral, Φ had to be measured separately. Corrections also had to be made for the rather large impurity content of the crystals used. Nevertheless, the results on Si and calcite are in excellent agreement.

Henins and Bearden (Bearden, 1965a; Henins and Bearden, 1964) have also reevaluated and updated the data of Smakula and coworkers (Smakula and Kalnajs, 1957; 1955; Smakula and Sils, 1955; Smakula, Kalnajs, and Sils, 1955) obtained on Al, CaF_2, CsI, Ge, TlCl, TlBr, and Si. Using the definition $\text{Cu}K\alpha_1 = 1.537400$ kxu and the atomic weights given by Cameron and Wichers (1962), they find*

$$N\Lambda^3 = 6.06002(39) \times 10^{26} \text{ kmole}^{-1} \text{ (65 ppm)}.$$

A similar reevaluation of data obtained by several different workers with $\text{Mo}K\alpha_1$ radiation on calcite, diamond, and quartz gives*

$$N\Lambda^3 = 6.05966(53) \times 10^{26} \text{ kmole}^{-1} \text{ (87 ppm)}.$$

It is apparent that all of the measurements agree within their assigned uncertainties, but of course these are quite large. In fact, the uncertainties assigned these last two values are so large compared with that assigned the Si result of Henins and Bearden that we shall not use them in our adjustments.

The observational equation for the quantity $N\Lambda^3$ is obviously

$$(\alpha^{-1})^0 e^0 K^0 N^1 \Lambda^3 = N\Lambda^3, \tag{86}$$

that is, $N\Lambda^3 = N\Lambda^3$.

(*c*) *Measurements of* Λ. There have been several direct measurements of Λ over the years, but these experiments are very difficult and cannot be considered reliable. Bearden (1931) carried out several determinations of Λ using Cu$K\beta_1\beta_3$, Cu$K\alpha_1\alpha_2$, Cr$K\beta_1\beta_3$, and Cr$K\alpha_1\alpha_2$ radiation. The technique employed was that of measuring the angles of incidence and diffraction of the x rays on a plane ruled grating. The absolute values of the x-ray wavelengths so obtained were then compared with their values in kx-units, thereby determining Λ. The mean of the four measurements on the Cu$K\alpha_1$ scale we are using is given by Henins and Bearden (1964) as

$$\Lambda = 1.002030 \pm 50 \text{ ppm}.$$

Henins and Bearden also point out that a similar measurement by Bearden (1935) using a large grating and a double crystal spectrometer is highly suspect on theoretical grounds and that a measurement by Bäcklin (1935) cannot be trusted because of the uncertain chemical condition of the aluminum emitter. These two measurements will therefore be ignored. [The values quoted (Henins and Bearden, 1964) for the two experiments are, respectively, $\Lambda = 1.00208$ and $\Lambda = 1.00199$; they differ by 90 ppm.]

Another direct determination of Λ should be mentioned. Tyrén (1938; 1940) measured the Al x-ray lines at about 8.3 Å using a 5-m concave grating spectrograph covering the range 20–100 Å. The reference wavelengths consisted of Lyman-series lines of the hydrogenlike spectra of highly ionized atoms. The wavelengths of these lines were not measured directly but calculated from the Dirac theory. However, they were not corrected for the then unknown Lamb shift.* Recently, Edlén and Svensson (1965) (Edlén, 1966) remeasured the plates of Tyrén and recalculated the wavelengths of the reference lines taking into account the Lamb shift. (Note that the reference lines themselves are doublets and the effective wavelength is the weighted average of the two components.) The result of this remeasurement is a value in angstroms for each of the wavelengths Al$K\alpha_1\alpha_2$, Al$K\alpha_3$, and $Al_2O_3K\alpha_1\alpha_2$. Edlén and Svensson compared these wavelengths with the corresponding wavelengths in kx-units as determined by Nordfors (1956) and obtained as a weighted average $\Lambda = 1.002060 \pm 23$ ppm, very close to the value $\Lambda = 1.002064 \pm 30$ ppm obtained from just the Al$K\alpha_1\alpha_2$ measurement. But the difficulty here is that there is a large degree of uncertainty associated with the kx-unit values of the Al wavelengths (even the kx-unit scale used is uncertain). Indeed, there are values in the literature which differ by as much as 100 ppm [see Cohen and DuMond (1965) for a further discussion]. If the values given by Bearden in his wavelength tables (Bearden, 1967) are used for Al$K\alpha_1$ and Al$K\alpha_2$ and if these two components are weighted in the usual 2:1 ratio in order to obtain Al$K\alpha_1\alpha_2$, then Λ turns out to be 1.002030, identical with the Bearden plane-grating result but 34 ppm less than the value 1.002064 implied by the work of Nordfors. (In making this calculation, we have converted the wavelengths given in Bearden's tables back to the scale on which Cu$K\alpha_1$= 1.537400 kxu by using his adopted value $\Lambda = 1.002056$.) It is quite clear that very little faith can presently be placed in the Tyrén–Edlén–Svensson result (or for that matter, in any of the direct determinations of Λ). For purposes of computation, we shall adopt the value implied by the wavelengths given in Bearden's tables,

* This was first pointed out by DuMond and Cohen (1956).

$$\Lambda = 1.002030 \pm 38 \text{ ppm},$$

since these tables represent the most consistent set of x-ray wavelengths presently available. The quoted uncertainty has been obtained by taking the RSS of the probable error of the composite Al$K\alpha_1\alpha_2$ line computed from the errors given by Bearden (Bearden, 1967) and the probable error of the Al$K\alpha_1\alpha_2$ line as given by Eldén and Svensson and multiplying by 1.48. The complexity of Tyrén's experiment is emphasized by the many problems which have arisen in a recent attempt by Kirkpatrick and co-workers to repeat it (Kirkpatrick, DuMond, and Cohen, 1968).

The observational equation for Λ is obviously just $\Lambda = \Lambda$ or

$$(\alpha^{-1})^0 e^0 K^0 N^0 \Lambda^1 = \Lambda. \tag{87}$$

(*d*) *Measurements of* λ_C. The Compton wavelength of the electron, $\lambda_C = h/m_e c$, can be obtained by measuring the wavelength of either of the two gamma rays emitted when an electron–positron pair having zero kinetic energy annihilate (the velocity of the center of mass must be zero). Such a situation nearly obtains when positrons annihilate with electrons in a solid since the probability for annihilation is large only when the relative velocity of the electron–positron pair is small. Under these circumstances, conservation of energy and momentum yield for the wavelength of the emitted γ rays, $2hc/\lambda = 2m_e c^2$ or $\lambda = h/m_e c \equiv \lambda_C$. In practice, λ_C is measured in kx-units rather than absolute units and the relation becomes

$$h/m_e c = 10^{-10} \Lambda (\lambda_C)_{\text{kxu}}. \tag{88}$$

There have been two relatively accurate measurements of λ_C by Knowles (1964; 1962a; 1962b) but, unfortunately, they are not in very good agreement. In Knowles' first experiment, the annihilation occurred in H_2O and the diffraction angle of the annihilation radiation was measured by a double plane crystal spectrometer. The wavelength of the radiation is related to the measured angle θ by the Bragg equation $\lambda = 2d_{211} \sin \theta_B$, where d_{211} is the lattice spacing of the calcite crystals employed and θ_B is one-half the measured angle. Knowles obtained $\theta = 7992.41 \pm 0.30$ μ rad. The lattice constant of the calcite used by Knowles was measured by Bearden and found to be 3.029463 kxu on our Cu$K\alpha_1$ scale (Knowles, private communication). The result of this experiment is therefore

$$\lambda_C = 24.21263(92) \times 10^{-3} \text{ kxu (38 ppm)},$$

where the quoted one-standard-deviation error is that given by Knowles.

In a subsequent experiment with the same type of spectrometer, the wavelength of the radiation resulting from positron annihilation in Ta was compared with a neutron-capture γ ray of ^{182}Ta at 171 keV using diffraction from higher-order crystal planes. In a separate experiment, the 171-keV γ ray was compared with W$K\alpha_1$ radiation generated by internal conversion in ^{182}W. The result of the experiment is the ratio $9\lambda_C/\lambda(\text{W}K\alpha_1) = 1.044811 \pm 15$ ppm. According to Bearden (Bearden, 1967; Henins and Bearden, 1964), on our present x-unit scale $\lambda(\text{W}K\alpha_1) = 0.2085811$ kxu. It therefore follows that

$$\lambda_C = 24.21421(36) \times 10^{-3} \text{ kxu (15 ppm)},$$

where the error is again that given by Knowles. This Ta result exceeds the value obtained for H_2O by 65 ppm, 1.6 times the 41-ppm standard deviation of the difference. The probability for this to occur by chance is about 11%. The origin of this discrepancy, if one may call it that, is presently unknown. It may result from systematic errors in the determination of the average value for d_{211} (Knowles, private communication). This was done by Bearden using millimeter-sized samples taken from one of the two 2 in.×2 in.×1 in. spectrometer crystals used by Knowles. Knowles and also DuMond (Knowles, private communication) have tried to explain the discrepancy as a difference in binding energy of the positron–electron pair in the different materials, but could only conclude that the uncertainty of pair binding is a basic limitation of the method.

The observational equation for $\lambda_C = h/m_e c$ can be readily derived by forming the ratio α^2/R_∞ [see Eq. (67)]; the result is $2\lambda_C = 2h/m_e c = \alpha^2/R_\infty$. If Eq. (88) is used to convert to an expression valid for λ_C expressed in kx-units one finally obtains

$$(\alpha^{-1})^{-2} e^0 K^0 N^0 \Lambda^{-1} = 2 \times 10^{-10} R_\infty (\lambda_C)_{\mathrm{kxu}}. \tag{89}$$

It is interesting to note that if λ_C could be measured accurately in absolute units, it would provide an additional means for obtaining the fine structure constant. Two possible methods for doing this using quantum phase coherence in superconductors will be discussed in Sec. VI.D.

7. *Summary of Stochastic Input Data*

Table XVI summarizes all of the stochastic input data we shall now consider for inclusion in our least-squares adjustment to obtain values of the fine structure constant and other physical constants without essential use of quantum electrodynamic theory. For convenience, each datum has been assigned an item number and will be referred to by this number, e.g., I4 means item 4, the measurement of K at NPL by Vigoureux. (Table XVI also contains additional data from QED experiments, I20, I21, and I22. These will be considered for use in the adjustment to obtain a final recommended set of constants, but will not be discussed or used here.) It will be noted that we have already eliminated from Table XVI some of the experimental results listed in Tables XII–XV. Throughout our discussion of the stochastic data, we have anticipated this expurgation by calling attention to instances in which we believed there was sufficient cause to exclude a particular result. However, because we wish to be quite explicit about the reasons for the absence of the discarded items, we shall briefly review each of the items here.

(*a*) *Measurements of K*. Two determinations of K listed in Table XII have been omitted from Table XVI, that of Driscoll in 1958 using the Pellat electrodynamometer, and the current-balance measurement of Curtiss *et al.* carried out in 1942 and revised by Driscoll and Cutkosky in 1958. The uncertainties quoted for both of these measurements are such that if they were included in an adjustment, they would carry comparable weight with the three other similar measurements. Nevertheless, we believe we are justified in excluding the revised result of Curtiss *et al.* because of its age and the concommitant added uncertainty in the electrical units in terms of which it was obtained. [A change in the definition of the ohm and volt in 1948 contributes significantly to this uncertainty (Driscoll and Cutkosky, 1958).] The omission of the 1958 dynamometer measurement is perhaps not so easy to justify and may be open to some criticism. However, we believe the many technical improvements incorporated in the apparatus since the 1958 experiments were completed make the 1968 measurements considerably more reliable, i.e., less likely to contain an unknown systematic error (see Sec. II.C.2). Indeed, the experimenters themselves recommend (Driscoll, private communication) discarding the older result for just this reason, and in our opinion, it is they who are best able to assess the relative merits of the two experiments. (We note that the weighted average of the three measurements of K listed in Table XVI is 1.0000088 ± 4.3 ppm. If the older dynamometer result is included, $K = 1.0000094 \pm 4.0$ ppm, a difference of only 0.6 ppm.)

(*b*) *Measurements of* γ_p. The value of γ_p' resulting from the NBS experiments (Table XIII) was derived in Sec. II.C.4 by averaging the mean of the two Fredericksburg measurements with the Washington and Gaithersburg results, but with each of the latter carrying twice as much weight as the former. This choice is primarily due to the poor control the experimenters had over the

TABLE XVI. Summary of the stochastic input data to be considered in the present paper for inclusion in the least-squares adjustments to obtain best values of the fundamental physical constants. Items I1–I19 come from experiments which do not require quantum electrodynamic theory for their analysis and are considered for use in obtaining the WQED values of the constants (Sec. III). Items I20–I22 involve QED in some way and will be considered for use in the adjustment to obtain the final recommended values of the constants (Sec. V).

Item number	Publication date and author	Quantity	Method	Value	Error (ppm)
1	1968, Parker, Langenberg, Denenstein, and Taylor	$2e/h$	Josephson effect	$4.835976(12)\times 10^{14}$ Hz/V_{NBS}	2.4
2	1968, Driscoll and Olsen	K	Pellat electrodynamometer	1.0000102(97)	9.7
3	1958, Driscoll and Cutkosky	K	NBS current balance	1.0000092(77)	7.7
4	1965, Vigoureux	K	NPL current balance	1.0000080(60)	6.0
5	1960, Craig, Hoffman, Law, and Hamer	F	Silver–perchloric acid coulometer	$9.648570(66)\times 10^{7}$ $A_{NBS}\cdot$sec kmole^{-1}	6.8
6	1958–1968, Driscoll, Olsen, and Bender	γ_p'	Low field	$2.6751525(99)\times 10^{8}$ Hz/T_{NBS}	3.7
7	1962, Vigoureux	γ_p'	Low field	$2.675144(16)\times 10^{8}$ Hz/T_{NBS}	5.8
8	1962–1966, Yagola, Zingerman, and Sepetyi	γ_p'	High field	$2.675105(20)\times 10^{8}$ Hz/T_{NBS}	7.4

9	1949–1951, Sommer, Thomas, and Hipple	μ_p'/μ_n	Omegatron	2.792690(30)	11
10	1957–1963, Sanders, Turberfield, Collington, and Dellis	μ_p'/μ_n	Inverse cyclotron	2.792701(73)	26
11	1961, Boyne and Franken	μ_p'/μ_n	Cyclotron	2.792832(55)	20
12	1965, Mamyrin and Frantsuzov	μ_p'/μ_n	Mass spectrometer	2.792794(17)	6.2
13	1967, Petley and Morris	μ_p'/μ_n	Omegatron	2.792746(52)	19
14	1964, Spijkerman and Bearden	hc/e	Short-wave length limit	12373.15(41) $V_{NBS}\cdot$kxu	33
15	1965, Bearden	$N\Lambda^3$	X-ray crystal density (calcite)	$6.05972(23)\times10^{26}$ kmole^{-1}	37
16	1964, Henins and Bearden	$N\Lambda^3$	X-ray crystal density (Si)	$6.059768(95)\times10^{26}$ kmole^{-1}	16
17	1931, Bearden 1938–1940, Tyrén [revised, Edlén and Svensson (1956)]	Λ	Ruled gratings	1.002030(38)	38
18	1962, Knowles	λ_C	Electron–positron annihilation—H_2O	$24.21263(92)\times10^{-3}$ kxu	38

TABLE XVI (*Continued*)

Item number	Publication date and author	Quantity	Method	Value	Error (ppm)
19	1964, Knowles	λ_C	Electron–positron annihilation—Ta	$24.21421(36)\times 10^{-3}$ kxu	15
20	1966, Vessot *et al.*	α^{-1}	ν_{Hhfs} via hydrogen maser, plus theory	137.03591(35)	2.6
21	1968, Metcalf, Brandenberger, and Baird	α^{-1}	Fine-structure splitting ΔE_{H} in H, $n=2$, plus theory	137.03545(59)	4.3
22a	1953–1968, Kaufman, Lamb, Lea, and Leventhal; Triebwasser, Dayhoff, and Lamb; Robiscoe	α^{-1}	$\Delta E_{\mathrm{H}}-\mathcal{S}_{\mathrm{H}}$ splitting in H, $n=2$, plus $\mathcal{S}_{\mathrm{H}}$, plus theory	137.03505(32)	2.4
22b	1968, Kaufman, Lamb, Lea, and Leventhal	α^{-1}	$\Delta E_{\mathrm{H}}-\mathcal{S}_{\mathrm{H}}$ splitting in H, plus theory	137.03673(25)	1.8

electrical standards at Fredericksburg. The average NBS value so obtained exceeds the mean of the two Fredericksburg values by 2.3 ppm, is essentially identical with the Gaithersburg value, and is 1.1 ppm less than the average Washington value. The differences are well within the assigned error of 3.7 ppm. We choose not to ignore entirely the Fredericksburg work because it is not outstandingly discrepant and because there are at present no really clear-cut experimental reasons for doing so.

As explained in Sec. II.C.4, the uncertainty of the NPL low-field measurement of γ_p' has been expanded so that the NPL result will carry a weight in the least-squares adjustments equal to that carried by either the NBS Washington or Gaithersburg results. Although this requires increasing the 2.6-ppm error originally assigned the NPL value to 5.8 ppm, we believe this error expansion is justified in view of the single location and limited time period associated with the NPL measurements as compared with those carried out at NBS.

The low-field measurement of γ_p' carried out at ETL by Hara and coworkers has been omitted because of possible systematic errors arising from several experimental sources. These include the blowing of cooling air through the solenoid, an observed thermal instability in the diameter of the solenoid, and an observed shift of γ_p' with the direction of the current in the prepolarizing coil (see Sec. II.C.4). Since little is known about these effects, there is no way of reliably estimating how much they might contribute to γ_p'; 5–10 ppm does not seem unreasonable. We therefore believe it best to censor the ETL result until these questions are resolved.

The determination of γ_p' in low fields carried out at VNIIM has been omitted because of the difficulty of assigning a meaningful uncertainty to the final experimental result. Studentsov *et al.* quote 1.7 ppm as the statistical uncertainty in their final value but give no estimates of the systematic errors. The scatter in the data, which is of order 15–25 ppm, as well as the history of the measurements, indicate that the true uncertainty may be as high as 10 ppm (see Sec. II.C.4). Until much more information about the errors associated with the experiment becomes available, we cannot justify its inclusion as a stochastic input datum.

We should emphasize that discarding the VNIIM and ETL values, expanding the error of the NPL value, and giving only half as much weight to the Fredericksburg measurement as compared with those carried out at Washington and Gaithersburg, do not result in a mean value of γ_p' which is drastically different from the value obtained if all the various measurements are treated on an equal footing. If the Fredericksburg result were given equal weight with the other two NBS measurements, the average NBS value for γ_p' would be (in units of 10^8 rad/sec$\cdot$$T_{NBS}$) $\gamma_p'=2.6751515$, only 0.4 ppm less than our adopted NBS average value, I6. If this "new" NBS average value is averaged with the NPL, ETL, and VNIIM values (all four carrying equal weight), then $(\gamma_p')_{Av}=2.6751423$. This result is only 2.9 ppm less than the weighted mean of our adopted NBS average value and the NPL value with its expanded uncertainty, I6 and I7; the 2.9-ppm difference is within the 3.1-ppm uncertainty of the weighted average of these two items.

The low-field determination of Wilhelmy has been discarded because it is not a high-precision measurement. This is apparent from the assigned uncertainty of 45 ppm, which is more than 10 times larger than that assigned to the NBS average value. Discarding this result is therefore in keeping with our general philosophy that it is usually incorrect to average data together which differ in uncertainty by large amounts. Since the high-field measurement of Capptuller has an uncertainty of 37 ppm, similar considerations also apply, and it too has been discarded. (Both of these results are also somewhat inconsistent with their high-accuracy counterparts.) However, the high-field experiment of Thomas, Driscoll, and Hipple (TDH) is rather different. It was apparently carried out with great care and, in principle, is a high-accuracy measurement.

Nevertheless, it exceeds by 47 ppm the precise high-field measurement of γ_p' by Yagola, Zingerman, and Sepetyi (YZS). This latter measurement is in excellent agreement with the high-field value of γ_p' implied by combining [via Eq. (47c)] the weighted mean of the three direct measurements of K, I2, I3, and I4, with the weighted mean of the two low-field values of γ_p', I6 and I7:

$$\gamma_p'(\text{high}) = 2.675103(24) \times 10^8 \,\text{rad/sec}\cdot\text{T}_{\text{NBS}}\ (9.1\ \text{ppm}). \tag{90}$$

The measurement of Yagola *et al.* exceeds this indirect value by only 0.7 ppm. In comparing the Thomas, Driscoll, and Hipple result with the directly measured value of Yagola *et al.*, I8, and the indirect value given in Eq. (90), we find that the standard deviations of the differences are 12 and 13 ppm, respectively, or about one-quarter of the approximate 50-ppm differences. The probability for this to occur by chance is ~0.02%, clearly implying that the measurement of Thomas *et al.* probably contains a systematic error and should be discarded. [The possibility that there is a fundamental difference in γ_p' as measured by the low- and high-field methods or by the free-precession and resonant-absorption technique would appear to be eliminated by the result of Yagola *et al.* and by the recent results of Driscoll and Olsen at Gaithersburg, in which both the free-precession and resonant-absorption techniques gave equivalent results in low fields (Driscoll and Olsen, 1968; Driscoll, private communication; see also Huggins and Sanders, 1965.] While we believe such a clear-cut discrepancy is in itself sufficient grounds for censoring an input datum, it is desirable to support the decision by appealing to experimental evidence. DuMond has visited the laboratory of Yagola *et al.* and has given an eyewitness account (Cohen and DuMond, 1965) of their attempts to examine carefully all possible sources of systematic error in a high-field experiment. DuMond reports that Yagola *et al.* demonstrated the necessity of having the faces of the pole pieces of the electromagnet very accurately plane and parallel, strictly vertical, and coaxial. If the pole pieces were not quite parallel and coaxial, the magnetic field in the gap was found to be quite inhomogeneous, thereby making it extremely difficult to center properly the coil of the Cotton balance. Departures from these rather strict requirements in the experiment of Thomas *et al.* (not realized at the time) could perhaps explain the 50-ppm discrepancy.*

Expurgating the measurement of Thomas, Driscoll, and Hipple is in keeping with the general philosophy (Cohen and DuMond, 1965) that it is incorrect to include in a least-squares adjustment (or for that matter, in a simple weighted average or one variable analog of the many-variable least-squares "average") two values of the same stochastic quantity which are in significant disagreement with each other, i.e., which differ by several times the standard deviation of their difference. This is because such a discrepancy implies that either one or both of the error estimates are incorrect and should not be used as a weight in an adjustment, or that the error estimates are realistic but that one or both of the measurements probably contain an unknown systematic error and should therefore be discarded. In such cases, it is desirable (if possible) to see if there is any experimental justification for expurgating the suspect datum since, if such justification can be found, a bit of the arbitrariness which is always present in a decision to expurgate will be removed. The present situation is such a case, for the measurement of γ_p' by Sommer, Thomas, and Hipple is in extreme disagreement with other accurate and consistent values of the same quantity, and there is some experimental evidence for the possible existence of a systematic error.

---

* On the other hand, Thomsen (private communication) has estimated the vertical magnetic-field component in the Thomas *et al.* experiment from their published field distribution curves. He concludes that the magnitude of the component is too small by an order of magnitude to account for the 50-ppm discrepancy.

(*c*) *Measurements of* μ_p/μ_n. We retain all of the measurements of μ_p'/μ_n listed in Table XIV for initial consideration as stochastic input data except those of Trigger and Marion and Winkler. Trigger's revised result for μ_p'/μ_n has been eliminated mainly because of its large uncertainty in comparison with the Sommer, Thomas, and Hipple measurement; the 36-ppm error of Trigger's value exceeds the error assigned the value of Sommer *et al.* by a factor of 3.4, significantly larger than the factor of 3 indicated by our rule of thumb (see Sec. II.A.3). More important, the theory for determining the cyclotron resonance frequency from the observed resonance curves is sufficiently uncertain in the measurements of Trigger that one cannot readily say whether the assigned 36-ppm error should be interpreted as a probable error or, as we have done, a standard deviation. The value for μ_p'/μ_n given by Marion and Winkler has also been censored because of its large (45 ppm) uncertainty, which exceeds by a factor of 4.2 the error assigned the value of Sommer *et al.* We do note that the values of μ_p'/μ_n given by Trigger and by Marion and Winkler are in good agreement with that given by Sommer *et al.*; the weighted mean of all three values is only 2.2 ppm less than the result of Sommer *et al.* or 1/5 of its 11-ppm uncertainty.

Examination of the five measurements of μ_p'/μ_n given in Table XVI, I9–I13, shows that they fall naturally into two groups (see also Fig. 3). The lowest three values I9, I10, and I13, are reasonably consistent among themselves, as are the three highest values, I11, I12, and I13. (Item 13 lies midway between the high and low values and may be considered as the highest of the low values or the lowest of the high values.) However, the means of the two groups differ by about 32 ppm. Since there are no salient experimental reasons to prefer one group over the other, we initially retain all five values in order to see how they agree (or disagree) with the remaining stochastic input data. This will be done in Sec. III.

(*d*) *Measurements of hc/e, N,* Λ *and* λ_C. The measurement of Hagström *et al.* (see Table XV) has been discarded because of its very large uncertainty and the fact that no attention was paid to the electrical units in terms of which it was measured. The two low-accuracy measurements of $N\Lambda^3$ have been omitted because their uncertainties are 4 to 5 times larger than the uncertainty of the Henins and Bearden measurement on Si. We do note that all of the $N\Lambda^3$ measurements are in quite good agreement; the weighted mean of the two relatively accurate values we have retained (I15, I16) has an uncertainty of 14 ppm, but is only 1.6 ppm less than the weighted mean of all four values. We retain only one of the two values of Λ listed in Table XV since they are both identical. However, other values of Λ, in particular those of Bearden (large grating) and Bäcklin, have already been discarded for both theoretical and experimental reasons, and the value of Λ implied by Tyrén's measurement (as revised by Edlén and Svensson) depends critically on the values used for the Al x-ray wavelengths expressed in kx-units. Literally, the only conclusion which can be drawn from the direct measurements of Λ is that Λ probably lies somewhere between 1.00200 and 1.00210!

As noted previously, the two measurements of λ_C by Knowles are in disagreement. They differ by 65 ppm, 1.6 times the standard deviation of the difference, which is only 41 ppm. The probability of this occurring by chance is 11%. Since this is not entirely unreasonable and since there are few experimental reasons for preferring one value over the other, both will be retained.

III. Least – Squares Adjustment to Obtain Values of the Constants without QED Theory

A. Preliminary Search for Discrepant Data

1. Inconsistencies among Data of the Same Kind

Before we can carry out a least-squares adjustment to obtain best WQED values of the fine structure constant and other fundamental constants, we must critically examine the over-all compatibility of the data listed in Table XVI. This is necessary because, as emphasized in Sec. II.C.7, it is incorrect to include inconsistent data in an adjustment since the inconsistencies imply either falacious error estimates or the presence of unknown systematic errors. One method for testing whether two measurements of the same quantity are compatible has already been given and simply involves comparing their difference with the standard deviation of their difference. Another test is the so-called Birge ratio. Birge (1932) was the first to emphasize that there are really two ways of assigning an uncertainty to the weighted average of a quantity, that obtained by the criterion of internal consistency and that obtained by the criterion of external consistency. The error determined by internal consistency, σ_I, is the usual uncertainty computed via Eq. (10), that is,

$$\sigma_I^2 = \sigma_m^2 = [\sum_{i=1}^{N} (1/\sigma_i^2)]^{-1}.$$

The error determined by external consistency, σ_E, is defined as

$$\sigma_E^2 = [\sum_{i=1}^{N} (X_i - \bar{X})^2/\sigma_i^2][(N-1) \sum_{i=1}^{N} (1/\sigma_i^2)]^{-1}, \quad (91)$$

where $\bar{X}$ is the weighted average computed in the usual way from the X_i and σ_i [Eq. (10)]. It thus follows that σ_I is the expected error in the mean as determined by the individual *a priori* assigned errors σ_i, while σ_E is the expected error as determined by how much each X_i actually deviates from the mean in comparison with its *a priori* uncertainty σ_i. If the data are actually statistically distributed according to the individual *a priori* errors (as would be expected if these errors are true indications of the experimental uncertainty), then σ_E should be equal to σ_I and the Birge ratio $R \equiv \sigma_E/\sigma_I$ should be one. If R is significantly larger than one, the data are suspect and there is a strong possibility that some of the σ_i have been underestimated or that some of the data contain systematic errors. On the other hand, if R is much less than one, the data are highly compatible, implying that the individual *a priori* error estimates may well be too large. These conclusions follow from the fact that if R is calculated for a set of data and each σ_i is then multiplied by R and the Birge ratio recomputed, the new value of R will be exactly equal to one. Thus, an initial Birge ratio of, for example, 1.2 implies that all of the individual uncertainties have been underestimated by 20%. Similarly, if $R = 0.80$, then the uncertainties have been overestimated by 20%. [This assumes of course that the same "mistake" has been made for each quantity, that there are no systematic errors, and that the deviation of R from one is not just a statistical fluctuation, i.e., due to the statistical uncertainties in σ_E and σ_I – see Birge (1932).]

We have computed the Birge ratios for all of the stochastic data of each kind listed in Table XVI. We find that (1) $R = 0.14$ for the three different measurements of K, I2, I3 and I4, (2) $R = 0.46$ for the NBS and NPL values of γ_p', I6 and I7, and (3) $R = 0.20$ for the two measurements of $N\Lambda^3$, I15 and I16. (For all four values of $N\Lambda^3$ given in Table XV, $R = 0.41$.) In each case, the Birge ratio is significantly less than unity and the data are highly compatible. However, as expected from the discussion in Secs. II.C.7.c and II.C.7.d, the

TABLE XVII. Least-squares investigation of the compatibility of the μ_p'/μ_n measurements. Since the adjustment involves only the single unknown μ_p'/μ_n, the adjusted value given in the table is simply the weighted mean of the items remaining after the indicated items are deleted. The initial set of data consisted of the five items I9–I13. The number of degrees of freedom is therefore $5-1=4$, minus the number of items deleted. All uncertainties are computed on the basis of internal consistency as is done throughout the present paper.

Item(s) deleted	R	χ^2	Degrees of freedom $N-1$	μ_p'/μ_n	Error (ppm)
1. None	1.68	11.34	4	2.792768(14)	4.9
2. I12: Mamyrin and Frantsuzov (MF)	1.34	5.39	3	2.792725(22)	8.0
3. I9: Sommer, Thomas, and Hipple (STH)	0.97	2.79	3	2.792789(15)	5.5
4. I11, I12: Boyne and Franken (BF) and MF	0.44	0.87	2	2.792704(24)	8.8
5. I9, I10: STH and Sanders and Turberfield (ST)	0.81	1.30	2	2.792793(16)	5.6
6. I11, I12, I13: BF, MF, Petley and Morris (PM)	0.14	0.02	1	2.792692(28)	9.9
7. I9, I10, I13: STH, ST, and PM	0.65	0.43	1	2.792797(16)	5.9

Birge ratio is larger than one for both the μ_p'/μ_n and λ_C data; the latter yield $R=1.59$, and the former, $R=1.68$. The compatability of the μ_p'/μ_n data may be further examined by deleting various measurements and recomputing R. [This is a form of "analysis of variance" as discussed in Cohen and DuMond (1965).] The results are summarized in Table XVII. As might have been anticipated, deleting either the most accurate high value of μ_p'/μ_n, that of Mamyrin and Frantsuzov, or the most accurate low value, that of Sommer, Thomas, and Hipple, reduces the incompatability of the data. Eliminating the *two* most accurate high values or the *two* most accurate low values reduces the incompatability to the point where R is significantly less than one. Thus, Table XVII clearly shows the division of the μ_p'/μ_n data into the two separate self-consistent groups of high and low values indicated in Sec. II.C.7.c.

Also listed in Table XVII is the familiar statistic "chi squared" (χ^2) [see, for example, Kendall (1952)] which is in fact closely related to R since

$$\chi^2 = \sum_{i=1}^{N} (X_i - \bar{X})^2/\sigma_i^2 ,$$

and therefore $\chi^2 = (N-1)R^2$. The expectation value of χ^2 is simply $N-1$, so that $\chi^2 > (N-1)$ has the same implication as $R>1$ and similarly for $\chi^2 < (N-1)$. Probability tables for χ^2 are available (Abramowitz and Stegun, 1964) for different "degrees of freedom" ν, where for the present case, ν is simply $N-1$. As an example, for 4 degrees of freedom, the probability that χ^2 will exceed 11.3 is only about 2%. Since this case corresponds to Line 1 of Table XVII, the incompatability of the μ_p'/μ_n data is clearly demonstrated.

2. *Inconsistencies among Dissimilar Data*

We now search for discrepancies between measurements of dissimilar quantities. This can of course be done by carrying out numerous least-squares adjustments involving all of the data (or a specific subset), systematically deleting various items, and studying how the different statistical measures of compatability, for example, χ^2 or R, depend on the deleted item or items. We shall make such a study, or analysis of variance in the next section, but we believe an over-all picture of the compatability of the data can be obtained more simply using straightforward calculations involving a few readily derived equations. For example, using Eq. (47c), we may compare the high-field measurement of γ_p', I8, with the two low-field measurements, I6 and I7. Using I8 and the different measurements of K indicated, we find (in units of 10^8 rad/sec$\cdot$T$_{NBS}$)

I2(NBS68): $\gamma_p' = 2.675159(56)$ (21 ppm),

I3(NBS58): $\gamma_p' = 2.675154(46)$ (17 ppm),

I4(NPL): $\gamma_p' = 2.675148(38)$ (14 ppm),

I2, I3, I4: (wt av) $\gamma_p' = 2.675152(30)$ (11 ppm).

(All uncertainties quoted here and in similar lists are the RSS of the individual *a priori* errors.) Each of these indirect low-field values of γ_p' is obviously in excellent agreement with both I6 and I7. Of course, Eqs. (47c) can be used equally well to compute values of K from I8 and I6 and I7. We find

I6(NBS): $K = 1.0000089 \pm 4.1$ ppm,

I7(NPL): $K = 1.0000073 \pm 4.7$ ppm,

I6, I7: (wt av) $K = 1.0000085 \pm 4.0$ ppm,

all in good agreement with the directly measured values I2, I3, and I4, but with considerably smaller uncertainties. This means that in a least-squares adjustment, the indirect value of K determined from the high- and low-field measurements of γ_p' will carry significantly more weight than the individual current-balance values and will therefore be a prime factor in determining the final adjusted value of K. (Note that in a least-squares adjustment, all possible routes for obtaining the adjustable constants are automatically considered and appropriately weighted.)

$$(\alpha^{-1})^{-1}e^{-1}K^{1}N^{0}\Lambda^{0} = \frac{\mu_o}{4}\,\frac{c\Omega_{ABS}}{\Omega_{NBS}}\left(\frac{2e}{h}\right)_{NBS} \quad (7)$$

$$(\alpha^{-1})^{0}e^{0}K^{1}N^{0}\Lambda^{0} = K \quad (38)$$

$$(\alpha^{-1})^{0}e^{1}K^{-1}N^{1}\Lambda^{0} = F_{NBS} \quad (46)$$

$$(\alpha^{-1})^{-3}e^{-1}K^{1}N^{0}\Lambda^{0} = \frac{\mu_o R_\infty}{(\mu_p'/\mu_B)}\,[\gamma_p'(\text{low})]_{NBS} \quad (69)$$

$$(\alpha^{-1})^{-3}e^{-1}K^{-1}N^{0}\Lambda^{0} = \frac{\mu_o R_\infty}{(\mu_p'/\mu_B)}\,[\gamma_p'(\text{high})]_{NBS} \quad (70)$$

$$(\alpha^{-1})^{-3}e^{-2}K^{0}N^{-1}\Lambda^{0} = \frac{\mu_o R_\infty}{(\mu_p'/\mu_B)}\,\frac{1}{M_p^*}\left(\frac{\mu_p'}{\mu_n}\right) \quad (82)$$

$$(\alpha^{-1})^{1}e^{1}K^{-1}N^{0}\Lambda^{-1} = \frac{2}{\mu_o c^2}\,\frac{10^{-10}}{\Omega_{ABS}/\Omega_{NBS}}\left(\frac{hc}{e}\right)_{NBS,kxu} \quad (85)$$

$$(\alpha^{-1})^{0}e^{0}K^{0}N^{1}\Lambda^{3} = (N\Lambda^3)_{kxu} \quad (86)$$

$$(\alpha^{-1})^{0}e^{0}K^{0}N^{0}\Lambda^{1} = (\Lambda)_{kxu} \quad (87)$$

$$(\alpha^{-1})^{-2}e^{0}K^{0}N^{0}\Lambda^{-1} = 2\times 10^{-10}\cdot R_\infty(\lambda_C)_{kxu} \quad (89)$$

$$(\alpha^{-1})^{1}e^{0}K^{0}N^{0}\Lambda^{0} = \alpha^{-1}$$

$$(\alpha^{-1})^{-1}e^{-1}K^{1}N^{0}\Lambda^{0} = \frac{\mu_o c}{4}\,(\Phi_o^{-1})_{NBS}$$

$$(\alpha^{-1})^{-3}e^{-1}K^{1}N^{0}\Lambda^{0} = 2\mu_o R_\infty[\Omega_o/(B_o)_{NBS}]$$

$$(\alpha^{-1})^{-2}e^{0}K^{0}N^{0}\Lambda^{0} = 2R_\infty\lambda_C$$

$$(\alpha^{-1})^{1}e^{2}N^{1}K^{0}\Lambda^{0} = \frac{2M_{He}^*g}{\mu_o c}\left(\frac{h}{gM_{He}}\right)$$

FIG. 4. Summary of the observational equations of interest in the present work. The numbers in parentheses correspond to the equation numbers used in the text. The last four equations relate to possible future measurements (see Sec. VI.D).

The other equations we shall require for our search for discrepant data are also on hand since they follow directly from the various observational equations derived throughout the paper. (For the convenience of the reader, these are summarized in Fig. 4). For example, by dividing the observational equation for $2e/h$, Eq. (7), by the similar equation for γ_p' (low field), Eq. (69), we obtain an equation for α^{-1}:

$$\alpha^{-1}=\left[\frac{1}{4R_\infty}\frac{c\Omega_{\mathrm{ABS}}}{\Omega_{\mathrm{NBS}}}\frac{\mu_p'}{\mu_B}\frac{2e/h}{\gamma_p'}\right]^{1/2}. \tag{92}$$

[If γ_p' is obtained from a high-field measurement, then γ_p' in Eq. (92) must be replaced by $\gamma_p'K^2$.] Taking the auxiliary constants as listed in Table XI, the Josephson-effect measurement of $2e/h$, I1, and the measurements of γ_p' and K as indicated, Eq. (92) gives

I6(NBS): $\alpha^{-1}=137.03601(30)$ (2.2 ppm),

I7(NPL): $\alpha^{-1}=137.03623(43)$ (3.1 ppm),

I6, I7: (wt av) $\alpha^{-1}=137.03608(27)$ (2.0 ppm),

I8(YZS) with I2, I3, I4: (wt av) $\alpha^{-1}=137.03603(79)$ (5.8 ppm). (93)

As anticipated from the high compatability of the different values of γ_p' and K, all four of these α^{-1} values are in excellent agreement.

We may obtain another equation for α^{-1} by combining the observational equations for μ_p'/μ_n, F, and $2e/h$, Eqs. (82), (46), and (7):

$$\alpha^{-1}=\left[\frac{M_p^*}{4R_\infty}\frac{c\Omega_{\mathrm{ABS}}}{\Omega_{\mathrm{NBS}}}\frac{\mu_p'}{\mu_B}\frac{2e/h}{K^2F(\mu_p'/\mu_n)}\right]^{1/2} \tag{94}$$

Using the auxiliary constants given in Table XI, the NBS measurement of F, I5, and the weighted average of the three measurements of K, I2, I3 and I4, Eq. (94) gives, for the values of μ_p'/μ_n indicated,

I9(STH): $\alpha^{-1}=137.0366(11)$ (7.8 ppm),

I10(ST): $\alpha^{-1}=137.0364(20)$ (14 ppm),

I11(BF): $\alpha^{-1} = 137.0332(15)$ (11 ppm),

I12(MF): $\alpha^{-1} = 137.03409(88)$ (6.4 ppm),

I13(PM): $\alpha^{-1} = 137.0353(15)$ (11 ppm),

I9, I10, I13: (wt av) $\alpha^{-1} = 137.03630(97)$ (7.1 ppm),

I11, I12, I13: (wt av) $\alpha^{-1} = 137.03412(86)$ (6.3 ppm).

Discrepancies among these calculated values of α^{-1} due to discrepancies among the μ_p'/μ_n data are present, as expected. Moreover, comparing these values with those computed from γ_p', it is immediately apparent that the two high values of μ_p'/μ_n, I11 and I12, give values of α^{-1} which are not compatible with those computed from γ_p'. On the other hand, the remaining data give values for α^{-1} which are in general agreement with those computed from γ_p'. Thus, assuming the measurements of $2e/h$, K, and γ_p' are reliable (as all experimental evidence would seem to indicate), we may conclude either that (1) the measurements of μ_p'/μ_n by Boyne and Franken and by Mamyrin and Frantsuzov are in error or that (2) both the measurement of the Faraday at NBS and μ_p'/μ_n by Sommer, Thomas, and Hipple are in error. (We are tacitly assuming of course that the quoted uncertainties are true estimates of the accuracy of the experiments.) We saw previously that there was some experimental evidence for distrusting the measurement of Boyne and Franken, but no such evidence presently exists for the measurement of Mamyrin and Frantsuzov. On the other hand, the NBS determination of the Faraday was carried out with great care and would seem to be quite reliable. The measurement of μ_p'/μ_n by Sommer, Thomas, and Hipple also appears to be reliable and, furthermore, it has the support of the measurements of Trigger, Sanders and Turberfield and also Marion and Winkler (see Table XIV).

We might also point out here that the uncertainties in the α^{-1} values derived from μ_p'/μ_n, F, K_{Av}, and $2e/h$ are significantly larger than the uncertainties in the α^{-1}

values derived from γ_p' and $2e/h$. This means that these α^{-1} values will carry very little weight in a least-squares adjustment and that α^{-1} will be determined primarily by γ_p' and $2e/h$.

There is another way to compare the various measurements of γ_p' and μ_p'/μ_n; equating Eqs. (92) and (94) yields

$$\mu_p'/\mu_n = M_p^* \gamma_p'/FK^2. \tag{95}$$

(If γ_p' is obtained from a high-field measurement, K^2 must be omitted.) Using the NBS measurement of the Faraday, I5, the weighted average of I2, I3, and I4 for K (where necessary), and the measurements of γ_p' indicated, Eq. (95) gives

I8(YZS): $\mu_p'/\mu_n = 2.792715(28)$ (10 ppm),

I6(NBS): $\mu_p'/\mu_n = 2.792715(32)$ (12 ppm),

I7(NPL): $\mu_p'/\mu_n = 2.792707(35)$ (12 ppm),

I6, I7: (wt av) $\mu_p'/\mu_n = 2.792713(32)$ (11 ppm).

As expected, these are in much better agreement with the lower values of μ_p'/μ_n, I9 and I10, than with the higher values, I11 and I12. (See Fig. 3; the indirect value shown is nearly identical to the last of the above values and is given exactly in Sec. III.B.2.) We also note that the uncertainties of these indirect values of μ_p'/μ_n are comparable with those assigned the direct measurements and will therefore carry considerable weight in an adjustment. Equation (95) clearly shows the critical role played by the Faraday constant, and the misfortune of having only one high-accuracy measurement of F. In this connection, we note that most of the experimental uncertainties in the determination of the electrochemical equivalent of Ag are such that if the present value of F is significantly in error, it is probably too large rather than too small. A smaller value of F would tend to favor the higher values of μ_p'/μ_n as obtained by Mamyrin and Frantsuzov and by Boyne and Franken. But F would have to be in error by some 30 ppm (over 4 times its assigned uncertainty) in order to give a value of μ_p'/μ_n which agrees with the higher experimental values. This can be seen by deriving indirect values of F from Eq. (95) and comparing them with I5. Taking the weighted average of I2, I3, and I4 for K, and the mean of I6 and I7 for γ_p', we find for the two weighted averages of μ_p'/μ_n indicated (in units of 10^3 $A_{NBS}\cdot$sec/kmole)

I9, I10, I13: (wt av) $F = 96\,486.01 \pm 1.22$ (13 ppm),

I11, I12, I13: (wt av) $F = 96\,482.94 \pm 1.03$ (11 ppm),

NBS value, I5: $F = 96\,485.70 \pm 0.66$ (6.8 ppm).

The direct value of F, I5, is only 3.2 ppm less than the indirect value calculated using the weighted average of the three low measurements of μ_p'/μ_n while it exceeds by 29 ppm the indirect value obtained from the high measurements.

Equation (95) can also be used to derive values of K. Using the weighted average of I6 and I7 for γ_p' and the NBS value for the Faraday, I5, we obtain from the measurements of μ_p'/μ_n indicated

I9(STH): $K = 1 + (12.8 \pm 6.5)$ ppm,

I10(ST): $K = 1 + (10.9 \pm 13.7)$ ppm,

I11(BF): $K = 1 - (12.4 \pm 10.5)$ ppm,

I12(MF): $K = 1 - (5.7 \pm 4.8)$ ppm,

I13(PM): $K = 1 + (2.8 \pm 10.0)$ ppm,

I9, I10, I13: (wt av) $K = 1 + (10.4 \pm 5.8)$ ppm,

I11, I12, I13: (wt av) $K = 1 - (5.5 \pm 4.7)$ ppm.

Again, the disagreement of the high values of μ_p'/μ_n with the other data is evident. It is also apparent that the uncertainties of some of these indirect values of K are comparable with those of the directly measured values, I2, I3, and I4. This means that such indirect values will carry similar weight in an adjustment, thereby playing an important role in determining K. (This situation is similar to that which obtains for the value of K arising from the high- and low-field measurements of γ_p'.)

In an attempt to obtain further indications as to which of the two groups of μ_p'/μ_n data is more nearly correct, we now examine the x-ray results. The observational equations for μ_p'/μ_n, γ_p', and $2e/h$, Eqs. (82), (69), and (7), may be combined to give an equation relating μ_p'/μ_n to the x-ray measurements of $N\Lambda^3$ and Λ:

$$\mu_p'/\mu_n = \frac{\Lambda^3}{(N\Lambda^3)}\left[\frac{M_p^{*2}\mu_0^2}{64R_\infty}\left(\frac{c\Omega_{\mathrm{ABS}}}{\Omega_{\mathrm{NBS}}}\right)^3\frac{\mu_p'}{\mu_B}\right]^{1/2}\left[\frac{\gamma_p'(2e/h)^3}{K^4}\right]^{1/2}. \tag{96}$$

(If γ_p' is obtained from a high-field measurement, then K^4 must be replaced by K^2.) Equation (96) may be evaluated using the auxiliary constants as given in Table XI, the Josephson-effect value of $2e/h$, I1, the high-field measurement of γ_p', I8 (we use this value because it results in a slightly lower uncertainty), the weighted average of I2, I3, and I4 for K, and the weighted average of the two values of $N\Lambda^3$, I15 and I16 (previously shown to be quite compatible). Equation (96) also requires values of Λ for its evaluation. These may be obtained from (1) the direct grating measurement of Λ, I17, (2) the measurement of hc/e by Spijkerman and Bearden, I14, and (3) the two measurements of λ_C by Knowles, I18 and I19. With the aid of the observational equations for hc/e, $2e/h$, and λ_C, Eqs. (85), (7), and (89), the Josephson-effect value of $2e/h$, I1, the weighted average of I2, I3, and I4 for K, and $\alpha^{-1}=137.03608(27)$ (2.0 ppm) as implied by Eq. (93), we find

$$\begin{aligned} &\text{I14(SB)}: && \Lambda = 1.002043 \pm 34\ \text{ppm},\\ &\text{I17(BT)}: && \Lambda = 1.002030 \pm 38\ \text{ppm},\\ &\text{I18(Kn, H}_2\text{O)}: && \Lambda = 1.002083 \pm 38\ \text{ppm},\\ &\text{I19(Kn, Ta)}: && \Lambda = 1.002018 \pm 15\ \text{ppm}, \end{aligned} \tag{97}$$

in not unreasonable agreement. The four values of μ_p'/μ_n implied by these Λ values and Eq. (96) are

I14(SB): $\mu_p'/\mu_n = 2.79243(28)$ (100 ppm),

I17(BT): $\mu_p'/\mu_n = 2.79232(32)$ (115 ppm),

I18(Kn, H_2O): $\mu_p'/\mu_n = 2.79277(32)$ (115 ppm),

I19(Kn, Ta): $\mu_p'/\mu_n = 2.79223(13)$ (48 ppm).

In comparing these indirect values of μ_p'/μ_n with those measured directly (Tables XIV and XVI), it is immediately evident that because of the large uncertainties of the x-ray data, very few meaningful conclusions may be drawn. However, these indirect values do appear to be more compatible with the lower values of μ_p'/μ_n. Thus, the x-ray data, like the combined measurements of γ_p', K, and F, point a finger of suspicion at the high values of μ_p'/μ_n.

For completeness, we point out that an independent value of μ_p'/μ_n can, in principle, be derived from measurements of the difference between components of the Hα and Dα lines in the spectra of H and D. If $\bar{\nu}_D$ is the wave number of a component of the Dα line, and $\bar{\nu}_H$ is the wave number of a component of the Hα line, then it can be shown that $\bar{\nu}_D - \bar{\nu}_H \equiv \Delta\bar{\nu} = R_\infty(1+m_e/M_d)U_D - R_\infty(1+m_e/M_p)U_H$. Here, U_D and U_H are *theoretical* numbers obtainable from the calculations of $\bar{\nu}_D$ and $\bar{\nu}_H$ by, for example, Garcia and Mack (1965). They are equal to $\bar{\nu}_D/R_D$ and $\bar{\nu}_H/R_H$, respectively, where R_D and R_H are the Rydberg constants for D and H. [If the simple Bohr theory were correct, then U_D and U_H would be the same and exactly equal to $(1/2)^2-(1/3)^2=5/36$. They actually differ by about $2/10^{10}$ and exceed 5/36 by several parts per million, depending on the component.] Using the fact that $M_p/m_e = (\mu_p'/\mu_n)(\mu_p'/\mu_B)^{-1}$ and $M_d/m_e = (M_D^*/M_H^*)(1+M_p/m_e)-1$ (see Sec. II.B.7), we obtain for μ_p'/μ_n

$$\frac{\mu_p'}{\mu_n} = \frac{\mu_p'}{\mu_B}\,\frac{U_D(1-M_H^*/M_D^*)-\Delta\bar{\nu}/R_\infty}{(\Delta\bar{\nu}/R_\infty)+(U_H-U_D)}.$$

Unfortunately, the available measurements of $\Delta\nu$ are few in number, inconsistent, and of large uncertainty [see Cohen (1952) for a summary]. But to demonstrate the method, we use the value of $\Delta\bar{\nu}$ obtained by Williams (Cohen, 1952; Williams, 1938) for the difference between the $D\alpha_2$ and $H\alpha_2$ lines: $\Delta\bar{\nu}=4.14691\pm 0.0006\ \text{cm}^{-1}$. (This measurement is probably the best of its type.) Using the values of $D\alpha_2$ and $H\alpha_2$ as derived in Table VIII from the calculations of Garcia and Mack, and therefore R_D and R_H as used by these workers, we find $U_D=0.13888952$ and $U_H-U_D=1.759\times 10^{-10}$ (note that U_D exceeds 5/36 by 4.5 ppm). The final value for μ_p'/μ_n is 2.7914(4) (145 ppm). Although the quoted uncertainty is much too large for this result to be of any use, we see that if $\Delta\bar{\nu}$ could be measured to $0.00005\ \text{cm}^{-1}$ (as might be possible with modern methods), then a value of μ_p'/μ_n accurate to about 12 ppm would result.

We can summarize the results of this section on the over-all compatibility of the data by stating that, with the exception of the two measurements of μ_p'/μ_n by Boyne and Franken and by Mamyrin and Frantsuzov, all of the non-x-ray data listed in Table XVI are highly compatible. Furthermore, the x-ray measurements, although of large uncertainty, also tend to favor the low values of μ_p'/μ_n. We shall decide how to handle this situation in the next section after summarizing the results of a rather complete analysis of variance.

B. Least-Squares Search for Discrepant Data

1. Summary of Least-Squares Procedure

In the next section, we report the salient results of an extensive analysis of variance of the WQED stochastic data given in Table XVI. In this section, we very briefly review the least-squares procedure. [For a particularly clear and concise discussion of the method, see Bearden and Thomsen (1957). See also Cohen and DuMond (1965) and Cohen, Crowe, and DuMond (1957).] We have initially N observational equations (one for each stochastic input datum) involving J dependent variables or adjustable constants, with $N>J$. The form of these observational equations is that given by Eq. (8). These equations are linearized around arbitrary fiducial values of the variables, Z_{j0}, which are chosen to be approximately equal to the expected output values. The result is a new set of N linearized observational equations of the form

$$\sum_{j=1}^{J} Y_{ji}z_j = B_i,$$

where $z_j = [(Z_j - Z_{j0})/Z_{j0}] \times 10^6$, $B_i = [(A_i - A_{i0})/A_{i0}] \times 10^6$, and A_{i0} is the value of the left-hand side of the original observational equation [Eq. (8)] evaluated with the Z_{j0}. The normalized residuals r_i are given by

$$\left(\frac{1}{\sigma_i}\right)\left[\sum_{j=1}^{J} Y_{ji}z_j - B_i\right]$$

with σ_i in parts per million. The least-squares solution is obtained by minimizing the sum of the squares of the normalized residuals with respect to z_j, i.e., by setting

$$\partial\left(\sum_{i=1}^{N} r_i^2\right)/\partial z_j = 0.$$

This leads to a set of J linear equations (known as the normal equations) given by

$$\sum_{k=1}^{J} G_{jk}z_k = D_j, \qquad j = 1 \cdots J,$$

where

$$G_{jk} = \sum_{i=1}^{N} \left(\frac{Y_{ji}Y_{ki}}{\sigma_i^2}\right)$$

and

$$D_j = \sum_{i=1}^{N} \left(\frac{Y_{ji}B_i}{\sigma_i^2}\right).$$

These equations are then solved for the J unknowns, z_k, and the adjusted values of the original variables computed from $Z_k = Z_{k0}(1 + z_k \times 10^{-6})$. It can be shown that the standard deviation uncertainty ϵ_k in Z_k is related to the matrix **G** by $\epsilon_k^2 = (\mathbf{G}^{-1})_{kk}$ or $\epsilon_k = (\mathbf{G}^{-1})_{kk}^{1/2}$. ($\mathbf{G}^{-1}$ is called the error matrix.)

The statistic χ^2 is defined by

$$\chi^2 = \sum_{i=1}^{N} [(\prod_{j=1}^{J} Z_j^{Y_{ji}} - A_i)^2/\sigma_i^2]$$

which upon linearization becomes simply

$$\chi^2 = \sum_{i=1}^{N} r_i^2.$$

(These expressions are to be evaluated with the adjusted values of the variables.) Thus, χ^2 is the sum of the squares of the normalized residuals, a normalized residual being the deviation of a stochastic input datum from its least-squares adjusted value divided by (or normalized to) the *a priori* uncertainty of the datum. The expectation value of χ^2 is just $N-J$, where $N-J$ is the number of degrees of freedom. A generalized Birge ratio may be defined by $R=[\chi^2/(N-J)]^{1/2}$ and, as in the one variable case, is a measure of the error computed on the basis of external consistency relative to that computed on the basis of internal consistency. (The ϵ_k above is based on *internal* consistency and will be so used throughout.)

While χ^2 and R indicate the over-all consistency of the adjustment and therefore the general compatibility of the data, the normalized residual of a particular observational equation is a measure of the compatibility of the particular stochastic datum contained therein with the remaining stochastic data. If the experimental data were all normally distributed (Gaussian), each normalized residual would be characterized by a Gaussian distribution with mean zero and variance less than one. The sum of the variances and the expectation value of χ^2 is $N-J$, and the average contribution of each residual to the expected value of χ^2 is $(N-J)/N$. (For $N=J$, the variances and the normalized residuals would be all identically zero.) Thus, we may state as a rule of thumb that if the normalized residual of an observational equation is significantly greater than unity, the stochastic datum it contains is somewhat incompatible with the other data. (Henceforth, we shall simply refer to "normalized residuals" as "residuals.") On the other hand, if a residual is significantly less than one, then the stochastic datum is quite consistent with the other data. But it should be noted that eliminating one datum may change the least-squares values of the adjustable constants, Z_j, by amounts so large that the residual of a second datum changes from greater than unity (incompatible), to less than unity (compatible), or vice versa. In a way, the residual is a measure of how much a particular datum contributes to the over-all "strain" of the system. Removing a datum with a large residual (one causing a lot of strain) may enable the system to relax to a state of lower strain, i.e., one in which the remaining residuals will be smaller. (See Cohen and DuMond, 1965 for a further discussion of this point of view.) Note that the above considerations are invalid if the datum is disjoint, i.e., enters the adjustment in such a way that other data play no role in determining its adjusted value. In this case, the residual will be identically zero.

It should now be clear why deleting different items and studying the variation of the residuals and χ^2 or R are useful techniques for investigating the compatibility of data. Indeed, such a study was actually done in a limited way in Sec. III.A.1 with μ_p'/μ_n (Table XVII). Although the adjustment was in one unknown only (i.e., a weighted average), all of the above discussion applies. For example, for all five values of μ_p'/μ_n (Line 1 of Table XVII), the residuals were $r9=2.60$, $r10=0.92$, $r11=-1.15$, $r12=-1.49$, and $r13=0.42$. Deleting I11 and I12 gave $r9=0.46$, $r10=0.04$, and $r13=-0.82$. Clearly, removing these two items enables the system to relax to a state of lower strain.

Before concluding this section on the least-squares procedure, we point out that, in general, the uncertainties of the adjusted constants Z_j will be *correlated*. (This is one of the few disadvantages of the least-squares method.) As a result, when numerical values of other quantities are computed from two or more of the adjusted variables, the generalized law of error propagation must be used in order to obtain the correct uncertainties to be assigned the derived quantities. This generalized law requires not only the variances of the individual adjusted constants, but also the covariances associated with pairs of adjusted constants. The variance v_{kk} of the kth adjustable constant is the square of its standard deviation, $v_{kk}=\epsilon_k{}^2=(\mathbf{G}^{-1})_{kk}$, and the covariance of the jth and kth adjustable constants is simply $v_{jk}=(\mathbf{G}^{-1})_{jk}$. We thus require the off-diagonal elements of the error matrix as well as the diagonal elements. In Appendix A, we give the variance–covariance or error matrix for our final adjustment to obtain WQED values of the constants, the similar matrix for our final adjustment to obtain a set of recommended or best values of the constants, and a brief description (with an example) of how the matrices are used.

2. *Analysis of Variance*

In the course of investigating the over-all compatibility of the stochastic input data, we have carried out several hundred separate least-squares adjustments, including many involving particular subsets of the data, e.g., $\gamma_p{}'$ and K. Since our conclusions are the same as those reached in our preliminary (and considerably more transparent) search for discrepant data in the previous section, we shall discuss here only a few of the most significant features of the results.

Table XVIII summarizes five of the more important adjustments involving the non-x-ray input data. Adjustment No. 1 includes I1–I13 and has a Birge ratio of $R=1.30$ and $\chi^2=15.27$. The probability that χ^2 will exceed this value for 9 degrees of freedom is about 8%. An examination of the residuals shows that I5, I9, I11, and I12 are the main contributors to the incompatibility of the data. This is as expected since the indirect values of $\mu_p{}'/\mu_n$ derived from I5 as well as the

Table XVIII. Results of five least-squares adjustments involving the non-x-ray data. The number of unknowns or adjustable constants is four (α^{-1}, e, K, N), and the initial set of data consisted of I1–I13. The number of degrees of freedom, $N-J$, is therefore $13-4=9$, minus the number of items deleted. The quantity e is in units of 10^{-19} C, and N is in units of 10^{26} kmole^{-1}. Also given are the Birge ratio R, χ^2, and the residual of each stochastic input datum, r.

	Adjustment number and items deleted				
	1. None	2. I9–I13	3. I9, I10	4. I5, I9, I10	5. I11, I12
α^{-1}	137.03602(26)	137.03607(26)	137.03601(26)	137.03607(26)	137.03608(26)
e	1.6021858(80)	1.6021896(82)	1.6021845(80)	1.6021896(82)	1.6021901(81)
K	1.0000059(26)	1.0000086(29)	1.0000049(26)	1.0000086(29)	1.0000090(27)
N	6.022112(38)	6.022167(47)	6.022093(38)	6.021997(52)	6.022174(41)
R	1.30	0.25	1.12	0.51	0.42
χ^2	15.27	0.26	8.74	1.56	1.21
$N-J$	9	4	7	6	7

*r*1	0.00	0.00	0.00	0.00	0.00
*r*2	−0.44	−0.16	−0.54	−0.16	−0.12
*r*3	−0.44	−0.08	−0.56	−0.08	−0.04
*r*4	−0.35	0.11	−0.51	0.11	0.16
*r*5	−1.29	0.00	−1.73	Deleted	0.16
*r*6	−0.04	−0.23	0.03	−0.23	−0.26
*r*7	0.52	0.40	0.57	0.40	0.38
*r*8	0.81	−0.04	1.10	−0.04	−0.14
*r*9	2.18	Deleted	Deleted	Deleted	0.62
*r*10	0.74	Deleted	Deleted	Deleted	0.11
*r*11	−1.38	Deleted	−1.12	−0.71	Deleted
*r*12	−2.22	Deleted	−1.37	−0.07	Deleted
*r*13	0.18	Deleted	0.46	0.89	−0.72

two low values of μ_p'/μ_n, I9 and I10, disagree with the high values, I11 and I12. We note also that the residual of I1, the Josephson-effect value of $2e/h$, is identically zero because it is disjoint; there are no other values of $2e/h$, either directly measured or derivable from any of the other stochastic input data. Thus, since the Josephson-effect measurement of $2e/h$ determines α from γ_p (as well as F, K, and μ_p/μ_n), it may not be deleted until its disjointness is removed. This will occur when the x-ray data and direct measurements of α are included.

In adjustment No. 2, all of the direct measurements of μ_p'/μ_n have been deleted, and χ^2 and R become extremely small, indicating high compatibility. Again, this is as expected since we saw in Sec. III.A that all of the data, with the exception of those involving μ_p'/μ_n, are quite consistent. We also note that deleting I9–I13 makes the Faraday measurement, I5, disjoint and that the adjusted value of α^{-1} does not change significantly from the value obtained in adjustment No. 1. This constancy of the fine structure constant follows from the nature of the present data; α is primarily determined by $2e/h$ and γ_p, there is only one value of $2e/h$, and all of the γ_p data are highly consistent. Thus, whatever decisions are made concerning the μ_p'/μ_n measurements, α will be essentially unchanged.

In adjustment No. 3, we delete the two low values of μ_p'/μ_n, I9 and I10. The reduction in the incompatibility of the data (or the lowering of the strain in the system) over that present in the first adjustment is evidenced by the lower values of R and χ^2. The probability that χ^2 will equal or exceed 8.74 by chance for 7 degrees of freedom is about 27%. However, significant strain still remains due to the measurement of the Faraday, I5. If this last item is deleted, (adjustment No. 4) the strain is greatly reduced and the remaining data are seen to be highly compatible; the Birge ratio is only 0.51. Such consistency may also be achieved by deleting the two high values of μ_p'/μ_n, I11 and I12, and retaining *all* of the remaining data, *including* I5. This is shown by the fifth adjustment.

It is perhaps also worthwhile to summarize briefly the results of several least-squares adjustments involving the directly measured values of μ_p'/μ_n, I9–I13, and the indirect value of μ_p'/μ_n obtained from a least-squares adjustment which includes all of the data except I9–I19 (i.e., the value of μ_p'/μ_n resulting from adjustment No. 2 in Table XVIII.) Such adjustments are similar to those presented in Table XVII except that the indirect value of μ_p'/μ_n has now been included. This indirect value, which is really determined by the Faraday via Eq. (95), turns out to be equal to 2.792714(25) (8.9 ppm). (This is the value shown in Fig. 3 and labeled "indirect.") If a least-squares adjustment is carried out using these six values of μ_p'/μ_n, we find that $\mu_p'/\mu_n=2.792756(12)$ (4.3 ppm), $R=1.73$, and $\chi^2=15.02$. The probability that χ^2 will equal or exceed this value for 5 degrees of freedom is only 1%. Clearly, the data are rather inconsistent. If we now delete I12, the Mamyrin and Frantsuzov high value of μ_p'/μ_n, we find $\mu_p'/\mu_n=2.792720(17)$ (6.0 ppm), $R=1.17$, and $\chi^2=5.51$. For 4 degrees of freedom, the chance probability for χ^2 to equal or exceed this value is 24%. The over-all compatability of the data has thus improved considerably. If both the high values of μ_p'/μ_n, I11 and I12, are deleted, we obtain $\mu_p'/\mu_n=2.792709(17)$ (6.2 ppm), $R=0.56$, and $\chi^2=0.96$. Thus, the remaining data, i.e., I9, I10, I13, and the indirect value are highly consistent as expected. For the least-squares adjustment in which only I9, the Sommer, Thomas, and Hipple low value of μ_p'/μ_n is deleted, we find $\mu_p'/\mu_n=2.792767(13)$ (4.6 ppm), $R=1.53$, and $\chi^2=9.35$. The chance probability for χ^2 to equal or exceed this value for 4 degrees of freedom is 5%. The primary reason for this large value of χ^2 is of course the presence of the low *indirect* value of μ_p'/μ_n. If this is deleted along with I9, then the quite compatable adjustment No. 3 of Table XVII results.

Table XIX summarizes five least-squares adjustments involving the stochastic data I1–I19. The purpose of these adjustments is to investigate which group of μ_p'/μ_n measurements is more consistent with

TABLE XIX. Results of five least-squares adjustments including the x-ray data. The number of unknowns or adjustable constants is five ($\alpha^{-1}, e, K, N, \Lambda$), and the initial set of data consisted of I1–I19. The number of degrees of freedom, $N-J$, is therefore $19-5=14$, minus the number of items deleted. The quantity e is in units of 10^{-19} C, and N is in units of 10^{26} kmole^{-1}, R is the Birge ratio and r is the residual of the stochastic input datum indicated.

	Adjustment number and items deleted				
	6. I11, I12	7. I5, I9, I10	8. I11, I12, I19	9. I5, I9, I10, I19	10. I1, I11, I12
α^{-1}	137.03593(26)	137.03588(26)	137.03609(26)	137.03608(26)	137.03135(176)
e	1.6021929(80)	1.6021914(82)	1.6021898(81)	1.6021890(82)	1.6023553(621)
K	1.0000091(27)	1.0000076(29)	1.0000088(27)	1.0000082(29)	1.0000089(27)
N	6.022182(41)	6.022028(52)	6.022178(41)	6.022007(52)	6.021560(239)
Λ	1.0020697(50)	1.0020771(52)	1.0020747(52)	1.0020837(55)	1.0021076(152)
R	1.16	1.40	0.58	0.73	0.91
χ^2	16.05	21.50	3.69	5.28	9.09
$N-J$	12	11	11	10	11
$r1$	−0.24	−0.29	0.01	0.00	Deleted
$r2$	−0.11	−0.26	−0.14	−0.20	−0.13

r3	−0.02	−0.21	−0.05	−0.13	−0.04
r4	0.18	−0.06	0.14	0.04	0.15
r5	0.60	Deleted	0.24	Deleted	0.34
r6	0.17	0.34	−0.29	−0.24	−0.15
r7	0.65	0.76	0.36	0.40	0.45
r8	0.04	0.52	−0.12	0.07	−0.07
r9	0.47	Deleted	0.58	Deleted	0.55
r10	0.04	Deleted	0.09	Deleted	0.08
r11	Deleted	−0.87	Deleted	−0.75	Deleted
r12	Deleted	−0.59	Deleted	−0.20	Deleted
r13	−0.81	0.72	−0.74	0.85	−0.76
r14	−0.78	−0.99	−0.94	−1.21	0.13
r15	−0.30	−0.39	0.08	0.04	−0.03
r16	−1.23	−1.44	−0.32	−0.41	−0.57
r17	1.06	1.25	1.19	1.43	2.07
r18	0.42	0.24	0.23	−0.01	1.18
r19	−3.28	−3.73	Deleted	Deleted	−1.35

the x-ray results. In adjustment No. 6, the high values of μ_p'/μ_n have been deleted and the over-all compatibility of the data is fairly reasonable. The probability that χ^2 will equal or exceed 16.05 for 12 degrees of freedom is about 19%. However, we note that about two-thirds of the strain in the system is coming from $r19$, the Knowles Ta value of λ_C. In adjustment No. 7, the low values of μ_p'/μ_n have been deleted as well as the measurement of the Faraday, I5. Clearly the over-all compatibility of the data is considerably worse than that for adjustment No. 6; the probability of χ^2 equalling or exceeding 21.50 for 11 degrees of freedom is only 3%. We therefore conclude that the x-ray data are in somewhat better agreement with the low values of μ_p'/μ_n than with the high values. On the other hand, since I19 also accounts for nearly two-thirds of the strain in adjustment No. 7, it is of interest to see what happens when I19 is deleted. This is done in adjustments Nos. 8 and 9. The values of χ^2 become rather small as expected, the over-all compatability of the data being comparable with that obtained in the corresponding adjustments Nos. 5 and 4, Table XVIII, in which the x-ray data were completely excluded. However, in comparing adjustments Nos. 8 and 9 with each other, in particular, the residuals of I14–I18, we see that in the former, the compatibility of x-ray data with the other data is somewhat better than in the latter. Indeed, the sum of the squares of the residuals $r14$–$r18$ for adjustment No. 8 is 2.5, while for adjustment No. 7, it is 3.7. Our conclusion concerning the compatibility of the x-ray data with the lower values of μ_p'/μ_n would seem to remain. In adjustment No. 10, we have repeated adjustment No. 6 but with I1 deleted in order to see what values of the adjusted constants would result. Although of much larger uncertainty, they are not overwhelmingly inconsistent with the output values of adjustment No. 6; the difference between each of the corresponding adjusted values is within 2.7 times the standard deviation of the difference. (If I19 had been deleted as well, then each difference would be within a small fraction of its standard deviation.)

In view of the general compatibility of the x-ray measurements with the other data, one may well ask why we have said throughout that none of the x-ray results will be used in the adjustments to obtain best values of the constants. There are several reasons for this: (1) The direct measurements of Λ via ruled gratings are generally in very poor agreement with each other and cannot be trusted (see Secs. II.C.6 and II.C.7.d). (2) The two measurements of λ_C by Knowles, I18 and I19, are in significant disagreement with one another, and there is no obvious experimental explanation for the discrepancy. The more accurate value I19 seems to be the more incompatible of the two. (3) The history of the measurements of hc/e via the short-wavelength limit of the continuous x-ray spectrum shows the difficulty in obtaining reproducible and consistent results in this type of experiment (Bearden and Thomsen, 1957; Cohen and DuMond, 1965; Cohen, Crowe, and DuMond, 1957). The work of Spijkerman and Bearden, I14, although undoubtedly the best such measurement, must still be viewed cautiously. Furthermore, when this result with its 33-ppm error is included in a least-squares adjustment having as input data the reliable and accurate measurements of $N\Lambda^3$ by Bearden, and Henins and Bearden, I15 and I16, the effect is equivalent to introducing a value of $2e/h$ into the adjustment with an error of 33 ppm, nearly 14 times the error of the Josephson-effect value of $2e/h$. (This follows from the fact that the mean value of $N\Lambda^3$ in combination with the adjusted value of N gives a value of Λ with only a 5.3-ppm error.) Clearly, the Spijkerman and Bearden short-wavelength-limit result has much too large an uncertainty to be included in the same adjustment with the Josephson-effect value. (A similar situation also occurs with the indirect values of α^{-1} implied by Knowles' measurements of λ_C.) The only x-ray data we might therefore consider for inclusion are the two determinations of $N\Lambda^3$. However, if all of the other x-ray data are eliminated, then the two $N\Lambda^3$ measurements together become disjoint and need not be included in the adjustment to obtain Λ, i.e., Λ may be obtained later from the adjusted value of N and the weighted mean of the two experimental values of $N\Lambda^3$. That is the procedure followed in the present paper.

It is clear from the discussion in Sec. II.A and this section, as well as Tables XVIII and XIX, that we are faced with a dilemma concerning the measurements of μ_p'/μ_n. There are several possible solutions to this dilemma including: (1) We might retain all of the data, I1–I13, but expand the errors of the five direct μ_p'/μ_n measurements by a factor which will make them compatible, i.e., make the Birge ratio of their weighted average unity or less. (From Table XVII, Line 1, we see that a factor of 1.68 would give a value for R of unity.) In a similar vein, we might consider expanding the errors of the five direct values as well as the Faraday by a factor such that the direct *and* indirect values of μ_p'/μ_n would be compatible. (An error expansion of 1.73 would give a unity Birge ratio for the weighted average of the direct and indirect values.) (2) We might discard the two low values of μ_p'/μ_n, I9 and I10, and the Faraday measurement I5 (adjustment No. 4, Table XVIII). (3) We might discard the two high values of μ_p'/μ_n, I11 and I12, and keep everything else (adjustment No. 5, Table XVIII). After carefully weighing all of the available evidence which bears on the problem, we have decided upon this last solution. Our reasons are as follows: (1) The measurement of Boyne and Franken, I11, is experimentally suspect due to the slope–intercept correlation effect (see Sec. II.C.5). (2) The indirect value of μ_p/μ_n derived from Eq. (95) and the measurements of γ_p', F, and K is of low uncertainty and in excellent agreement with the directly measured low values of μ_p'/μ_n, I9 and I10, but in clear disagreement with the high values, I11 and I12 (see Sec. III.A.2 and this section). The NBS value of F, which is probably the most suspect quantity which enters Eq. (95), is, we believe, quite trustworthy in view of the great care taken by the experimenters and the over-all consistency of the measurements which were carried out under a wide variety of experimental conditions. Since the origins of this indirect value of μ_p'/μ_n are completely different than the directly measured values, we believe it should not be regarded as just another μ_p'/μ_n measurement but should have considerably more influence on our final decision regarding the handling of the μ_p'/μ_n problem. In particular, we believe there is absolutely no reason for expanding its error. (We note that including the expurgated values of γ_p' would tend to *lower* the mean indirect value of μ_p'/μ_n, thus favoring the lower direct values still more.) (3) Although they have relatively large uncertainties, Trigger's and Marion and Winkler's expurgated values of μ_p'/μ_n are in better agreement with I9 and I10 than with I11 and I12. Similarly, the x-ray data tend to support the lower values. We have not chosen the safe, conservative solution to the problem, that of expanding the errors assigned some or all of the experimental quantities, because we believe it amounts to sweeping the problem under the rug; such a procedure would throw away information and there are no convincing reasons for doing so. In essence, we feel it better to focus attention on the problem by discarding the questionable data rather than to cover the problem up by expanding errors in a manner which is difficult to justify.

The major flaw in our solution to the μ_p/μ_n problem is the fact that we have deleted what might be considered the best determination of μ_p/μ_n, that of Mamyrin and Frantsuzov. The nature of the experiment, the care with which it was carried out, and the thorough search made for systematic errors would seem to make it highly reliable. On the other hand, history clearly shows that systematic errors may be present even when experiments appear to be well under control, e.g., the well-known discrepancies in the measurements of the velocity of light carried out prior to 1950.* We believe that the evidence in favor of the low values of μ_p/μ_n indicates that such a systematic error may be present in the work of Mamyrin and Frantsuzov, even though

* See Table IV, and the footnote on the bottom of page 30 .

there are at present no obvious experimental reasons for believing that such an error actually exists. The decision to expurgate this measurement is therefore based solely on its incompatibility with the other data. Hopefully, the future will not prove this decision to be incorrect. But let us again emphasize that as far as our best WQED value of α is concerned (and its implications for QED), the particular decision made is irrelevant; the value of α is primarily determined by the Josephson-effect measurement of $2e/h$ and by the measurements of γ_p and is almost independent of μ_p/μ_n. Unfortunately this is not true for some of the other constants, in particular, μ_p'/μ_n and Avogadro's number. We therefore note that if we had in fact decided to expand the errors of the directly measured values of μ_p'/μ_n by the aforementioned factor of 1.68, then the adjusted value of μ_p'/μ_n using all the data I1–I13 would be 2.792743(17) (6.0 ppm), an increase of 12 ppm over the WQED value of 2.792709(17) (6.2 ppm) given in the next section. Similarly, the WQED value of Avogadro's number would decrease from $6.022174(41)\times 10^{26}$ kmole^{-1} (6.8 ppm) to $6.022128(41)\times 10^{26}$ kmole^{-1} (6.8 ppm), a change of 7.5 ppm. Increasing the individual errors of the Mamyrin and Frantsuzov and Boyne and Franken measurements by a still larger factor, perhaps justifiable because of the lack of detailed information regarding the former experiment and the experimental problems associated with the latter, would result in values of μ_p/μ_n and N intermediate between our WQED values and those just given above. (For values of the adjusted constants which would result if some of the other possible solutions to the μ_p/μ_n problem had been adopted, see Table XVIII.)

C. Final Adjustment to Obtain Best Values of the Constants without QED Theory

We present here the results of our final least-squares adjustment to obtain a best WQED value of α as well as a best or recommended set of WQED fundamental constants. As discussed in the previous section, we have decided to delete all of the x-ray data, I14–I19 (Table XVI), as well as the measurements of μ_p'/μ_n by Boyne and Franken and by Mamyrin and Frantsuzov, I11 and I12. The results of this adjustment (identical to adjustment No. 5, Table XVIII), together with a fairly complete set of physical constants derived from the adjusted quantities α^{-1}, e, K, and N (and the auxiliary constants as required), are given in Tables XX–XXIII. The standard-deviation errors quoted in these tables have been computed from the variance-covariance matrix presented and discussed in Appendix A. Although similar tables containing a final set of best or recommended constants based on QED as well as WQED data will be presented in Sec. V, we include these extensive tables of WQED constants here in order to facilitate evaluation of theoretical expressions with values of the constants which are free of uncertainties about the validity of QED (see Sec. IV).

From Table XX, the best WQED value of α^{-1} is seen to be

$$(\alpha^{-1})_{\mathrm{WQED}}=137.03608(26)\ (1.9\ \mathrm{ppm}). \qquad (98)$$

This value is 1.3 ppm higher than the value $\alpha^{-1}=$ 137.0359 derived by Parker, Taylor, and Langenberg (1967). There are two reasons for this. First, the value of $2e/h$ used here is the final value reported by Parker, Langenberg, Denenstein, and Taylor (1969); it exceeds the preliminary value reported by Parker, Taylor, and Langenberg, (1967) by about 1 ppm. Second, Parker, Taylor, and Langenberg calculated α using the 1963 adjusted value of γ_p (Cohen and DuMond, 1965) which is about 1.6 ppm larger than the mean value resulting from the present adjustment.

Several comments concerning some of the derived constants given in Tables XX through XXIII are in order:

Table XX. (1) The atomic mass of the proton has been calculated as in Sec. II.B.7 using the adjusted

TABLE XX.[a] A list of physical constants derived from WQED data. (See Table XXXII for our final recommended values.) The stochastic input data include I1–I13 with the exception of I11 and I12. (This is adjustment No. 5 of Table XVIII.) $R=0.416$, $\chi^2=1.212$, and $N-J=7$. The numbers in parentheses are the standard-deviation uncertainties in the last digits of the quoted value, computed on the basis of internal consistency.

Quantity	Symbol	Value	Error (ppm)	Units	
				SI	cgs
Velocity of light	c	2.9979250(10)	0.33	10^8 m sec^{-1}	10^{10} cm sec^{-1}
Fine-structure constant,	α	7.297348(14)	1.9	10^{-3}	10^{-3}
$[\mu_0 c^2/4\pi](e^2/\hbar c)$	α^{-1}	137.03608(26)	1.9		
Electron charge	e	1.6021901(81)	5.0	10^{-19} C	10^{-20} emu
		4.803246(24)	5.0		10^{-10} esu
Planck's constant	h	6.626186(57)	8.5	10^{-34} J·sec	10^{-27} erg·sec
	$\hbar=h/2\pi$	1.0545903(90)	8.5	10^{-34} J·sec	10^{-27} erg·sec
Avogadro's number	N	6.022174(41)	6.8	10^{26} kmole^{-1}	10^{23} mole^{-1}
Atomic mass unit	amu	1.660530(11)	6.8	10^{-27} kg	10^{-24} g
Electron rest mass	m_e	9.109553(56)	6.2	10^{-31} kg	10^{-28} g
	m_e^*	5.485931(34)	6.2	10^{-4} amu	10^{-4} amu
Proton rest mass	M_p	1.672613(11)	6.8	10^{-27} kg	10^{-24} g
	M_p^*	1.00727661(8)	0.08	amu	amu

Neutron rest mass	M_n	1.674919(11)	6.8	10^{-27} kg	10^{-24} g
	M_n^*	1.00866520(10)	0.10	amu	amu
Ratio of proton mass to electron mass	M_p/m_e	1836.109(11)	6.2		
Electron charge to mass ratio	e/m_e	1.7588022(56)	3.2	10^{11} C kg^{-1}	10^7 emu g^{-1}
		5.272757(17)	3.2		10^{17} esu g^{-1}
Magnetic flux quantum, $[c]^{-1}(hc/2e)$	Φ_0	2.0678527(74)	3.6	10^{-15} T·m^2	10^{-7} G·cm^2
	h/e	4.135705(15)	3.6	10^{-15} J·sec C^{-1}	10^{-7} erg·sec emu^{-1}
		1.3795227(50)	3.6		10^{-17} erg·sec esu^{-1}
Quantum of circulation	$h/2m_e$	3.636944(14)	3.8	10^{-4} J·sec kg^{-1}	erg·sec g^{-1}
	h/m_e	7.273888(28)	3.8	10^{-4} J·sec kg^{-1}	erg·sec g^{-1}
Faraday constant, Ne	F	9.648667(54)	5.6	10^7 C kmole^{-1}	10^3 emu mole^{-1}
		2.892598(16)	5.6		10^{14} esu mole^{-1}
Rydberg constant, $[\mu_0 c^2/4\pi]^2(m_e e^4/4\pi\hbar^3 c)$	R_∞	1.09737312(11)	0.10	10^7 m^{-1}	10^5 cm^{-1}
Bohr radius, $[\mu_0 c^2/4\pi]^{-1}(\hbar^2/m_e e^2)=\alpha/4\pi R_\infty$	a_0	5.291769(10)	1.9	10^{-11} m	10^{-9} cm
Classical electron radius $[\mu_0 c^2/4\pi](e^2/m_e c^2)=\alpha^3/4\pi R_\infty$	r_0	2.817935(16)	5.7	10^{-15} m	10^{-13} cm
Electron magnetic moment in Bohr magnetons	μ_e/μ_B	1.0011596384(33)	0.0033		

Table XX (*Continued*)

Quantity	Symbol	Value	Error (ppm)	Units	
				SI	cgs
Bohr magneton, $[c](e\hbar/2m_ec)$	μ_B	9.274079(79)	8.5	10^{-24} J T^{-1}	10^{-21} erg G^{-1}
Electron magnetic moment	μ_e	9.284833(79)	8.5	10^{-24} J T^{-1}	10^{-21} erg G^{-1}
Gyromagnetic ratio of protons in H_2O	γ_p' $\gamma_p'/2\pi$	2.6751260(86) 4.257595(14)	3.2 3.2	10^8 rad $sec^{-1} \cdot T^{-1}$ 10^7 Hz T^{-1}	10^4 rad $sec^{-1} \cdot G^{-1}$ 10^3 Hz G^{-1}
γ_p' corrected for diamagnetism of H_2O	γ_p $\gamma_p/2\pi$	2.6751956(86) 4.257706(14)	3.2 3.2	10^8 rad $sec^{-1} \cdot T^{-1}$ 10^7 Hz T^{-1}	10^4 rad $sec^{-1} \cdot G^{-1}$ 10^3 Hz G^{-1}
Magnetic moment of protons in H_2O in Bohr magnetons	μ_p'/μ_B	1.52099312(10)	0.066	10^{-3}	10^{-3}
Proton magnetic moment in Bohr magnetons	μ_p/μ_B	1.52103264(46)	0.30	10^{-3}	10^{-3}
Proton magnetic moment	μ_p	1.410618(12)	8.5	10^{-26} J T^{-1}	10^{-23} erg G^{-1}
Magnetic moment of protons in H_2O in nuclear magnetons	μ_p'/μ_n	2.792709(17)	6.2		
μ_p'/μ_n corrected for diamagnetism of H_2O	μ_p/μ_n	2.792782(17)	6.2		

Nuclear magneton, $[c](e\hbar/2M_pc)$	μ_n	5.050942(55)	11	10^{-27} J T^{-1}	10^{-24} erg G^{-1}
Compton wavelength of the electron, h/m_ec	λ_C $\lambda_C/2\pi$	2.4263075(93) 3.861588(15)	3.8 3.8	10^{-12} m 10^{-13} m	10^{-10} cm 10^{-11} cm
Compton wavelength of the proton, h/M_pc	$\lambda_{C,p}$ $\lambda_{C,p}/2\pi$	1.3214399(93) 2.103137(15)	7.1 7.1	10^{-15} m 10^{-16} m	10^{-13} cm 10^{-14} cm
Compton wavelength of the neutron, h/M_nc	$\lambda_{C,n}$ $\lambda_{C,n}/2\pi$	1.3196207(93) 2.100242(15)	7.1 7.1	10^{-15} m 10^{-16} m	10^{-13} cm 10^{-14} cm
Gas constant	R_0	8.31434(35)	42	10^3 J $kmole^{-1} \cdot K^{-1}$	10^7 erg $mole^{-1} \cdot K^{-1}$
Boltzman's constant, R_0/N	k	1.380621(59)	43	10^{-23} J K^{-1}	10^{-16} erg K^{-1}
Stefan–Boltzman constant, $\pi^2k^4/60\hbar^3c^2$	σ	5.66962(96)	170	10^{-8} W m^{-2} K^4	10^{-5} erg sec^{-1} $cm^{-2} \cdot K^{-4}$
First radiation constant, $8\pi hc$	c_1	4.992571(43)	8.5	10^{-24} J·m	10^{-15} erg·cm
Second radiation constant, hc/k	c_2	1.438831(61)	43	10^{-2} m·K	cm·K
Gravitational constant	G	6.6732(31)	460	10^{-11} N·m^2 kg^{-2}	10^{-8} dyn·cm^2 g^{-2}

[a] Note that the unified atomic mass scale $^{12}C \equiv 12$ has been used throughout, that amu = atomic mass unit, C = coulomb, G = gauss, Hz = hertz = cycles/sec, J = joule, K = kelvin (degrees kelvin), T = tesla (10^4 G), V = volt, and W = watt. In cases where formulas for constants are given (e.g., R_∞), the relations are written as the product of two factors. The second factor, in parentheses, is the expression to be used when all quantities are expressed in cgs units, with the electron charge in electrostatic units. The first factor, in brackets, is to be included only if all quantities are expressed in SI units. We remind the reader that with the exception of the auxiliary constants which have been taken to be exact, the uncertainties of these constants are correlated, and therefore the general law of error propagation must be used in calculating additional quantities requiring two or more of these constants. (See Appendix A; for further comments on the table, see text.)

TABLE XXI. The WQED energy conversion factors. (See text for discussion and Table XXXIII for our final recommended values.)

Quantity	Value	Unit	Error (ppm)
1 kg	5.609543(28)	10^{29} MeV	5.0
1 amu	931.4814(52)	MeV	5.6
Electron mass	0.5110043(16)	MeV	3.2
Proton mass	938.2595(52)	MeV	5.6
Neutron mass	939.5529(52)	MeV	5.6
1 electron volt	1.6021901(81)	10^{-19} J	5.0
		10^{-12} erg	
	2.4179671(87)	10^{14} Hz	3.6
	8.065469(29)	10^{5} m^{-1}	3.6
		10^{3} cm^{-1}	
	1.160485(49)	10^{4} K	42
Energy–wavelength conversion	1.2398535(45)	10^{-6} eV·m	3.6
		10^{-4} eV·cm	

Rydberg constant, R_∞	2.179911(19)	10^{-18} J	8.5
		10^{-11} erg	
	13.605819(49)	eV	3.6
	3.2898423(11)	10^{15} Hz	0.35
	1.578935(67)	10^{5} K	43
Bohr magneton, μ_B	5.788376(22)	10^{-5} eV T^{-1}	3.8
	1.3996103(45)	10^{10} Hz T^{-1}	3.2
	46.68597(15)	$m^{-1} \cdot T^{-1}$	3.2
		10^{-2} $cm^{-1} \cdot T^{-1}$	
	0.671732(29)	K T^{-1}	43
Nuclear magneton, μ_n	3.152523(22)	10^{-8} eV T^{-1}	7.1
	7.622698(43)	10^{6} Hz T^{-1}	5.6
	2.542658(14)	10^{-2} $m^{-1} \cdot T^{-1}$	5.6
		10^{-4} $cm^{-1} \cdot T^{-1}$	
	3.65846(16)	10^{-4} K T^{-1}	44
Gas constant, R_0	8.20562(35)	10^{-2} $m^3 \cdot$ atm $kmole^{-1} \cdot K^{-1}$	42
Standard volume of ideal gas, V_0	22.4136	m^3 $kmole^{-1}$	

TABLE XXII. The WQED values of various quantities involving as-maintained electrical units. (See text for further explanation and Table XXXIV for our final recommended values.)

Quantity	Symbol	Value (Prior to 1 Jan. 1969)	Value (After 1 Jan. 1969)	Error (ppm)	Units
Ratio of NBS ampere to absolute ampere	$K \equiv A_{NBS}/A_{ABS}$	1.0000090(27)	1.0000006(27)	2.7	
Ratio of NBS volt to absolute volt	V_{NBS}/V_{ABS}	1.0000086(27)	1.0000002(27)	2.7	
Ratio of BIPM ampere to absolute ampere	A_{BIPM}/A_{ABS}	1.0000114(27)	1.0000004(27)	2.7	
Ratio of BIPM volt to absolute volt	V_{BIPM}/V_{ABS}	1.0000112(27)	1.0000002(27)	2.7	

Ratio absolute ohm to NBS ohm	$\Omega_{ABS}/\Omega_{NBS}$	1.00000036(70)	1.00000036(70)	0.7	
Ratio absolute ohm to BIPM ohm	$\Omega_{ABS}/\Omega_{BIPM}$	1.00000017(70)	1.00000017(70)	0.7	
Josephson frequency–voltage ratio	$2e/h$	4.835976(12)	4.835935(12)	2.4	10^{14} Hz V_{NBS}^{-1}
Faraday constant	F	9.648581(55)	9.648662(55)	5.7	10^{7} $A_{NBS}\cdot$sec kmole^{-1}
Gyromagnetic ratio (low field) of protons in H_2O	γ_p' $\gamma_p'/2\pi$	2.6751500(80) 4.257633(13)	2.6751275(80) 4.257598(13)	3.0 3.0	10^{8} rad sec$^{-1}\cdot T_{NBS}^{-1}$ 10^{7} Hz T_{NBS}^{-1}
γ_p' (low field) corrected for diamagnetism of H_2O	γ_p $\gamma_p/2\pi$	2.6752195(80) 4.257744(13)	2.6751971(80) 4.257708(13)	3.0 3.0	10^{8} rad sec$^{-1}\cdot T_{NBS}^{-1}$ 10^{7} Hz T_{NBS}^{-1}
Voltage–wavelength conversion, $V\lambda$	hc/e	1.2398428(30)	1.2398532(30)	2.4	10^{-6} $V_{NBS}\cdot$m

values of the constants, but the result is identical to that given in Table XI. The atomic mass of the neutron is that given by Mattauch *et al.* (1965).

(2) The value of μ_e/μ_B has been calculated from the theory [Eq. (19)] using the adjusted value of α, and it has been assumed that the uncertainty in the coefficient of the $(\alpha/\pi)^3$ term is ± 0.20. The values of μ_p/μ_B and μ_p'/μ_B are those given in Table XI; the difference between the present value of μ_e/μ_B and the value originally used to calculate these quantities is sufficiently small to leave them unchanged.

(3) The prime on γ_p and μ_p means, as before, "for protons in a spherical sample of H_2O." In correcting for the diamagnetism of H_2O, we have assumed a diamagnetic shielding constant of 26.0 ppm (see Sec. II.B.6).

(4) The gas constant is that derived by Cohen and DuMond (1965) from several different measurements.

(5) The gravitational constant is the value determined by Heyl and Chrzanowski (1942) [see also Cook (1968)].

Table XXI: (1) Note that (a) the thermodynamic scale is defined by 273.16 kelvin (K) = the triple point of H_2O; (b) one normal atmosphere equals 101325 N/m^2; (c) we choose to ignore the calorie because of the arbitrariness of the unit. For the convenience of the reader, we note that one thermochemical calorie equals 4.184 J, and one international steam table calorie equals 4.1868 J. We also note that the liter has recently been redefined by the International Committee of Weights and Measures (CIPM) to be exactly 10^{-3} m^3 (Page, private communication).

(2) The value of V_0 is taken from Cohen and DuMond (1965).

Table XXII. (1) Two values for $K \equiv A_{NBS}/A_{ABS}$ are given. The first corresponds to the units as-maintained by NBS prior to 1 January 1969. The second corresponds to the units as maintained after this date. The difference is due to the fact that the CIPM, in their 1968 session, approved a recommendation of the BIPM Advisory Committee on Electricity that the BIPM as-maintained volt should be adjusted downwards (effective 1 January 1969) by 11 ppm to bring it into better agreement with the absolute volt (Page, private communication; Terrien, private communication). The BIPM as-maintained ohm is to remain unchanged since, to within experimental error, it is equal to the absolute ohm (see Sec. II.B.3). Other national laboratories are also to adjust their as-maintained volts so that they too are in better agreement with the absolute volt. The exact correction is to be determined from the 1967 comparisons. Thus, since in 1967, $V_{NBS} = V_{BIPM} - 2.58\ \mu V$ (see Table I), the NBS volt will be corrected by exactly $(11.0-2.6)$ ppm$=8.4$ ppm. Since the NBS as-maintained ohm will not be changed, this is the correction we have applied to the value of K obtained from our least-squares adjustment in order to calculate a value of K applicable to the post-1 January 1969 units. A similar correction of 8.4 ppm has been applied to the other quantities which have been expressed in terms of these new units.

(2) In addition to K, we also give a value for V_{NBS}/V_{ABS}. It has been calculated from K using the Thompson result [Eq. (16b)], $\Omega_{ABS}/\Omega_{NBS} = 1.00000036$. The required relation is

$$\frac{V_{NBS}}{V_{ABS}} = \frac{A_{NBS}\Omega_{NBS}}{A_{ABS}\Omega_{ABS}} = \frac{K}{\Omega_{ABS}/\Omega_{NBS}}.$$

Since $\Omega_{ABS}/\Omega_{NBS}$ has been taken throughout the present work to be an auxiliary constant, we have assumed it to be exact in obtaining the quantities given in Table XXII. Using the results of the 1967 comparisons, Tables I and II, we have also calculated A_{BIPM}/A_{ABS} and V_{BIPM}/V_{ABS} from the corresponding NBS quantities.

Table XXIII. (1) We believe the most reliable x-ray measurements from which $\Lambda = \lambda(\text{Å})/\lambda(\text{kxu})$ may be obtained are the two determinations of $N\Lambda^3$ by Bearden and Henins and Bearden, I15 and I16. Their weighted mean is

TABLE XXIII. The WQED values of various quantities involving the kilo-x-unit and angstrom-star x-ray scales. (See text for a detailed discussion and Table XXXV for our final recommended values.)

Quantity	Symbol	Value	Error (ppm)	Units
kx-unit-to-angstrom conversion factor, $\Lambda = \lambda(\text{Å})/\lambda(\text{kxu})$; $\lambda(\text{Cu}K\alpha_1) \equiv 1.537400$ kxu	Λ	1.0020762(53)	5.3	
Å*-to-angstrom conversion factor, $\Lambda = \lambda(\text{A})/\lambda(\text{Å}^*)$; $\lambda(\text{W}K\alpha_1) \equiv 0.2090100$ Å*	Λ^*	1.0000195(56)	5.6	
Voltage–wavelength conversion product, $V\lambda = hc/e$	$V\lambda(\text{kxu})$	1.2372851(70)	5.6	10^4 V·kxu
	$V\lambda(\text{Å}^*)$	1.2398298(73)	5.9	10^4 V·Å*
$V\lambda$ in NBS units prior to 1 Jan. 1969	$V_{NBS}\lambda(\text{kxu})$	1.2372740(66)	5.3	10^4 V_{NBS}·kxu
	$V_{NBS}\lambda(\text{Å}^*)$	1.2398187(70)	5.6	10^4 V_{NBS}·Å*
$V\lambda$ in NBS units after 1 Jan. 1969	$V_{NBS}\lambda(\text{kxu})$	1.2372844(66)	5.3	10^4 V_{NBS}·kxu
	$V_{NBS}\lambda(\text{Å}^*)$	1.2398291(70)	5.6	10^4 V_{NBS}·Å*
Compton wavelength of the electron	$\lambda_C(\text{kxu})$	2.421280(14)	5.9	10^{-2} kxu
	$\lambda_C(\text{Å}^*)$	2.426260(15)	6.2	10^{-2} Å*

$$(N\Lambda^3)_{\text{Av}} = 6.059761(88) \times 10^{26} \text{ kmole}^{-1} \text{ (14.5 ppm)}. \tag{99}$$

When combined with the adjusted value of N in Table **XX**, it yields

$$(\Lambda)_{\text{WQED}} = 1.0020762(53) \text{ (5.3 ppm)}. \tag{100}$$

This is the value given in Table **XXIII**. Of course, values of the WQED adjusted constants can be combined with the remaining x-ray measurements, I14, I18, and I19, to obtain other values of Λ, as was done to derive Eq. (97). The results are

$$\begin{aligned} &\text{I14(SB)}: && \Lambda = 1.002043(34) \text{ (34 ppm)}, \\ &\text{I17(B, T)}: && \Lambda = 1.002030(38) \text{ (38 ppm)}, \\ &\text{I18(Kn, } H_2O\text{)}: && \Lambda = 1.002083(38) \text{ (38 ppm)}, \\ &\text{I19(Kn, Ta)}: && \Lambda = 1.002018(16) \text{ (15.5 ppm)}. \end{aligned} \tag{101}$$

With the exception of Knowles' measurement of λ_C using Ta, these values agree reasonably well with our recommended value, Eq. (100). We choose not to give as a recommended WQED value a weighted average of the different Λ's because of the much greater accuracy and reliability of the value derived from the $N\Lambda^3$ results.

(2) We also give in Table **XXIII** a value for $\Lambda^* = \lambda(\text{Å})/\lambda(\text{Å}^*)$, the "angstrom star"-to-angstrom conversion factor (see Sec. II.C.6). This has been obtained from the relation

$$\Lambda^* = \Lambda/(1.00205667 \pm 1.8 \text{ ppm}), \tag{102}$$

which follows from the definition of the kx-unit scale we have been using, $\lambda(\text{Cu}K\alpha_1) \equiv 1.537400$ kxu, the ratio $\lambda(\text{Cu}K\alpha_1)/\lambda(\text{W}K\alpha_1) = 7.370757 \pm 1.2$ ppm (P.E.) as given by Bearden (Bearden, 1967), and the definition of the Å* scale, $\lambda(\text{W}K\alpha_1) \equiv 0.2090100$ Å* (Bearden, 1967).

IV. Implications for Quantum Electrodynamics

A. g Factors

1. Free Electron and Positron

We shall now discuss the implications of the new WQED value of α derived in Sec. III for quantum electrodynamics.* We begin with the g factors for the free electron and positron. The deviation of these g factors from 2 is a purely QED effect arising from the virtual emission and absorption of photons and the polarization of the vacuum with electron–positron pairs. Thus, calculations of these so-called radiative corrections and comparison of the theoretical value of g with experimental values provide a critical test of QED. Some discussion of g_s^-, the free-electron g factor, has already been presented in Sec. II.B.5. Indeed, we gave there [Eq. (19)] the present "best" QED expression for μ_e/μ_B, from which we now write the electron g factor anomaly defined by $a_e^- \equiv (g_s^- - 2)/2$:

$$a_e^- = \tfrac{1}{2}(\alpha/\pi) - 0.3285(\alpha/\pi)^2 + 0.13(\alpha/\pi)^3. \qquad (103)$$

The coefficient of the sixth-order term $[(\alpha/\pi)^3]$ is that given by Parsons (1968) as calculated using dispersion theory and is estimated only within reasonable bounds. The quantity "0.3285" stands of course for the closed form expression for the coefficient of the fourth-order term $[(\alpha/\pi)^2]$, Eq. (18). It is believed to be exact and has been obtained independently by several workers using rather different methods (Sommerfield,

* See also Drell (1969) and Hughes (1969).

1957; 1958; Petermann, 1957a; 1958a). [See also Smrz and Ulehla (1960) and Terent'ev (1962).] [An additional contribution to a_e of order $(\alpha/\pi)^2$ has recently been calculated and will be given in the next section. It depends on m_e/m_μ, where m_μ is the mass of the muon, but is so small that it may be neglected; it contributes less than 0.003 ppm to a_e.] The g factor for the positron, g_s^+, is predicted to be identical to g_s^- by CPT invariance which requires that the masses and magnetic moments of single particle states of a particle–antiparticle set be identical [see, for example, Sakurai (1964)].

Using $(\alpha^{-1})_{\mathrm{WQED}} = 137.03608(26)$ (1.9 ppm), our least-squares-adjusted, WQED result, Eq. (103) gives for the g factor anomaly,

$$a_e^- = a_e^+ = 0.0011596384(33) \ (2.9 \text{ ppm}), \qquad (104)$$

where it has been assumed that the uncertainty in the sixth-order coefficient is ± 0.20.* We note that: (1) The entire sixth-order term contributes only 1.6×10^{-9} to a_e or 16 in the last two places (+1.4 ppm); (2) the 1.9-ppm uncertainty in α corresponds to $0.18(\alpha/\pi)^3 = 2.2\times 10^{-9}$ or 22 in the last two places and is therefore larger than the entire sixth-order term; (3) the "5" in 0.3285, the coefficient of the fourth-order term, must be included since it corresponds to $-0.22(\alpha/\pi)^3 = -2.7\times 10^{-9}$ or 27 in the last two places (−2.3 ppm; we have actually used −0.32847897 throughout our calculations).

As indicated in Sec. II.B.5, the only really accurate measurement of a_e^- is that of Wilkinson and Crane (WC) (1963), who found $a_e^- = 0.001159622(27)$ (23 ppm). In their experiment, a_e^- is determined by measuring directly the difference between the spin precession frequency and the cyclotron frequency for electrons in a magnetic field. Electrons are initially polarized by right-angle Mott scattering from a gold foil, allowed to enter a magnetic bottle where they are

* See asterisk footnote, page 48.

trapped with the aid of an electric field for a measured time, released from the trap, and finally Mott-scattered from a second gold foil into a detector. The number of electrons scattered into the detector is a function of the final direction of the electron polarization which in turn depends on the length of time spent in the magnetic field. Thus, a plot of detector output vs trapping time is a cosine curve whose frequency is the difference between the cyclotron or orbital frequency and the spin precession frequency and is directly related to a_e^-.

As noted in Sec. II.B.5, Rich (R) (1968a; 1968b; 1968c; see also Henry and Silver, 1969) has recently corrected the original value reported by Wilkinson and Crane for relativistic effects not originally included and has also carefully reconsidered the calculation of the mean value of the magnetic field. He finds that the additional relativistic corrections decrease a_e^- by about 25 ppm and that reevaluation of the magnetic field leads to an additional decrease of about 40 ppm. This last change results from an improved numerical integration technique which takes into account in a precise way the relatively large amount of time the electrons spend near their turning points in the magnetic bottle. The final result is

$$(a_e^-)_{\mathrm{WC,R}} = 0.001159549(30) \ (26 \text{ ppm}), \quad (105)$$

63 ppm less than the original value.

A comparison of this experimental result with the theoretical result, Eq. (104), shows that they are in poor agreement:

$$a_e(\text{exptl}) - a_e(\text{theory}) = (-89 \pm 30) \times 10^{-9} [(-77 \pm 26) \text{ ppm}].$$

The difference is 3.0 times the standard deviation of the difference, and the probability for this to occur by chance is about 0.3%. The disagreement between theory and experiment can be shown in an alternate way by writing the experimental result in the form

$$(a_e^-)_{\mathrm{WC,R}} = \frac{1}{2}\left(\frac{\alpha}{\pi}\right) - (0.345 \pm 0.006)\left(\frac{\alpha}{\pi}\right)^2 + 0.13\left(\frac{\alpha}{\pi}\right)^3,$$

using α_{WQED}. The magnitude of the "experimental" coefficient of the $(\alpha/\pi)^2$ term exceeds the 0.3285 derived from QED by 0.016, well outside the experimental error. It is also of interest to calculate an experimental value for the coefficient of the sixth-order term using α_{WQED} and assuming the theoretical values for the coefficients of the second- and fourth-order terms. The result is

$$(a_e^-)_{\mathrm{WC,R}} = \tfrac{1}{2}(\alpha/\pi) - 0.3285(\alpha/\pi)^2 - (7.0 \pm 2.4)(\alpha/\pi)^3.$$

The experimental coefficient differs in sign from the +0.13 predicted by QED, is about 54 times as large, and exceeds in magnitude the coefficient of the fourth-order term by over a factor of 20. This seems rather unreasonable. It therefore appears that we are faced with a clear-cut discrepancy between QED and experiment.* However, this discrepancy should perhaps not be taken overly seriously at present since it arises primarily from the corrections applied to the original Wilkinson and Crane result. The introspective reader will appreciate that any worker's motivation to carry a complex calculation to still higher order or to ferret out obscure sources of systematic error in a difficult experiment drops off rapidly as soon as the calculation agrees with experiment or the experiment agrees with theory. Any experimental result which originally agrees with theory and then disagrees after *a posteriori* correction or because the theory is modified should be viewed with some caution, since the correction may leave untouched some source of error in the original experiment whose presence might have been detected were it not for the

---

* The possibility that the discrepancy could be due to an electric dipole moment for the electron would appear to be ruled out by the recent work of Weisskopf, Carrico, Gould, Lipworth, and Stein (1968). These workers showed that such a moment, if it exists, must be less than 3×10^{-24} $e \cdot$cm. See also Carrico *et al.* (1968).

TABLE XXIV. Summary of measurements of the electron g-factor anomaly a_e. Those experiments which measure the difference between the electron spin precession frequency and cyclotron frequency, $\omega_s - \omega_c$, determine a_e directly, while those which measure the ratio ω_s/ω_c determine $(1+a_e) = g_s/2$.

Year of publication and authors	Value
1968, Gräff, Major, Roeder, and Werth	1.001159(2)
1968, Klein plus Lambe (1959)	1.00115980(49)
1966, Rich and Crane (positron)	0.001168(11)
1963, Wilkinson and Crane [corrected by Rich (1968); see also Henry and Silver (1969)]	0.001159549(30)
1963, Farago, Gardner, Muir, and Rae	0.001153(23)
1961, Schupp, Pidd, and Crane, (uncorrected)	0.0011609(24)
Theory, $\alpha_{WQED}^{-1} = 137.03608(26)$ (1.9 ppm)	0.0011596384(33)

unfortunate initial agreement between experiment and theory. (The original result of Wilkinson and Crane agreed well with the theoretical value of a_e^- calculated using the *then* accepted value of α, and it would agree to within its assigned error with our newer value.)

There are other values of g_s in the literature besides that of Wilkinson and Crane but they are of relatively large uncertainty and therefore of limited usefulness (see Table XXIV). A value for a_e^- was reported by the Michigan group in 1961 (Schupp, Pidd, and Crane, 1961) using an earlier version of the apparatus of Wilkinson and Crane. They found $a_e^- = 0.0011609(24)$, but corrections like those applied to the Wilkinson and Crane result by Rich are probably required. Farago, Gardiner, Muir, and Rae (1963) [see also Galbraith and Gardiner (1968)] obtained the value $a_e^- = 0.001153(23)$ using a technique similar to that of Wilkinson and Crane. The most recent measurement is that of Gräff, Major, Roeder, and Werth (1968), who measured the ratio of the spin resonance frequency ω_s to the cyclotron frequency ω_c of free electrons confined in a small volume; ω_s was observed through spin-dependent collision processes with polarized Na atoms and ω_e, by the rapid loss of electrons when the cyclotron motion was excited by an rf field. They found $\omega_s/\omega_e = g_s/2 = 1.001159(2)$. A value of g_s of some interest may be derived from the measurement of $\omega_p(H_2O)/\omega_e = \mu_p'/\mu_B$ by Klein in combination with the Lambe measurement of μ_e/μ_p' (see Sec. II.B.6). Noting that $g_s/2 = \mu_e/\mu_B = (\mu_p'/\mu_B)(\mu_e/\mu_p')$, we obtain from Eqs. (27) and (24) $g_s/2 = 1.00115980(49)$ (0.49 ppm) or $a_e^- = 0.00115980(49)$ (420 ppm). This result agrees with that predicted by theory, as can be seen by comparing it directly with Eq. (104) or by writing

$$(a_e^-)_{\mathrm{K,L}} = \tfrac{1}{2}(\alpha/\pi) - (0.30 \pm 0.09)(\alpha/\pi)^2.$$

However, the error of this Klein–Lambe value is so large that it is in equally good agreement with the experimental result of Wilkinson and Crane (as revised by Rich).

The only measurement of the g factor anomaly for free positrons is that of Rich and Crane (1966). They give

$$a_e^+ = 0.001168 \pm 0.0000055.$$

(Note that the original error of ± 0.000011 quoted by Rich and Crane corresponds to two standard deviations.) The method used was very similar to that employed by Wilkinson and Crane for electrons but no initial polarization by Mott scattering was necessary since the positrons were already polarized upon emission from the ^{58}Co source. The direction of polarization of the positrons after being trapped in the magnetic field for a given time was determined by detecting gamma quanta from the annihilation of triplet-state positronium formed when the positrons were stopped in a plastic scintillator in a 1.3-T magnetic field. (The fraction of triplet positronium formed is different for positron spins parallel and antiparallel to the field.) The measured value of a_e^+ exceeds the theoretical value, Eq. (104), by 1.5 times the standard deviation of the difference, but the experimental error is quite large.

Experiment and theory may also be compared by writing

$$(a_e^+)_{\mathrm{RC}}=\tfrac{1}{2}(\alpha/\pi)+(1.2\pm1.0)(\alpha/\pi)^2,$$

where the experimental coefficient of the $(\alpha/\pi)^2$ term is to be compared with the theoretical coefficient of -0.3285. It is also instructive to compare a_e^+ and a_e^- directly. Using the revised value of Wilkinson and Crane, Eq. (105), the result is

$$(a_e^+-a_e^-)_{\mathrm{RC,WC,R}}=(8.4\pm5.5)\times10^{-6},$$

in only fair agreement with the expected theoretical value of zero. However, this is obviously not a particularly critical test of CPT in view of the large uncertainty associated with a_e^+. We will have to wait for the completion of several e^+ and e^- g-factor experiments presently underway before a more stringent test can be made and before any firm conclusions can be drawn about the possible inadequacy of the present QED expression for g_s. (See also Notes Added in Proof.)

2. Free Muon

The present point of view concerning the muon is that it is simply a heavy electron, i.e., an elementary particle with a mass m_μ some 207 times larger than m_e but one which interacts with the electromagnetic field, has weak interactions, *etc.*, in exactly the same manner as does the electron. A critical comparison of the experimental and theoretical values of g_μ, the g factor of the muon, is therefore of great importance because it may indicate that there are some basic differences in the interaction properties of muons and electrons, or the existence of new couplings (Kinoshita, 1968). This comparison also provides an important test for possible deficiencies in the present formulation of QED. Any such breakdown of QED might be observable in g_μ since the muon "energy scale" is m_μ/m_e times that of the electron and therefore tests QED at shorter distances.

The theoretical QED expression for g_μ or the muon g-factor anomaly a_μ [$g_\mu=2(1+a_\mu)$] may be written in a power series involving α in the same manner as for the electron. Since the positive and negative muons are antiparticles, we have

$$a_\mu{}^+ = a_\mu{}^- = B_1(\alpha/\pi) + B_2(\alpha/\pi)^2 + B_3(\alpha/\pi)^3 + \cdots . \quad (107)$$

Because the muon is believed to interact with the electromagnetic field in the same manner as the electron, the second-order radiative correction should be the same as for the electron since it is independent of mass. Thus, $B_1 = \frac{1}{2}$ as for the electron (see Sec. II.B.5). Similarly, the fourth-order radiative corrections must contain a contribution equal to the $-0.3285(\alpha/\pi)^2$ characteristic of the electron since it too is mass independent. [Throughout, -0.3285 will stand for the exact expression, Eq. (18)]. However, for the muon there is an additional fourth-order contribution due to the fact that a virtual photon emitted by a muon may polarize the vacuum with a virtual electron–positron pair. This contribution has been calculated by Elend (1966a; 1966b), who gives

$$\frac{1}{3}\left(\frac{\alpha}{\pi}\right)^2 \left\{ \ln \frac{m_\mu}{m_e} - \frac{25}{12} + \frac{3\pi^2}{4}\frac{m_e}{m_\mu} + 3\left(3 - 4\ln\frac{m_\mu}{m_e}\right)\left(\frac{m_e}{m_\mu}\right)^2 + O\left(\frac{m_e}{m_\mu}\right)^3 \right\} \quad (108)$$

[see also Suura and Wichmann (1957) and Petermann, (1957b; 1958b)]. There is actually a similar fourth-order contribution to the electron moment due to the polarization of the vacuum with muon pairs, but it is not usually mentioned since it is so small. It has been calculated recently by Lautrup and de Rafael (1968) and is

$$\left\{ \frac{1}{45}\left(\frac{m_e}{m_\mu}\right)^2 + O\left[\left(\frac{m_e}{m_\mu}\right)^4 \ln \frac{m_\mu}{m_e}\right] \right\} \left(\frac{\alpha}{\pi}\right)^2 = (5.20\times 10^{-7}) \left(\frac{\alpha}{\pi}\right)^2, \quad (108a)$$

using the value of m_μ/m_e to be derived below.* This implies that the numerical value of the coefficient of the $(\alpha/\pi)^2$ term [Eq. (18)] in the theoretical expression for a_e is no longer $-0.32847897\cdots$, but more correctly, $-0.32847845\cdots$. The change is of course entirely negligible, i.e., ≈ 0.002 ppm in a_e.

The required mass ratio m_μ/m_e can best be obtained from measurements of μ_μ/μ_p', μ_e/μ_p', g_s, and g_μ via the equation

$$\frac{m_\mu}{m_e}=\frac{g_\mu}{g_s}\frac{\mu_e/\mu_p'}{\mu_\mu/\mu_p'}. \tag{109}$$

This follows from the fact that $g_\mu/2=\mu_\mu(e\hbar/2m_\mu)^{-1}$, where μ_μ is the muon magnetic moment, and $g_s/2=\mu_e(e\hbar/2m_e)^{-1}=\mu_e/\mu_B$.† The quantity μ_μ/μ_p' is derivable from the measurements of Hutchinson, Menes, Shapiro, and Patlach (1963a; 1963b). These workers measured the ratio of the spin precession frequency of positive muons stopped in water and aqueous HCl solutions to that of protons in water in the same magnetic field. They obtained

$$\omega_\mu'/\omega_p'=\mu_\mu'/\mu_p'=3.18338(4)\ (13\ \text{ppm}). \tag{110}$$

The prime on ω_μ and μ_μ means as obtained for muons in water. Similar experiments by Bingham (1963) gave

---

* Recently, Erickson and Liu (1968) have carried out an exact evaluation (in terms of Spence functions) of these vacuum polarization contributions to the magnetic moment of both the electron and muon. The analytical form of the contribution is the same in both cases and reduces to Eq. (108) or Eq. (108a) in the appropriate mass ratio limit (i.e., the mass of the virtual particle much greater than or much less than the mass of the particle under consideration). Erickson and Liu give $(1/70)\,[(9/280)-\ln(m_\mu/m_e)]$. as the coefficient of the $(m_e/m_\mu)^4$ factor in Eq. (108a).

† It will be assumed throughout not only that $e_\mu=e$, but also that $e_\mu^+=e_\mu^-$, $m_\mu^+=m_\mu^-$, and $g_\mu^+=g_\mu^+$. The present level of accuracy of the muon experiments does not warrant distinguishing between measurements involving positive muons and negative muons. See Feinberg and Lederman (1963) for a review of the physics of muons and related work.

essentially the same results as those of Hutchinson, Menes, Shapiro, and Patlach but with a slightly larger uncertainty. (Both Hutchinson *et al.* and Bingham were able to demonstrate that the magnetic moments of positive and negative muons are the same to at least $3/10^4$.)

To use Eq. (110) in Eq. (109) requires the application of a diamagnetic shielding correction to μ_μ'. Until recently, this correction was believed to be identical to that for protons in water, i.e., 26 ppm (see Sec. II.B.6). However, Ruderman (1966) has shown that because of the muon's smaller mass, a positive muon can form a bond with water molecules considerably stronger than the usual hydrogen bond and therefore the muon does not simply replace a proton in a water molecule. As a result, the diamagnetic shielding (or chemical shift) for μ^+ is estimated to be only 10 ppm rather than 26 ppm. Applying a 10-ppm correction gives

$$\mu_\mu/\mu_p' = 3.18341(4) \ (13 \text{ ppm}). \tag{111}$$

Using this value of μ_μ/μ_p', g_μ as measured at CERN (to be discussed below), $\mu_e/\mu_B = g_s/2$ as given in Table XX, and μ_p'/μ_e as given in Table XI, we obtain from Eq. (109)

$$m_\mu/m_e = 206.769(3) \ (13 \text{ ppm}). \tag{112}$$

This value is in excellent agreement with the value $m_\mu/m_e = 206.76(2)$ (96 ppm) obtained from measurements of muonic x rays in phosphorous.†

We should point out that a_μ is really rather insensitive to the value of m_μ/m_e used; a relative change in m_μ/m_e gives rise to a relative change in a_μ, 1.5×10^{-3} times as large. Thus, as far as a_μ is concerned, it makes little difference whether we apply a correction to μ_μ' of 26 or 10 ppm. However, the theoretical equation for the hyperfine splitting in muonium depends directly on μ_μ/μ_p' and, when discussing that quantity, we shall retain two values of μ_μ/μ_p', one with and one without

† See Feinberg and Lederman (1963).

the Ruderman correction since this correction may be open to some question.

With the above value for m_μ/m_e, Eq. (108) yields

$$[1.094261(4)](\alpha/\pi)^2,$$

where the error quoted is that due to the 13-ppm uncertainty in m_μ/m_e. Combining this with the exactly known mass-independent contribution $-0.3285(\alpha/\pi)^2$, Eq. (18), we finally obtain for the fourth-order term in a_μ

$$[0.765782(4)](\alpha/\pi)^2, \tag{113}$$

For all practical purposes, we may take $B_2 \equiv 0.76578$ since a 10^{-4} error in B_2 will give rise to an error in a_μ of less than 0.05 ppm. This is entirely negligible at the present levels of experimental (and theoretical) accuracy for a_μ.

We now turn our attention to the sixth-order radiative corrections. The dominant contributions to the difference between the sixth-order term for muons and electrons have been calculated by Kinoshita (1968; 1967), Drell and Trefil (1967), and by Lautrup and de Rafael (1968) and may be written as

$$(a_\mu - a_e)^{(6)} = \left(\frac{\alpha}{\pi}\right)^3 \left\{\tfrac{1}{4} \ln\left(\frac{m_\mu}{m_e}\right) + \tfrac{1}{2}\zeta(3) - \tfrac{5}{12} - 0.3285(\tfrac{4}{3})\right.$$

$$\times\left[\ln\left(\frac{m_\mu}{m_e}\right) - \frac{25}{12}\right] + \frac{2}{9}\left[\ln^2\left(\frac{m_\mu}{m_e}\right) - \tfrac{25}{6}\ln\left(\frac{m_\mu}{m_e}\right)\right.$$

$$\left.\left. + \tfrac{317}{72} + \tfrac{1}{6}\pi^2\right] + O\left(\frac{m_e}{m_\mu}\right)\right\} = (2.819 \pm 0.3)\left(\frac{\alpha}{\pi}\right)^3,$$

using the value of m_μ/m_e given in Eq. (112).* The quoted uncertainty is meant to take into account in an

---

* This expression does not include a contribution from the light by light-scattering diagram. Such a contribution may be of order $(\alpha/\pi)^3 \ln(m_\mu/m_e)$ and is presently being evaluated by Brodsky, Dufner, and Kinoshita (Brodsky, private communication).

approximate way the uncertainty due to the constant term $-0.3285(\frac{4}{3})(-\frac{25}{12})$ which has only been estimated (Kinoshita, 1968) and uncalculated terms; it is simply 10% of the numerical value of the coefficient. Using the Parsons value $a_e^{(6)}=0.13(\alpha/\pi)^3$ with an uncertainty of ±0.20 in the coefficient 0.13 yields

$$a_\mu^{(6)}=(2.95\pm0.35)(\alpha/\pi)^3. \qquad (114)$$

So far, only contributions to a_μ arising from the muon, electron, and photon interactions have been considered. But there are additional contributions due to the fact that a virtual photon emitted by a muon may polarize the vacuum with strongly interacting particles, for example, pion pairs, triples, *etc.* Usually these contributions can be ignored since they are surpressed by a factor $(m_\mu/M)^2$, where M is a mass characteristic of the intermediate state. However, if the particles interact strongly, their contribution can be enhanced considerably. The biggest effects are expected from the ρ meson resonance in the 2π intermediate state and the ω and ϕ resonances in the 3π intermediate state. Kinoshita and Oakes (1967) (see also Durand, 1962) have calculated these hadronic contributions to the muon magnetic moment and find

$$(a_\mu)_\rho=6.1\times10^{-8}=4.9(\alpha/\pi)^3,$$

$$(a_\mu)_\omega=0.72\times10^{-8}=0.58(\alpha/\pi)^3,$$

$$(a_\mu)_\phi=0.63\times10^{-8}=0.50(\alpha/\pi)^3,$$

$$(a_\mu)_{\text{strong}}=7.5\times10^{-8}=6.0(\alpha/\pi)^3,$$

where $(a_\mu)_{\text{strong}}\equiv(a_\mu)_{2\pi+3\pi}$ is the sum of the individual ρ, ω, and ϕ contributions. However, Bowcock (1968) has reconsidered the calculation of Kinoshita and Oakes, taking into account more recent experimental data on the pion form factor. His final result is

$$(a_\mu)_{\text{strong}}=3.4\times10^{-8}=2.7(\alpha/\pi)^3.$$

From the uncertainty in the experimental data, Bowcock estimates the uncertainty in $(a_\mu)_{\text{strong}}$ as

$(+1.9 \text{ or } -0.9)\times 10^{-8}$ or $(+1.5 \text{ or } -0.7)(\alpha/\pi)^3$. Similarly, Gourdin and de Rafael (1969) have obtained a value of $(a_\mu)_{\text{strong}}$ from the recently reported results of the Orsay colliding beam experiments. Their result is

$$(a_\mu)_{\text{strong}} = (6.5\pm 0.5)\times 10^{-8} = (5.2\pm 0.4)(\alpha/\pi)^3, \quad (115)$$

which is in only fair agreement with the Bowcock result. Recently, Terazawa (1968) has estimated all the hadronic contributions to a_μ (i.e., contributions from other intermediate states such as 4π, 5π, $K\bar{K}$, $\Sigma\bar{\Sigma}$, $2N2\bar{N}$, etc.). Assuming the correctness of a certain hypothesis concerning the hadronic electromagnetic current operator to be correct, he finds

$$0 < (a_\mu)_{\text{all hadrons}} - (a_\mu)_{2\pi+3\pi} < 14\times 10^{-8}. \quad (116)$$

Using the Gourdin and de Rafael value for $(a_\mu)_{2\pi+3\pi}$, Eq. (115), this inequality becomes

$$6.5\times 10^{-8} < (a_\mu)_{\text{all hadrons}} < 20.5\times 10^{-8}$$

or

$$5.2(\alpha/\pi)^3 < (a_\mu)_{\text{all hadrons}} < 16.4(\alpha/\pi)^3.$$

(We use the Gourdin and de Rafael result rather than that of Bowcock because the former is based on more extensive data. Although the two results differ by nearly a factor of 2, it is relatively unimportant in relationship to the present experimental error in a_μ.) Gathering the terms of order $(\alpha/\pi)^3$, Eqs. (114) and (115), gives for the α^3 term

$$[8.1(+11 \text{ or } -0.5)](\alpha/\pi)^3. \quad (117)$$

The quoted uncertainty was obtained by adding RSS the uncertainties of the individual terms including that implied by Eq. (116). Although it may be closer to a limit of error than to a standard deviation, it is still relatively small in comparison with the error in the best experimental value of a_μ. We also note that Burnett and Levine (1967) and also Brodsky and Sullivan (1967) have recently estimated that if the intermediate vector boson (the W boson) exists, then it should contribute $\approx -2\times10^{-8}$ or $-1.6(\alpha/\pi)^3$ to a_μ. However, we shall ignore this contribution since it is so uncertain and because it is small compared with the present experimental uncertainty in a_μ.

The theoretical expression for a_μ may finally be written as

$$a_\mu = \tfrac{1}{2}(\alpha/\pi) + 0.76578(\alpha/\pi)^2 + [8.1(+11 \text{ or } -0.5)](\alpha/\pi)^3$$
$$= 0.001165643(+140 \text{ or } -7)\ (+120 \text{ or } -6 \text{ ppm}), \qquad (118)$$

using α_{WQED}. Note that the 1.9-ppm error in α_{WQED} contributes only a small fraction of the over-all uncertainty.

The most accurate measurement of a_μ is that carried out by a group at CERN using the CERN muon storage ring (Bailey, Bartl, von Bochmann, Brown, Farley, Jöstlein, Picasso, and Williams, 1968).* They report

$$a_\mu = 0.00116616(31)\ (270 \text{ ppm}). \qquad (119)$$

This value includes measurements on both μ^+ and μ^-: $a_{\mu^+} = 0.00116575(71)$, and $a_{\mu^-} = 0.00116625(24)$, where the errors are only statistical. This implies $a_{\mu^-} - a_{\mu^+} = (5.0\pm7.5)\times10^{-7}$. The principle of the $g-2$ experiment for muons is exactly the same as that for electrons as carried out by Wilkinson and Crane, i.e., the magnetic-moment anomaly is determined directly by measuring the difference between the spin precession frequency and orbital or cyclotron frequency. However, the experiments differ considerably in detail. In the muon experiment, spin polarized muons are produced by the in-flight decay of pions produced when a target is struck by 10.5-GeV protons from the CERN proton synchrotron. The muons are then trapped in the muon storage

* This value differs somewhat from an earlier value $a_\mu = 0.00116656(27)$ reported by Farley (1968a; private communication). See also Farley (1968b), and Bailey, Bartl, *et al.* (1967).

ring. The spin direction of the muons is followed in time by using the fact that the muons decay in flight by electron emission, and the electron angular distribution has its maximum in the direction of the muon spin. The spin direction or forward decay electrons have the highest energy and may be detected with counters which accept only electrons of a certain minimum energy. Thus, since the muon spin is precessing and only forward-direction decay is detected, the counting rate for an initial muon bunch will be a modulated exponential. The quantity a_μ is directly related to the modulation frequency ω_a and the magnetic field B via the equation $\omega_a = a_\mu(e/m_\mu)B$. The magnetic field is measured in terms of ω_p', the precession frequency of protons in water, so that this equation becomes

$$\omega_a = [a_\mu/(1+a_\mu)](\mu_\mu/\mu_p')\omega_p',$$

using the definitions $\hbar\omega_p' = 2\mu_p'B$, and $g_\mu/2 = (1+a_\mu) = \mu_\mu(e\hbar/2m_\mu)^{-1}$ [see Eqs. (109) and (110)]. The value for a_μ obtained by the CERN group and given in Eq. (119) was calculated using the Ruderman corrected value for μ_μ/μ_p'. However, the uncertainty in a_μ due to the uncertainty in this correction (even if it is assumed to be as large as 20 ppm) is negligible in comparison with the 270-ppm experimental uncertainty, since a relative change in μ_μ/μ_p' results in essentially the same relative change in a_μ (but of opposite sign). A major experimental difficulty was the determination of the mean effective orbit radius in the storage ring for muons which gave detectable decay electrons. (The magnetic field seen by the muons depends on the orbit radius because of a radial magnetic-field gradient which is necessary for vertical focusing.) The experimental uncertainty in the radius was ± 3 mm corresponding to an uncertainty in a_μ of about 160 ppm. The statistical error in the experiment was about 210 ppm, and the error quoted by the CERN group is the RSS of these two uncertainties. We note that the value quoted in Eq. (119) is in agreement with an earlier but less accurate measurement by Charpak, Farley, Garwin,

Muller, Sens, and Zichichi (1965) at CERN on positive muons using a shaped magnetic field. They found $a_\mu{}^+=0.001162(5)$. An earlier CERN muon-storage-ring value of $a_\mu{}^-=0.001165(3)$ has also been reported (Farley, Bailey, *et al.*, 1966).

In comparing the theoretical and experimental results, Eqs. (118) and (119), we see that there is somewhat of a discrepancy:

$$a_\mu(\text{exptl})-a_\mu(\text{theory})$$

$$=52(+31 \text{ or } -34)\times 10^{-8}[440(+270 \text{ or } -290)] \text{ ppm}.$$

The difference exceeds the standard deviation of the difference by about a factor of 1.6. The probability for this to occur by chance is $\approx 11\%$. [Note that this disagreement is *opposite* to that found for a_e since a_e(theory) exceeded a_e(exptl) by (77 ± 26) ppm.] The origin of the discrepancy, if one may call it that, is presently unknown.

The theoretical and experimental values of a_μ may also be compared by taking the theoretical coefficient of the $(\alpha/\pi)^2$ term to be correct and calculating the coefficient of the $(\alpha/\pi)^3$ term implied by the experimental value. The result is $B_3(\text{exptl})=49\pm 25$ as compared with $B_3(\text{theory})=8.1(+11 \text{ or } -0.5)$. Similarly, assuming B_3(theory) is correct, including the error estimates, we find $B_2(\text{exptl})=0.86(+0.03 \text{ or } -0.06)$ as compared with $B_2(\text{theory})=0.76578$.

The possible implications of the above discrepancy in a_μ [which was larger with the results reported earlier (Farley, 1968a; 1968b; Bailey, Bartl, *et al.* 1967)] have been considered recently by Brodsky and de Rafael (1968) who calculated the second-order contribution to a_μ due to the coupling of hypothetical scalar and vector bosons to lepton pairs. From the difference between a_μ(theory) and a_μ(exptl), they were able to place limits on the quantity $(f^2/4\pi)M^{-2}$, where f is the coupling constant and M is the mass. Nieto (1968) has shown that the existence of such a boson is consistent with the direct production of high-energy muons in cosmic rays and the observation of certain cosmic-ray events.

In concluding this section, it may be of interest to compare the theoretical and experimental values for the difference between the anomalous moment of the electron and muon, $\Delta a \equiv a_\mu - a_e$. We find

$$\Delta a(\text{theory}) = 600(+14 \text{ or } -1) \times 10^{-8},$$

$$\Delta a(\text{exptl}) = 661 \pm 31 \times 10^{-8},$$

$$\Delta a(\text{exptl}) - \Delta a(\text{theory}) = 61(+31 \text{ or } -34) \times 10^{-8},$$

where the experimental values are those given in Eqs. (119) and (105). The agreement is not particularly good, the difference exceeding the standard deviation of the difference by ~ 1.8. The chance probability for this is about 7%.

B. Ground-State Hyperfine Splittings

1. Atomic Hydrogen

The QED corrections to the theoretical expression for the hyperfine splitting (hfs) in the ground state of atomic hydrogen amount to only 60 ppm in addition to the radiative corrections to the free-electron magnetic moment. Consequently, the hfs in H is perhaps not as good a testing ground for QED as is the g factor of the electron and muon. On the other hand, the hfs can be measured with extreme accuracy, and Drell and Sullivan (1967) have pointed out that a comparison between theory and experiment is of great interest because it (1) provides an important test of relativistic bound-state methods and (2) provides a link between precision atomic physics and high-energy electron-scattering physics since the hfs in H is sensitive to the structure of the proton.

The theoretical expression for the hfs in H (not including nuclear structure corrections) has recently been systematically reexamined by Brodsky and Erickson (1966). [See also Grotch and Yennie (1969;

1967) and Guérin (1967a; 1967b). For a more descriptive discussion of the hfs, see Iddings (1969).] They not only verified previous calculations of order α, $\alpha(Z\alpha)$, $\alpha(Z\alpha)^2 \ln^2 (Z\alpha)^{-2}$, and $\alpha(Z\alpha)^2 \ln (Z\alpha)^{-2}$ relative to the lowest-order Fermi splitting, but also obtained a result for the dominant contribution to order $\alpha(Z\alpha)^2$. Their final result is

$$\nu_{\text{Hhfs}} = \tfrac{16}{3}\alpha^2 R_\infty c \frac{\mu_p}{\mu_B}\left(1+\frac{m_e}{M_p}\right)^{-3} \{1+a_e+\tfrac{3}{2}(Z\alpha)^2$$
$$+\alpha(Z\alpha)(-\tfrac{5}{2}+\ln 2)+(\alpha/\pi)(Z\alpha)^2[-\tfrac{2}{3}\ln^2 (Z\alpha)^{-2}$$
$$+(\tfrac{37}{72}+\tfrac{4}{15}-\tfrac{8}{3}\ln 2)\ln (Z\alpha)^{-2}+(18.36\pm 5)]+\delta_N\}, \tag{120}$$

where a_e is the theoretical expression for the electron magnetic-moment anomaly, Eq. (103). The term 18.36 is the newly calculated contribution and the ± 5 is the estimated error for uncalculated contributions to order $(\alpha/\pi)(Z\alpha)^2$ and terms quadratic in the field strength (uncertainties of ± 4 and ± 1, respectively). The reduced-mass term describes the main effect of nuclear motion or recoil due to the finite mass of the proton. The third term in the braces is the Breit relativistic correction, and the remaining terms are radiative corrections to the electron moment due to binding; the $\alpha(Z\alpha)$ term was first calculated by Kroll and Pollack and Karplus, Klein, and Schwinger, and the logarithmic terms, by Layzer and by Zwanziger [see Brodsky and Erickson (1966)]. The last term δ_N represents the effects of internal proton structure and finite proton size and mass aside from the purely kinematic reduced-mass correction. It may be conceptually written as the sum of two terms, $\delta_N = \delta_N^{(1)} + \delta_N^{(2)}$, where $\delta_N^{(1)}$ is the next-higher-order recoil correction of order $\alpha m_e/M_p$ due to the finite mass and size of the proton; the proton charge and magnetic moment are assumed to be extended, but the proton is taken to be in its ground state. The term $\delta_N^{(2)}$ is the proton polarizability contribution due to the various excited states or the internal structure of the proton. (Both terms arise mainly from processes in which two intermediate virtual photons are exchanged between the electron and proton.) Drell and Sullivan (1967) have pointed out that if the orbital electron in H were able to follow the instantaneous charge and magnetization position of the proton, then the proton would appear as a point and $\delta_N^{(2)}$ would cancel the finite-size correction $\delta_N^{(1)}$.

On the assumption that the proton is simply a point mass with an anomalous point magnetic moment $\kappa = (\mu_p/\mu_n) - 1$, Arnowitt (1953) and Newcomb and Salpeter (1955) have calculated $\delta_N^{(1)}$. They find

$$[\delta_N^{(1)}]_{\text{point}} = -\frac{\alpha m_e}{\pi M_p(1+\kappa)} \times \left((3 - \tfrac{3}{4}\kappa^2) \ln \frac{M_p}{m_e} - \tfrac{1}{8}\kappa^2 + \tfrac{9}{4}\kappa^2 \ln \frac{\Lambda}{M_p} \right), \quad (121)$$

where the cutoff $\Lambda \approx 2M_p$ is introduced to avoid the logarithmic infinity arising from the interaction of the anomalous moments of the electron and proton. (The cutoff corresponds to a spreading of the total proton moment about Λ^{-1} or twice its Compton wavelength.) Iddings (1965) [see also Iddings and Platzman (1959b) and Zemach (1956)] has treated the cutoff problem by using an experimental form factor for the charge and magnetic moment of the proton as obtained from high-energy electron–proton-scattering data. This was conveniently accomplished by calculating the difference between $\delta_N^{(1)}$ for a point proton and an extended one; the difference contains a logarithmic divergence which just cancels that in Eq. (121).* His result is

* Recently, F. Guérin (1967a; 1967b) and Grotch and Yennie (1967; 1969) obtained a slightly different result for Eq. (121) by introducing the cutoff Λ in a covariant way. Since the calculation of Iddings was made with the same assumptions regarding Λ as was the derivation of Eq. (121), we must use the older result. However, the final answer for $\delta_N^{(1)}$ will be independent of how Λ is introduced.

$$[\delta_N^{(1)}]_{\text{ext}}-[\delta_N^{(1)}]_{\text{point}}$$
$$=-\frac{\alpha m_e}{\pi M_p(1+\kappa)}\left(\delta\mathcal{E}'+\tfrac{17}{16}\kappa^2-\tfrac{9}{4}\kappa^2\ln\frac{\Lambda}{M_p}\right). \quad (122)$$

Defining $\delta\mathcal{E}=\delta\mathcal{E}'+17\kappa^2/16$, Iddings obtained values for $\delta\mathcal{E}$ varying between 67 and 73 for a number of reasonable form factors [see Table III of Iddings (1965)]. For three different approximations to the form factors which fit the fairly reliable elastic *e–p*-scattering data of Hofstadter and coworkers, he found $\delta\mathcal{E}=71.1$, 72.0, and 73.0 (cases III–V in his Table III). It would thus appear that $\delta\mathcal{E}=72\pm2$ is a reasonable best estimate for this quantity.† Using this value, and combining Eqs. (121) and (122) gives‡

$$\delta_N^{(1)}=-\frac{\alpha m_e}{\pi M_p(\mu_p/\mu_n)}(76.0\pm2)=-(34.4\pm0.9)\text{ ppm}. \quad (123)$$

(The coefficient of $\delta\mathcal{E}$ is 0.453×10^{-6} and thus the particular values of the constants used are not critical; we take those given in Table XX.) Although newer form-factor data exists [see, for example, Goitein, Dunning, and Wilson (1967)] and may well give rise to some small changes in the value of $\delta\mathcal{E}$ used here, such changes should not exceed the assigned error (Iddings, 1969).

The most comprehensive discussion of $\delta_N^{(2)}$ is probably that given recently by Drell and Sullivan (1967). They conclude that while nonrelativistic Schrödinger models of proton structure give sizable but model-

† We wish to thank C. K. Iddings for this suggestion.

‡ As noted by Iddings (1969), one may wonder why $\delta_N^{(1)}$ turns out to be so large when the form factors are introduced since this corresponds to a very large cutoff in Eq. (121). The reason for this (Iddings, 1969) is that the point-proton result, Eq. (121), is anomalously small because several large but finite terms tend to cancel to give ~2 ppm for the cutoff-independent part. When the form factors are introduced, they no longer cancel. Iddings thus concludes that most of $\delta_N^{(1)}$ comes from terms which are finite for a point proton.

dependent polarizability contributions, a relativistic dispersion-theory approach fails to provide any significant contributions (~2 ppm at most). While they find no clear candidate which might contribute as much as 10 ppm to the hfs, they feel that the calculations are sufficiently sensitive to various amplitudes which can only be approximately calculated that it cannot be said with any certainty that such a contribution does not exist. The general conclusions of Drell and Sullivan concerning the small size of $\delta_N^{(2)}$ are supported by the results of several authors: Iddings (1965), Iddings and Platzman (1959a), Verganelakis and Zwanziger (1965), and Guérin (1967b) all find model-dependent contributions on the order of $|1|$ to $|2|$ ppm or less. (Much of this work on the polarizability contribution has been concerned with the 33 resonance.) Fenster and Nambu (1965) [see also Fenster, Köberle and Nambu (1965) and Drell and Sullivan (1965)] have considered a quark model of the proton but their treatment is probably unrealistic [see Drell and Sullivan (1967)]. It would thus appear that $\delta_N^{(2)}$ is zero to within one or two ppm.* However, we shall retain it for the moment in order to compare theory and experiment. Thus, using α_{WQED} and the auxiliary constants given in Table XI, we find from Eq. (123) and Eq. (120) with the theoretical expression for a_e [Eq. (103)]

$$\nu_{\text{Hhfs}} = 1420.4023(1+\delta_N^{(2)}) \pm 0.0057 \text{ MHz } (4.0 \text{ ppm}). \tag{124}$$

The quoted uncertainty is the root sum square of the following errors: 2×1.9 ppm $=3.8$ ppm in α_{WQED}^2, 0.6 ppm due to the ± 5 uncertainty for uncalculated terms, and 0.9 ppm in $\delta_N^{(1)}$. (We assume as usual that the auxiliary constants are exactly known.)

The most precise experimental determinations of ν_{Hhfs} have been made using the hydrogen maser

* For a discussion of the possible effect of very virtual photons on the magnitude of δ_N, see Bjorken (1966).

(Kleppner, Goldenberg, and Ramsey, 1962). In this device atomic hydrogen obtained from an rf discharge source passes through an inhomogeneous state-selecting magnetic field. The field focuses atoms in the $(F=1, m_F=0)$ and $(F=1, m_F=1)$ states (see Fig. 2) onto an aperture in a Teflon coated quartz bulb located in the center of a cylindrical rf cavity. When the cavity is tuned to the $(F=1, m_F=0)\leftrightarrow(F=0, m_F=0)$ transition (1420 MHz), self-excited maser oscillations result. Contributing to the success of the maser are (1) a transition time >1 sec, (2) the fact that the hydrogen atoms spend most of their time in free space and are little perturbed by collisions with the Teflon coated walls, and (3) the great reduction of the effect of the first-order Doppler shift because the average velocity of the atoms in the bulb is nearly zero.

Crampton, Kleppner, and Ramsey (1963) reported the first ultrahigh-accuracy hydrogen-maser measurement of ν_{Hhfs}. They obtained $\nu_{\mathrm{Hhfs}}=1420.405751800(28)$ MHz $(2/10^{11})$. More accurate measurements made at NBS were reported in 1966 by Vessot and coworkers and represent the most precise determination of any physical quantity. Their result is

$$\nu_{\mathrm{Hhfs}}=1420.4057517864(17)\ \mathrm{MHz}\ (1.2/10^{12}), \qquad (125)$$

in excellent agreement with the Ramsey value. (It will probably be some time before theoretical physicists will be able to make use of this accuracy since they have over six orders of magnitude to go!) There have been other determinations of ν_{Hhfs} by the hydrogen-maser technique as well as by other methods. In Table XXV we summarize some of the more pertinent measurements.

Using Eqs. (124) and (125), we compare the theoretical and experimental values of ν_{Hhfs} and find

$$\frac{\nu_{\mathrm{Hhfs}}(\mathrm{exptl})-\nu_{\mathrm{Hhfs}}(\mathrm{theory})}{\nu_{\mathrm{Hhfs}}(\mathrm{exptl})}=(2.5\pm4.0)\ \mathrm{ppm}\ -\delta_N{}^{(2)},$$

which implies $\delta_N{}^{(2)}=(2.5\pm4.0)$ ppm. We may therefore conclude that (1) the various calculations which give a small (1 to 2 ppm) proton polarizability correction to ν_{Hhfs} are essentially correct and that the ideas of Drell and Sullivan concerning the magnitude of uncalculated contributions are probably too conservative and (2) the over-all calculation of the hydrogen hyperfine splitting is on firm ground including the higher-order recoil correction $\delta_N{}^{(1)}$ calculated by Iddings.* This agreement is in marked contrast to what was implied by the old value of α derived from the fine-structure splitting in deuterium, $\alpha^{-1}=137.0388(6)$ (4.5 ppm); it gave $[\nu(\text{exptl})-\nu(\text{theory})]/\nu(\text{exptl})=(42\pm9)$ ppm $-\delta_N{}^{(2)}$ or $\delta_N{}^{(2)}=(42\pm9)$ ppm, in complete disagreement with theory. Indeed, Fenster and Nambu (1965) claimed the discrepancy between the theoretical and experimental values of ν_{Hhfs} was perhaps the only major unsolved problem of QED at that time.

2. *Muonium*

The muonium atom consists of a muon and an electron (μ^+e^-) and is similar in many respects to the hydrogen atom, its 2.2-μsec lifetime notwithstanding. The QED corrections to the hyperfine splitting in the muonium ground state are $\sim$200 ppm (in addition to the radiative corrections to the free-electron moment), and hence the muonium hfs provides a reasonable testing ground for QED. More important, muonium is one of the simplest systems involving the muon and electron, and therefore a comparison of the theoretical and experimental values of the hfs may indicate whether there are any basic differences in the interaction properties of electrons and muons (see Sec. IV.A.2).

On the assumption that the muon is just a heavy electron, the theoretical equation for the hfs in muonium is exactly the same as that for the hfs in H as given by

---

*The theoretical expression for the ratio of the hfs in the $2S$ state of a hydrogenic atom to that in the $1S$ state is independent of δ_N. Thus, a comparison of the theoretical and experimental values of this ratio can be used as a check on the validity of some of the higher-order QED corrections to the hfs. [See Fortson, Major, and Dehmelt (1966), Sternheim (1963), and Zwanziger (1961).]

TABLE XXV. Summary of some measurements of the ground-state hyperfine splitting in hydrogen.

Publication date and authors	Method	Value (MHz)
1966, Vessot *et al.*	Hydrogen maser	1420.4057517864(17)
1965, Peters and Kartaschoff[a]	Hydrogen maser	1420.405751785(16)
1963, Crampton, Kleppner, and Ramsey	Hydrogen maser	1420.405751800(28)
1962, Pipkin and Lambert[b]	Optical pumping	1420.4057383(60)
1956, Wittke and Dicke[c]	Microwave absorption	1420.40572(4)
1955, Kusch[d]	Atomic beam magnetic resonance	1420.40573(5)

[a] H. E. Peters and P. Kartaschoff, Appl. Phys. Letters **6,** 35 (1965).
[b] F. M. Pipkin and R. H. Lambert, Phys. Rev. **127,** 787 (1962) (corrected to ^{133}Cs definition of the second).
[c] J. P. Wittke and R. H. Dicke, Phys. Rev. **103,** 620 (1956) (unextrapolated value).
[d] P. Kusch, Phys. Rev. **100,** 1188 (1955).

Brodsky and Erickson, Eq. (120), but with the proton parameters replaced by muon parameters, i.e., m_e/M_p must be replaced by m_e/m_μ, μ_p/μ_B by μ_μ/μ_B, and the numerical value for δ_N by that appropriate to muonium. Since the anomalous moment of the muon a_μ is $\sim\alpha/2\pi$, treating the muon as a point with zero anomalous moment is sufficient to obtain δ_N to order $\alpha m_e/m_\mu$. Thus, Eq. (121) of the previous section which was calculated for a point proton with anomalous moment κ becomes for the muon

$$\delta_N = -\frac{3\alpha}{\pi}\frac{m_e}{m_\mu}\ln\frac{m_\mu}{m_e} = -179.7 \text{ ppm}. \tag{126}$$

We may take $\delta_N^{(1)} = \delta_N$ because no polarizability contributions to δ_N are expected, in contrast to the more complex situation which exists for the proton. The numerical value quoted was calculated using m_μ/m_e as given in Eq. (112) which was derived from Eqs. (109) and (110) using the Ruderman diamagnetic shielding correction. However, δ_N would be the same to within 0.002 ppm if the standard 26-ppm correction had been used; this difference is entirely negligible.

To evaluate Eq. (120) for muonium requires writing $\mu_\mu/\mu_B = (\mu_\mu/\mu_p')(\mu_p'/\mu_B) = (1+\sigma_\mu)(\mu_\mu'/\mu_p')(\mu_p'/\mu_B)$ since μ_μ'/μ_p' is the experimentally measured quantity. Here, the diamagnetic shielding correction σ_μ for muons in water is written explicitly. Similarly, we may rewrite Eq. (109) as

$$\frac{m_\mu}{m_e} = \frac{(g_\mu/g_s)(\mu_e/\mu_p')}{(1+\sigma_\mu)(\mu_\mu'/\mu_p')}.$$

Using α_{WQED}, the auxiliary constants of Table XI, g_μ and μ_μ'/μ_p' as given in Sec. IV.A.2, we obtain from Eq. (126), Eq. (120) with the theoretical expression for a_e [Eq. (103)], and the above equations for m_μ/m_e and μ_μ/μ_B

$$\sigma_\mu = 10 \text{ ppm:} \quad \nu_{\text{Mhfs}} = 4463.272(61) \text{ MHz (14 ppm)},$$

$$\sigma_\mu = 26 \text{ ppm:} \quad \nu_{\text{Mhfs}} = 4463.342(61) \text{ MHz (14 ppm)}. \tag{127}$$

Here, $\sigma_\mu = 10$ ppm is the Ruderman correction and $\sigma_\mu = 26$ ppm is the standard correction for protons in water (see Sec. II.B.6). The quoted uncertainty is the RSS of the following errors: $2 \times 1.9 = 3.8$ ppm in α^2, 13 ppm in μ_μ'/μ_p, and 0.6 ppm due to the ± 5 uncertainty in the $(\alpha/\pi)(Z\alpha)^2$ term of Eq. (120).

Extensive measurements of ν_{Mhfs} have been made by Hughes and coworkers at Yale over the last several years. [For a summary of this work, see Hughes (1966; 1967). See also Cleland, Bailey, Eckhause, Hughes, Mobley, Prepost, and Rothberg (1964) and Thompson, Amato, Hughes, Mobley, and Rothberg, (1967).] The experiments may be classified according to whether they were carried out in a high magnetic field ($\sim$0.55 T), a weak field ($\sim 3 \times 10^{-4}$ T), or a very weak field ($\sim 10^{-6}$ T). In the high-field experiments, a spin polarized beam of μ^+ is obtained from the in-flight decay of a beam of π^+. The μ^+ are stopped in a high-pressure argon-gas target (10–65 atm), where they capture electrons and form muonium with the same direction of polarization. If a strong field is present in the polarization direction, then the muonium is formed with $m_I = \pm \frac{1}{2}$ (in state 1 or 4 of Fig. 2; the energy levels for muonium and hydrogen are the same except for scale, $\nu_{\mathrm{Mhfs}} \sim 4.4$ GHz while $\nu_{\mathrm{Hhfs}} \sim 1.4$ GHz.) In the absence of an external perturbation, the muonium atoms decay with the decay positrons preferentially emitted in the direction of polarization, in this case, the field direction. (Recall the g_μ experiment, Sec. IV.A.2.) If a microwave field is applied with frequency such that it can induce transitions between, say, states 1 and 2, $[(m_j = \frac{1}{2},\ m_I = \frac{1}{2}) \leftrightarrow (m_j = \frac{1}{2},\ m_I = -\frac{1}{2})]$, the number of decay positrons in the applied field direction decreases since in state 2 the muon spins point in the opposite direction. Hence, an induced transition can be detected by measuring the angular distribution of the decay positrons. The frequency of the transition and the magnetic field at which it occurs (measured in proton frequency units) are used in conjunction with the theory of the hfs energy levels to calculate ν_{Mhfs}. Although this calculation requires use of the quantity μ_μ/μ_p', it enters in such a way that its experimental uncertainty and the uncertainty as to which value of σ_μ should be used contribute less than 2 ppm to the uncertainty in ν_{Mhfs}. This is negligible compared with the 13-ppm experimental error in ν_{Mhfs} which arises mainly from statistics and magnetic-field uncertainties. The low-field measurements are very similar to those carried out at high field, except that the transition involved is, say, $(F=1, m_F=-1)\leftrightarrow(F=0, m_F=0)$. Furthermore, μ_μ/μ_p is even less important in the weak-field data analysis than it is in the high-field data analysis. One of the main difficulties in these experiments is the shift of ν_{Mhfs} with argon pressure due to collisions with argon atoms (~0.1 MHz/10 atm). To circumvent this problem, data are taken as a function of pressure and extrapolated to zero pressure. Since there is no evidence of any quadratic dependence of ν_{Mhfs} with pressure, the extrapolation is done linearly. The results of the high- and weak-field work are (Thompson *et al.*, 1967; Hughes, 1966; Cleland *et al.*, 1964)

$$\text{high field:}\quad \nu_{\mathrm{Mhfs}}=4463.15(6)\ \mathrm{MHz}\ (13\ \mathrm{ppm}), \tag{128a}$$

$$\text{weak field:}\quad \nu_{\mathrm{Mhfs}}=4463.21(6)\ \mathrm{MHz}\ (13\ \mathrm{ppm}). \tag{128b}$$

Since the weak-field data consisted of but a single series of measurements taken at 35 atm of argon, the pressure-shift correction used was that obtained from the high-field work. The quoted errors correspond to one standard deviation and for the high-field result includes the statistical counting error and uncertainties in the magnetic-field measurements. For the low-field result, the uncertainty in the pressure shift as obtained from the high-field measurements has been included along with the statistical counting error. The two measurements are clearly consistent with each other.

Recently, Brown and Pipkin (1968) reported new, highly accurate measurements of the fractional pressure shifts in argon (0.06–0.3 atm) of the hfs in hydrogen and tritium using an optical pumping technique. They found a fractional pressure shift for ν_{Hhfs} of $-4.78\pm 0.03\times10^{-9}$ torr^{-1} at 0°C, in fair agreement with (but more accurate than) the value $-4.05\pm0.49\times10^{-9}$ torr^{-1} obtained for muonium. On the assumption that the pressure shift for ν_{Hhfs} and ν_{Mhfs} are the same, Brown and Pipkin have reanalyzed the Yale muonium data (including the low-field result) and obtained

$$\nu_{\mathrm{Mhfs}}(\mathrm{exptl}) = 4463.23(2)\ \mathrm{MHz}\ (5\ \mathrm{ppm}), \qquad (129)$$

as compared with Eqs. (128a) and (128b). Brown and Pipkin have given plausible arguments in general support of the three main assumptions which underlie Eq. (129): (1) the pressure shift is independent of isotopic mass and is therefore the same for H and muonium; (2) the dependence of ν_{Mhfs} on pressure is linear and therefore measurements made at 0.06–0.3 atm are usable for correcting data obtained at 10–65 atm; and (3) the muonium atoms thermalize and are at room temperature.

Still more recently, the Yale group has completed a new series of very weak-field ($\sim10^{-6}$ T) measurements (Thompson, Amato, Crane, Hughes, Mobley, zu Pulitz, and Rothberg, 1969). For these small fields, the two separate transitions $(F=1, m_F=1)\leftrightarrow(F=0, m_F=0)$ and $(F=1,\ m_F=-1)\leftrightarrow(F=0,\ m_F=0)$ are not separately resolved as for the weak-field case. As a result, the resonance line is broadened and the line-shape theory must take into account all three levels. In the new experiments, muons were stopped in both argon (62 and 32 atm) and krypton (42 and 21 atm). The fractional pressure shift of ν_{Mhfs} in Ar was found to be $(-4.07\pm0.25)\times10^{-9}$ torr^{-1} and in Kr $(-10.4\pm0.3)\times 10^{-9}$ torr^{-1}. The measured shift in Ar agrees with that obtained in the high-field measurements but disagrees with the value implied by the work of Brown and Pipkin. On the other hand, the pressure shift in Kr for muonium is identical to the value $(-10.4\pm0.2)\times10^{-9}$ torr^{-1} found by Ensberg and Morgan (1968) for hydrogen in Kr ($\sim$0.06 to 0.3 atm). This suggests an isotope dependence of the pressure shift or a nonlinear density dependence of ν_{Mhfs} (Thompson *et al.*, 1969). The values of ν_{Mhfs} obtained from the very-weak-field measurements are

$$\text{argon:} \quad \nu_{\mathrm{Mhfs}}=4463.302(27)\ \mathrm{MHz}\ (6.0\ \mathrm{ppm}), \tag{130a}$$

$$\text{krypton:} \quad \nu_{\mathrm{Mhfs}}=4463.220(33)\ \mathrm{MHz}\ (7.4\ \mathrm{ppm}), \tag{130b}$$

where it should be noted that the Ar result includes the earlier single weak-field measurement made at 35 atm of Ar. The quoted one-standard-deviation errors are statistical counting errors and are much larger than the estimated systematic errors which include: (1) uncertainties in the microwave power level in the cavity and the microwave frequency, 5 kHz and 0.1 kHz, respectively; (2) magnetic-field instabilities and inhomogeneities, 3 kHz; (3) uncertainties in gas pressure and temperature, 3 kHz; and (4) approximations in deriving the theoretical line shape, 3 kHz. The difference between the krypton and argon values is 0.082 ± 0.043 MHz [(18 ± 10) ppm] or 1.9 times the standard deviation of the difference; the chance probability for this is about 5%. This may be evidence for a nonlinearity in the Ar pressure shift. To arrive at a final value for ν_{Mhfs}, Thompson *et al.* took the weighted average of the high-field result, Eq. (128a), with the Ar and Kr very-weak-field results, Eqs. (130a) and (130b), to obtain $\nu_{\mathrm{Mhfs}}=4463.255(20)$ MHz. However, in order to allow for a possible nonlinear pressure shift, they doubled the uncertainty to 0.04 MHz since this corresponds to the standard deviation of the mean of the three measurements. Although it is possible that even this expanded error may not adequately reflect the uncertainties about nonlinear pressure shifts, we shall adopt it for the purpose of comparing theory with experiment. We thus take as the final result of the muonium work

$$\nu_{\mathrm{Mhfs}} = 4463.255(40) \text{ MHz } (9.0 \text{ ppm}). \quad (130c)$$

In adopting this result we have ignored the pressure-shift work of Brown and Pipkin, but we believe this is justified in view of the uncertainties in the assumptions made by Brown and Pipkin and the much improved statistics of the newer muonium measurements.

We compare theory and experiment by computing $\Delta\nu/\nu \equiv [\nu_{\mathrm{Mhfs}}(\text{exptl}) - \nu_{\mathrm{Mhfs}}(\text{theory})]/\nu_{\mathrm{Mhfs}}(\text{exptl})$ using Eqs. (127) and (130). The result is

$$\sigma_\mu = 10 \text{ ppm}: \quad \Delta\nu/\nu = (-3.7 \pm 16) \text{ ppm},$$

$$\sigma_\mu = 26 \text{ ppm}: \quad \Delta\nu/\nu = (-19 \pm 16) \text{ ppm}. \quad (131)$$

Clearly, theory and experiment are in excellent agreement if the Ruderman correction is used, and in reasonable agreement if the standard proton correction is used. The validity of treating the muon as a heavy Dirac particle is therefore further verified, and the Ruderman correction given some experimental support. Note also that the higher-order recoil term δ_N contributes some 180 ppm to ν_{Mhfs} and thus the agreement between experiment and theory indicates that it must be essentially correct.

It is also of interest to compare the experimental and theoretical values for the ratio $\nu_{\mathrm{Mhfs}}/\nu_{\mathrm{Hhfs}} \equiv R_{\mathrm{M/H}}$. From Eqs. (125) and (130c) we find

$$R_{\mathrm{M/H}}(\text{exptl}) = 3.142239(28) \ (9 \text{ ppm}). \quad (132)$$

From Eq. (120) it may be shown that

$$R_{\mathrm{M/H}}(\text{theory}) = \frac{\mu_\mu'}{\mu_p'}\left[\frac{1+\sigma_\mu}{1+\sigma(\mathrm{H_2O})}\right]\left(\frac{1+m_e/M_p}{1+m_e/m_\mu}\right)^3$$

$$\times\left[\frac{1+\delta_N(\mathrm{M})}{1+\delta_N^{(1)}(\mathrm{H})+\delta_N^{(2)}(\mathrm{H})}\right]$$

$$=3.142227(41)\left[\frac{1+\sigma_\mu}{1+\delta_N^{(2)}(\mathrm{H})}\right](13\ \mathrm{ppm}), \tag{133}$$

where $\sigma(H_2O)$ is, as before, the diamagnetic shielding constant for protons in a spherical sample of water; our adopted value for this quantity is 26.0 ppm (see Sec. II.B.6 and Table XI). The higher-order recoil correction for muonium, $\delta_N(\mathrm{M})$, is given by Eq. (126), and the similar correction for hydrogen, by Eq. (123). $\delta_N^{(2)}(\mathrm{H})$ is of course the proton polarizability correction to ν_{Hhfs}. In evaluating Eq. (133), we have used $1+m_e/M_p$ as given in Table XI, and m_μ/m_e as given in Eq. (112). [The fact that m_μ/m_e actually depends on σ_μ may be ignored because of the relatively small contribution σ_μ makes to $(1+m_e/m_\mu)$ and the large 13-ppm uncertainty of Eq. (133) due to the uncertainty in the experimental value of μ_μ'/μ_p', Eq. (110).] It should be noted that Eq. (133) is independent of α, and that the various QED radiative corrections have canceled. This means that any error in these terms and the omission of higher-order terms is unimportant as far as the ratio $R_{\mathrm{M/H}}$ is concerned.

Equations (132) and (133) may be used to calculate either a value for $\delta_N^{(2)}(\mathrm{H})$ or for σ_μ. This could be of some importance since these two quantities are the most questionable items which enter into the theoretical expressions for ν_{Hhfs} and ν_{Mhfs}, respectively. Assuming $\delta_N^{(2)}(\mathrm{H})=0$, as the majority of the theoretical calculations indicate, yields

$$\delta_N^{(2)}(\mathrm{H})=0\ \mathrm{ppm}:\quad \sigma_\mu=(4\pm16)\ \mathrm{ppm}. \tag{134}$$

On the other hand, if σ_μ is assumed to be equal to 10 ppm (the Ruderman correction) or 26 ppm (the standard proton correction), we find for $\delta_N^{(2)}(\mathrm{H})$

$$\sigma_\mu=10\ \mathrm{ppm}:\quad \delta_N^{(2)}(\mathrm{H})=(6\pm16)\ \mathrm{ppm},$$

$$\sigma_\mu=26\ \mathrm{ppm}:\quad \delta_N^{(2)}(\mathrm{H})=(22\pm16)\ \mathrm{ppm}. \tag{135}$$

Although the uncertainties in Eqs. (134) and (135) are quite large, zero proton polarizability and the Ruderman correction do tend to support each other.

Thompson, Amato *et al.* (1969) have also pointed out that one may use the experimental value of ν_{Mhfs}, the theoretical equation for ν_{Mhfs}, and a value of α to obtain an independent value of μ_μ/μ_p. Using α_{WQED} and Eq. (120) with μ_p/μ_B replaced by

$$\mu_\mu/\mu_B = (\mu_\mu/\mu_p)(\mu_p/\mu_B)$$

and m_e/M_p by

$$m_e/m_\mu = [(\mu_\mu/\mu_p)(g_s/g_\mu)](\mu_e/\mu_p)^{-1},$$

we find

$$\mu_\mu/\mu_p = 3.183319(30) \quad (9.4 \text{ ppm})$$

and thus $m_\mu/m_e = 206.770(2)$ (9.4 ppm). [For a discussion of muonic phosphorous, see Hughes, (1969).]

3. *Positronium*

Both the theoretical and experimental values for the hfs in positronium (e^+e^-) are considerably more uncertain than those for muonium and are included for completeness only. The present best theoretical equation for ν_{Phfs} is

$$\begin{aligned}\nu_{\mathrm{Phfs}} &= \alpha^2 c R_\infty[\tfrac{7}{6} - (\tfrac{16}{9} + \ln 2)(\alpha/\pi) + (2\pi^2 \pm 2\pi)(\alpha/\pi)^2] \\ &= 203.3996(59) \text{ GHz } (29 \text{ ppm}), \end{aligned} \tag{136}$$

using α_{WQED} and the auxiliary constants of Table XI. The first two terms of this expression were first calculated by Karplus and Klein (1952). The last term is a recent order-of-magnitude estimate by Erickson (1967; private communication) of the contributions of order α^2. It is equal to (92 ± 29) ppm or (0.0187 ± 0.0059) GHz.

The most recent measurement of ν_{Phfs} is that of Theriot, Beers, and Hughes (1967), who find

$$\nu_{\mathrm{Phfs}} = 203.403(12) \text{ GHz } (59 \text{ ppm}). \tag{137}$$

As in the muonium experiments, the positronium was

formed in argon (3–5 atm), and an extrapolation of $\nu_{\mathrm{Ph\,fs}}$ to zero pressure was required. The above value is in reasonable agreement with the older, less accurate results of Weinstein, Deutsch, and Brown (1954; 1955; Deutsch and Brown, 1952), $\nu_{\mathrm{Phfs}}=203.380(40)$ GHz (200 ppm), and Hughes, Marder, and Wu (1957), $\nu_{\mathrm{Ph\,fs}}=203.220(40)$ GHz (200 ppm), when the pressure-shift correction implied by the new measurements is made to the older data (Theriot, Beers, and Hughes, 1967).

In comparing theory with experiment, Eqs. (136) and (137), we find

$$\frac{\nu_{\mathrm{Ph\,fs}}(\mathrm{exptl})-\nu_{\mathrm{Phfs}}(\mathrm{theory})}{\nu_{\mathrm{Ph\,fs}}(\mathrm{exptl})}=(17\pm 66)\ \mathrm{ppm}.$$

Although the agreement between theory and experiment is excellent, Erickson's estimate will have to be replaced by an exact calculation and the error in the experiments significantly reduced before a critical test of QED can be made using the positronium hfs. [For a discussion of the annihilation rate of ortho- and parapositronium, see Hughes (1969).]

We conclude this section on hyperfine splittings by noting that accurate measurements of the ground-state hfs in relatively simple systems other than the ones we have discussed do exist, e.g., atomic deuterium, tritium, and ionized ^{3}He (see Crampton, Robinson, Kleppner, and Ramsey, 1966; Mathur, Crampton, Kleppner, and Ramsey, 1967; Fortson, Major, and Dehmelt, 1966). However, the present state of the theory of the hfs in these atoms is not yet sufficiently advanced to permit meaningful comparisons with experiment [but see * footnote on page 217 and Bethe and Salpeter (1957), pages 107–114].

C. Fine Structure of Hydrogenic Atoms

1. Lamb Shift in H and D, n=2

The fine structure of atomic hydrogen and hydrogen-like atoms such as deuterium and singly ionized helium

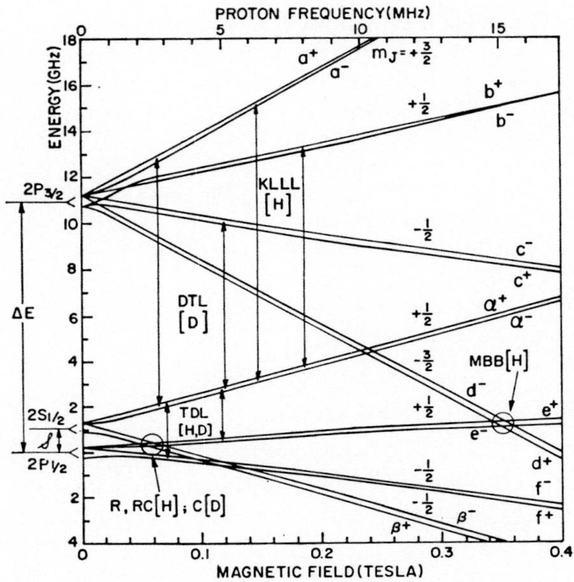

FIG. 5. Energy-level diagram of the fine-structure Zeeman levels of the $n=2$ state of atomic hydrogen [taken in part from Brandenberger (1968) and Brodsky and Parsons (1967)]. The hyperfine splitting of the levels is shown schematically and is not to scale. The states are labeled in the conventional manner (Lamb, 1952), and the + and − on the state labels correspond to $m_I=+\frac{1}{2}$ or $-\frac{1}{2}$, respectively. The Lamb shift $\mathcal{S}$ is the energy difference between the $2S_{1/2}$ and $2P_{1/2}$ levels, and the fine-structure splitting ΔE is the energy difference between the $2P_{3/2}$ and $2P_{1/2}$ levels. For the $n=2$ state in H, $\mathcal{S}\approx 1.058$ GHz, $\Delta E\approx 10.969$ GHz, and the hyperfine splittings of the $2S_{1/2}$, $2P_{1/2}$, and $2P_{3/2}$ states are ≈178, 59, and 24 MHz, respectively. The transitions and crossings which have yielded precise values of ΔE, $\mathcal{S}$, and $\Delta E-\mathcal{S}$ are denoted by the initials of the experimenters who carried out the work (see text). [H] and [D] indicate that the measurement was made in hydrogen or deuterium, respectively.

has proved to be an important testing ground for QED as well as a prime source of numerical values for the fine-structure constant. Figure 5 shows schematically the Zeeman splitting of the $n=2$ energy levels of atomic hydrogen. The principal quantities of interest here are the zero-field fine-structure (fs) splitting of the $2P_{3/2}$ and $2P_{1/2}$ levels, ΔE, and the splitting of the $2S_{1/2}$

and $2P_{1/2}$ levels or Lamb shift $\mathcal{S}$. The major portion of ΔE is due to spin–orbit coupling between the electron spin and orbital moments and can be accounted for within the Dirac theory; it is of order $\alpha^2 R_\infty$. The QED corrections other than those arising from the anomalous moment of the electron contribute only a few parts per million to ΔE (1.2 ppm in H for $n=2$). On the other hand, the Lamb shift is a purely QED effect since, within the Dirac theory, $nP_{1/2}$ and $nS_{1/2}$ levels are degenerate.* [$\mathcal{S}$ is of order $\alpha^3 R_\infty \ln (Z\alpha)^{-2}$, corresponding to about 10% of ΔE for $n=2$ in H.] Historically, ΔE has been of interest primarily as a source of information about α, and $\mathcal{S}$ as a testing ground for QED. Indeed, the discovery of the Lamb shift played a central role in the development of quantum electrodynamics.

A comprehensive theoretical discussion of the Lamb shift has been given recently by Erickson and Yennie (1965a; 1965b). These workers not only verified previous calculations of the terms of order $(Z\alpha)^4 \ln (Z\alpha)^{-2}$, $(Z\alpha)^4$, $(Z\alpha)^5$, and $(Z\alpha)^6 \ln^2 (Z\alpha)^{-2}$, but calculated the complete state dependence of the $(Z\alpha)^6 \ln (Z\alpha)^{-2}$ term as well as the dominant contribution to the term of order $(Z\alpha)^6$. In addition, they carefully estimated the size of uncalculated terms and were thus able to give estimates of the uncertainties to be assigned the theoretical values of $\mathcal{S}$. Their result for $\mathcal{S}$ in the nth level of a hydrogenic atom of nuclear mass M_i and charge Z may be written as

$$\mathcal{S}=\frac{8\alpha^3 R_\infty c Z^4}{3\pi n^3}\left\{\left(1+\frac{m_e}{M_i}\right)^{-3}\right.$$

$$\times\left(\ln\frac{1+(m_e/M_i)}{(Z\alpha)^2}+\frac{19}{30}+\ln\frac{K_0(n,1)}{K_0(n,0)}\right)+\frac{1}{8}\left(1+\frac{m_e}{M_i}\right)^{-2}$$

* Strictly speaking, this is not quite true. These levels would be slightly split by nuclear size corrections even if QED did not exist.

$$+\frac{\alpha}{\pi}\left[\tfrac{3}{2}m-0.3285-\tfrac{82}{81}(\tfrac{3}{4})\right]+\pi Z\alpha\left(\frac{427}{128}-\frac{3\ln 2}{2}\right)$$

$$+(Z\alpha)^2\Bigg[-\tfrac{3}{4}\ln^2(Z\alpha)^{-2}+C_n\ln(Z\alpha)^{-2}$$

$$-\left(\frac{4\pi^2}{3}+4+4\ln^2 2\right)\Bigg]+Z\frac{m_e}{M_i}\left[\frac{1}{4}\ln(Z\alpha)^{-2}\right.$$

$$\left.+2\ln\frac{K_0(n,1)}{K_0(n,0)}-\frac{1}{12}+D_n\right]\Bigg\}+\frac{4R_\infty cZ^4}{3n^3}\frac{r_n^2}{a_0^2}. \quad (138a)$$

[This equation also includes terms previously calculated by other workers; see Erickson and Yennie (1965a; 1965b).] In Eq. (138a), $\ln K_0(n, 0 \text{ or } 1)$ is the Bethe logarithmic excitation energy, m is the coefficient of the fourth-order radiative correction recently calculated exactly by Soto (1966) and numerically equal to 0.215296114, 0.3285 stands for the exact expression given in Eq. (18) for the coefficient of the fourth-order correction to the anomalous moment of the electron, and C_n and D_n are constants which depend on n:

$$C_n=7\ln 2-3\ln n+3\sum_{q=1}^{n}q^{-1}-\frac{757}{240}-\frac{4}{5n^2}, \quad (138b)$$

$$D_n=\frac{91}{24}+\frac{7}{2}\left(\ln\frac{2}{n}+\sum_{q=1}^{n}q^{-1}-\frac{1}{2n}\right). \quad (138c)$$

Although D_n was originally calculated exactly only for $n=2$, Erickson (1969; 1967) has recently shown it to be correct for all n. The last term of Eq. (138a) takes into account the effects of nuclear structure; r_n is the root-mean-square radius of the nuclear charge distribution and a_0 is the Bohr radius.

We have evaluated Eq. (138) for the $n=2$, 3, and 4 levels in H, D, and $^4He^+$ using the WQED adjusted values of α and a_0, the values of r_n given by Erickson and Yennie, the values of the Bethe logarithm given by Schwartz and Tiemann (1959) ($n=2$) and by Harriman (1956) ($n=3$, 4), and the auxiliary constants listed in

TABLE XXVI. Theoretical values of the Lamb shift for the $n=2$, 3, and 4 levels of H, D, and $^4He^+$ as calculated using the adjusted WQED value of α. The errors are *three standard deviations* (see text).

	Lamb shift $\mathcal{S}$, (MHz)		
n	H	D	$^4He^+$
$n=2$	1057.559 ± 0.086	1058.821 ± 0.145	$14\,038.93\pm4.11$
$n=3$	314.791 ± 0.025	315.165 ± 0.043	$4\,182.66\pm1.22$
$n=4$	133.041 ± 0.011	133.198 ± 0.018	$1\,768.34\pm0.51$
$n=2$	$\mathcal{S}_D-\mathcal{S}_H=1.262\pm0.077$		

Table XI. The results are presented in Table XXVI. The errors quoted are *extreme limits of error* (3 standard deviations) and for $n=2$ are as given by Erickson and Yennie (1965a) (their Table III) except for the following modifications: (1) the uncertainty in the size of the $\alpha(Z\alpha)^4$ term included by Erickson and Yennie has been eliminated as a result of Soto's recent exact calculation of this term, (2) the auxiliary constants R_∞ and c have been taken to be exact, and (3) the uncertainty in $\mathcal{S}$ due to the uncertainty in α has been reduced from 38 ppm (0.040 MHz for H and D, $n=2$) to 16 ppm (0.017 MHz) because of the smaller uncertainty in α_{WQED}. The quoted errors for $n=3$ and 4 have been obtained from the $n=2$ errors by using the multiplicative factors 8/27 and 8/64, respectively. This procedure takes into account the dependence of the various uncertainties on n since $\mathcal{S}\propto n^{-3}$.

The fine structure of the $n=2$ level of H and D was first studied experimentally with high accuracy by Lamb and coworkers and reported in a series of now classic papers (Lamb and Retherford, 1950; 1951; 1952; Lamb, 1952; Triebwasser, Dayhoff and Lamb, 1953; Dayhoff, Triebwasser and Lamb, 1953). In these experiments, microwave transitions between the $2S$ and $2P$ states were observed by making use of the metastability of the $2S$ state (its lifetime against decay to the ground or $1S$ state is $\sim\frac{1}{8}$ sec). A beam of atoms in this state was produced by thermal dissociation of molecular hydrogen followed by excitation of the resulting atomic hydrogen by electron bombardment. Atoms excited to the $2P$ states decayed very rapidly to the ground state, leaving a beam composed of atoms in the metastable $2S$ state. The beam passed through a microwave interaction region to a detector in which the metastable atoms ejected electrons from a metal surface, and the resulting electron current was measured by an electrometer. When microwave power was applied to the beam at a frequency corresponding to the difference in energy between a pair of $2S$ and $2P$ levels, transitions to the $2P$ state were induced, followed rapidly by decay to the ground state. The transitions were observed via the resulting reduction ("quenching") of the detected metastable beam intensity. The magnetic field at which the transition occurs for fixed (known) microwave frequency, together with an extrapolation along the Zeeman lines to zero field, determines $\mathcal{S}$, ΔE, or $\Delta E-\mathcal{S}$, as the case may be. The frequency width of the transitions is determined primarily by the radiative lifetime of the $2P$ states and is about 100 MHz. Since the fine-structure splitting and Lamb shift are of order 10 and 1 GHz, respectively, for H, D, $n=2$, a measurement of the former with an accuracy of 10 ppm and the latter with an accuracy of 100 ppm requires locating the centers of the transition resonance lines to within 10^{-3} of their widths. An extremely careful analysis of the many factors contributing to line shapes and shifts was therefore necessary, including a detailed understanding of the Zeeman structure of Fig. 5.

In 1953, Triebwasser, Dayhoff, and Lamb (TDL) reported their final results for the Lamb shift in the $n=2$ state of H and D. Using the αe transition at ≈ 0.1160 T ($\nu \approx 2195$ MHz), and the αf transition at ≈ 0.0704 T ($\nu \approx 2395$ MHz) (see Fig. 5), they found (in megahertz)

$$\begin{aligned} \mathrm{H}(\alpha e): &\quad \mathcal{S}_{\mathrm{H}} = 1057.752 \pm 0.095, \\ \mathrm{H}(\alpha f): &\quad \mathcal{S}_{\mathrm{H}} = 1057.795 \pm 0.089, \\ \text{average}: &\quad \mathcal{S}_{\mathrm{H}} = 1057.774 \pm 0.10, \end{aligned}$$

$$\begin{aligned} &\mathrm{D}(\alpha e): && \mathcal{S}_{\mathrm{D}}=1059.057\pm 0.074, \\ &\mathrm{D}(\alpha f): && \mathcal{S}_{\mathrm{D}}=1058.950\pm 0.033, \\ &\text{average}: && \mathcal{S}_{\mathrm{D}}=1059.004\pm 0.10. \end{aligned} \tag{139}$$

The quoted error in $\mathcal{S}$ for each transition is the average deviation of the several separate runs of which it is the mean, while the error of the average values obtained from both the αe and αf transitions is approximately 3 times the average deviation of the mean (all uncertainties and $\mathcal{S}$ values are as given by Triebwasser, Dayhoff, and Lamb). The mean value was obtained by taking a straight average rather than a weighted one because Triebwasser, Dayhoff, and Lamb felt the αe and αf measurements were of equal inherent accuracy, fluctuations in the individual deviations being of a normal random nature.

The values of $\mathcal{S}$ given above were calculated by Triebwasser, Dayhoff, and Lamb using magnetic fields obtained (indirectly) from proton resonance frequencies via the γ_p result of Thomas, Driscoll, and Hipple (1950a; 1950b) and other constants from Bearden and Watts (1951). These values of course differ from the currently accepted ones (recall that the Thomas, Driscoll, and Hipple value of γ_p is probably in error by some 50 ppm; see Sec. II.C.7). However, the Zeeman-theory relations used by Triebwasser, Dayhoff, and Lamb to analyze their data [Eqs. (164)–(172) of Lamb (1952)] may be rewritten in terms of proton resonance frequency. If this is done, the only constants which enter are g_s, $1-m_e/M_p$ (or $1-m_e/M_d$), and μ_p'/μ_B, all of which are known to better than 0.1 ppm. (Any difference between Triebwasser *et al.*'s cylindrical, doped-water sample and a spherical pure-water sample may be neglected.) Also required is ΔE, but $\mathcal{S}$ is not very sensitive to its value; a 1-MHz uncertainty in ΔE gives rise to a 0.03-MHz uncertainty in $\mathcal{S}(\alpha e)$ and 0.013 MHz in $\mathcal{S}(\alpha f)$. We have recalculated the results of Triebwasser, Dayhoff, and Lamb by reconverting their resonance magnetic fields to equivalent proton resonance frequencies using their conversion factor. With the aid of the auxiliary constants given in Table XI and the theoretical value of ΔE implied by α_{WQED} (see next section), we find (in megahertz)

$$\begin{aligned}
&\text{H}(\alpha e): && \mathcal{S}_{\text{H}}=1057.738\pm0.095,\\
&\text{H}(\alpha f): && \mathcal{S}_{\text{H}}=1057.807\pm0.089,\\
&\text{average}: && \mathcal{S}_{\text{H}}=1057.772\pm0.063\ (60\ \text{ppm}),\\
&\text{D}(\alpha e): && \mathcal{S}_{\text{D}}=1059.042\pm0.074,\\
&\text{D}(\alpha f): && \mathcal{S}_{\text{D}}=1058.950\pm0.033,\\
&\text{average}: && \mathcal{S}_{\text{D}}=1058.996\pm0.064\ (60\ \text{ppm}).
\end{aligned}\tag{140}$$

Although the individual values change slightly, the average values remain essentially the same [see Eq. (139)]. In principle, it should also be possible to recalculate $\mathcal{S}$ using the precise ($\sim$1 ppm or 0.001 MHz) Zeeman theory of Brodsky and Parsons (1967; 1968) [see also Brodsky and Primack (1968, and to be published)] which is based on diagonalization of an appropriate Hamiltonian rather than the perturbation theory of Lamb. But in view of the close quantitative agreement (within 0.02 MHz) between the two methods as found by Robiscoe in the analysis of his own Lamb-shift experiments (see below), we did not believe it necessary to undertake this rather complex calculation. [This agreement has also been confirmed by Kaufman (1968).]

The uncertainties quoted in Eq. (140) for the individual transitions are the same as in Eq. (139), but those given for the average values of $\mathcal{S}_{\text{H}}$ and $\mathcal{S}_{\text{D}}$ are our own estimates and are meant to represent one standard deviation, including both statistical and systematic errors. They were obtained in the following way. The statistical standard deviation of the mean value of $\mathcal{S}_{\text{H}}$ (or $\mathcal{S}_{\text{D}}$) as obtained from the two separate measurements (αe and αf) is $\sim$0.04 MHz when calculated according to Eq. (10). To this, we add an estimated systematic error of 0.05 MHz in the usual RSS manner. It includes ~0.015 MHz for uncertainties in the magnetic field (~20 ppm), ~0.01 MHz for magnetic-field inhomogeneities, ~0.02 MHz for possible differences between the perturbation theory and more exact theory of Brodsky and Parsons, and from 0.02 to 0.04 MHz for possible uncertainties in the applied corrections. We have also compared the experiment of Triebwasser, Dayhoff, and Lamb with the similar experiments of other workers (to be discussed below and in the next section). We conclude that in most cases, 0.05 MHz is a reasonable 70% confidence-level estimate of the systematic error, and this value will therefore be adopted for the purpose of comparing theory and experiment.

Recently, measurements of the Lamb shift in the $n=2$ state of H have been made by Robiscoe (R) (1965) and Robiscoe and Cosens (RC) (1966a) using a level-crossing technique. The method is generally similar to that used by Lamb and coworkers, except that the quenching of the metastable beam is induced by a static electric field at the crossing of the β and e levels which occurs at about 0.0570 T (see Fig. 5). The experiment does not require an applied microwave field and is essentially a zero-frequency Lamb–Retherford experiment. An important feature of this work was that the metastable beam could be prepared in either of the two hyperfine levels, i.e., β^+ or β^- ($\pm$ means $m_I=\pm\frac{1}{2}$). This allowed the individual level crossings β^+e^+ at 0.0538 T and β^-e^- at 0.0605 T to be observed separately, thereby considerably reducing the resonance linewidth and simplifying the analysis of the quenching line shape. [These crossings are referred to as crossing A or H(538) and crossing B or H(605), respectively.] This is in marked contrast to the Triebwasser, Dayhoff, and Lamb work in which two overlapping resonances were actually observed, one for each of the $2S$–$2P$ transitions with $\Delta m_I=0$. [Kaufman (1968) has shown using a matrix diagonalization method that higher-order terms must be included in the Lamb perturbation theory in order to account for the hyperfine levels adequately. However, very little error is introduced by ignoring these terms for a composite line as did Triebwasser, Dayhoff, and Lamb since the levels are shifted symmetrically.]

In 1965, Robiscoe reported $\mathcal{S}_H = 1058.07 \pm 0.10$ MHz as obtained from the H(605) crossing, and in 1966, Robiscoe and Cosens (1966a) reported $\mathcal{S}_H = 1058.04 \pm 0.10$ MHz as obtained from the H(538) crossing using a completely new apparatus. (The quoted errors were meant to be ~2 standard deviations.) Both of these results were obtained by extrapolating to zero field using the Lamb perturbation theory of the Zeeman levels. Recently, Robiscoe (1968) has reported that a contribution to the asymmetry in the observed resonance curve due to quenching of the metastable beam by motional electric fields had been overlooked. Correcting the above values for this asymmetry and using the Brodsky–Parsons (1967) theory rather than the Lamb perturbation theory, Robiscoe obtained (in megahertz)

$$\begin{aligned} &\text{H}(538): && \mathcal{S}_H = 1057.84 \pm 0.10, \\ &\text{H}(605): && \mathcal{S}_H = 1057.89 \pm 0.15, \\ &\text{average}: && \mathcal{S}_H = 1057.855 \pm 0.063 \text{ (60 ppm)}. \end{aligned} \qquad (141)$$

The quoted uncertainty for the weighted average of the two transitions is our own standard-deviation estimate and follows from our adopted systematic error of 0.05 MHz and the fact that 0.10 MHz is stated to be at least twice the standard deviation of the mean of the means of 10 independent runs comprising over 200 line-center measurements. [Robiscoe (1968) comments that all *known* systematic errors were calculated to an accuracy of better than 0.03 MHz.] In comparing Robiscoe's final result with that of Triebwasser, Dayhoff, and Lamb [Eq. (140)], we see that the two are in good agreement; their 0.083-MHz difference is 0.92 times the 0.090-MHz standard deviation of their difference. The probability for this to occur by chance is about 35%.

A fairly accurate value of $\mathcal{S}_H$ for $n=2$ can also be obtained from the separate measurements of $\Delta E_H - \mathcal{S}_H$ by Kaufman, Lamb, Lea, and Leventhal (KLLL), and of ΔE_H by Metcalf, Brandenberger, and Baird (MBB) [Eqs. (154) and (158), respectively; these experiments will be discussed in detail shortly]. The result is

$$\mathcal{S}_H = 1057.750 \pm 0.099 \text{ MHz (93 ppm)}, \qquad (142)$$

in excellent agreement with the Triebwasser, Dayhoff, and Lamb determination and in reasonable agreement with that of Robiscoe.

Recently Cosens (1968) reported a measurement of the Lamb shift in the $n=2$ state of deuterium using the same technique used by Robiscoe and Robiscoe and Cosens. (This is the final result of work reported earlier by Robiscoe and Cosens, 1966b.) However, since the spin of the deuteron is one, the β and e levels are composed of three hyperfine sublevels rather than two as in H. Thus, in D, there are three observable transitions with $\Delta m_I = 0$ at the βe crossing point. They are crossing A or D(564), crossing B or D(574), and crossing C or D(584) [the quantum numbers m_F, m_I, and m_J are for A: $\beta(+\frac{1}{2}, +1, -\frac{1}{2})$, $e(+\frac{3}{2}, +1, +\frac{1}{2})$; for B: $\beta(-\frac{1}{2}, 0, -\frac{1}{2})$, $e(+\frac{1}{2}, 0, +\frac{1}{2})$; for C: $\beta(-\frac{3}{2}, -1, -\frac{1}{2})$, $e(-\frac{1}{2}, -1, +\frac{1}{2})$]. Using the theory of Brodsky and Parsons and correcting for the effect of motional fields as did Robiscoe, Cosens finds (in megahertz),

$$\text{D(574)}: \quad \mathcal{S}_D = 1059.288 \pm 0.042,$$

$$\text{D(584)}: \quad \mathcal{S}_D = 1059.165 \pm 0.055,$$

$$\text{average}: \quad \mathcal{S}_D = 1059.244 \pm 0.064 \text{ (60 ppm)}. \qquad (143)$$

The errors quoted for the separate measurements are those given by Cosens and are intended to be one-standard-deviation uncertainties including estimates of systematic error. This would imply a one-standard-deviation uncertainty in the weighted average of only 0.033 MHz. We believe this to be unrealistically low in view of the many corrections required in the experiment and the fact that the two transitions give results which differ by 1.8 times the standard deviation of their difference. We shall therefore assume an error of 0.064 MHz (60 ppm) as in the experiments of Triebwasser, Dayhoff, and Lamb and of Robiscoe. (Corrections are necessary for the effect of Stark matrix-element variation, level curvature, Stark shift, other transitions, velocity distribution distortion, finite size of the quench region, hydrogen impurities, etc., and amount to −177 ppm for crossing B and −361 ppm for crossing C.)

In comparing the Cosens value of $\mathcal{S}_D$ with that of Triebwasser, Dayhoff, and Lamb [Eq. (140), average value], we see the former exceeds the latter by 0.248 ± 0.090 MHz or (235 ± 85) ppm. The difference thus exceeds its standard deviation by a factor of 2.8. The probability for this to occur by chance is $\sim$0.6%; this is somewhat surprising. The existence of a systematic difference between the level crossing and microwave resonance methods would seem unlikely in view of the good agreement between $\mathcal{S}_H$ obtained by Triebwasser, Dayhoff, and Lamb and $\mathcal{S}_H$ obtained by Robiscoe using the same level-crossing technique as Cosens. (Note however, that both level-crossing values are higher than the corresponding microwave-transition values.) On the other hand, the difficulties inherent in locating the center of a resonance line to better than one part in 10^3 of its width are formidable, and there are many opportunities for systematic errors to "rear their ugly heads." The history of the Lamb shift clearly shows that over the years, the uncertainties assigned both the theoretical and experimental values of $\mathcal{S}$ have usually been too small, e.g., the estimate of the fourth-order radiative correction prior to Soto's exact calculation* and Robiscoe's motional-field correction. If the error assigned the Triebwasser, Dayhoff, and Lamb and the Cosens $\mathcal{S}_D$ results is increased to 0.1 MHz, a value

* This was due to the fact that the error limits originally assigned were rigorous upper and lower bounds given by Weneser, Bersohn, and Kroll (1953) to certain integrals. The integrals have since been found to be incorrect by Soto (Erickson, private communication).

which may better reflect the true 70% confidence-level uncertainty in this type of experiment, their difference becomes 0.25±0.14 MHz. The probability for this to occur by chance is 8%, and is not unbelievable. We shall have more to say on the error problem at the end of this section.

We now compare with theory the various Lamb-shift measurements so far discussed. Defining $\Delta\mathcal{S}$ as $\mathcal{S}$(exptl) − $\mathcal{S}$(theory), we find from Table XXVI and the average values given in Eqs. (140)–(143) (in megahertz)

$$(\Delta\mathcal{S}_{\mathrm{H}})_{\mathrm{TDL}} = 0.213 \pm 0.070[(200 \pm 66)\ \mathrm{ppm}],$$

$$(\Delta\mathcal{S}_{\mathrm{H}})_{\mathrm{R}} = 0.296 \pm 0.070[(280 \pm 66)\ \mathrm{ppm}],$$

$$(\Delta\mathcal{S}_{\mathrm{H}})_{\mathrm{MBB,KLLL}} = 0.191 \pm 0.105[(180 \pm 97)\ \mathrm{ppm}],$$

$$(\Delta\mathcal{S}_{\mathrm{D}})_{\mathrm{TDL}} = 0.175 \pm 0.080[(165 \pm 75)\ \mathrm{ppm}],$$

$$(\Delta\mathcal{S}_{\mathrm{D}})_{\mathrm{C}} = 0.422 \pm 0.080[(400 \pm 75)\ \mathrm{ppm}]. \tag{144}$$

In computing the errors, we have divided the uncertainties of the theoretical values of $\mathcal{S}$ by 3 to convert them to standard deviations. (Recall that these errors were meant to be extreme limits of error or three standard deviations.) Taken at face value, Eq. (144) indicates that theory and experiment are in significant disagreement, the various differences exceeding their standard deviations by factors of 2 to more than 5. Even if the error assigned both $\mathcal{S}$(theory) and $\mathcal{S}$(exptl) were increased to the not-unreasonable value of 0.1 MHz, the discrepancies in Eq. (144) would still remain sizable; in all five comparisons, experiment would exceed theory by 1.2 to 3 standard deviations. Thus, it would appear that in addition to the electron magnetic-moment anomaly, we are faced with another clear-cut discrepancy between QED and experiment.

Recently, Barrett, Brodsky, Erickson, and Goldhaber (1968) have investigated the idea that the above discrepancies are in fact real and could perhaps be due to a "proton halo" of radius $\sim$8 F and positive charge $\approx 0.01e$.* They showed that such a halo would increase $\mathcal{S}$(theory) for both H and D by about 0.25 MHz, could explain the small discrepancy between muonic x-ray and electron-scattering measurements of the nuclear charge structure of ^{209}Bi, and would not be in disagreement with other experiments which might be affected by it, viz., the present good agreement between the theoretical and experimental values of the hfs in H, and electron–proton-scattering data [but see Anderson *et al.* (1969)]. This conjecture is presently being tested experimentally. Yennie and Farley (Yennie, 1967) have also speculated about the exchange of a new scalar particle between the electron and proton in order to explain the disagreement between $\mathcal{S}$(exptl) and $\mathcal{S}$(theory). Cochran and Franken (1968) have pointed out that their recent experiments establishing that the exponent in Coulomb's law is $2(1 \pm 4.6 \times 10^{-12})$ rule out the possibility that the discrepancy could be accounted for by a deviation from Coulomb's law [see also Bartlett and Phillips (1969)].

We next compare the experimental values of $\mathcal{S}_D - \mathcal{S}_H$ with the theoretical value given in Table XXVI. Since there are two measurements of both $\mathcal{S}_D$ and $\mathcal{S}_H$, there are four distinct values of $\mathcal{S}_D - \mathcal{S}_H \equiv \mathcal{S}_{D-H}$. [We ignore in this comparison the value of $\mathcal{S}_H$ obtained from the separate measurements of ΔE_H and $\Delta E_H - \mathcal{S}_H$, Eq. (142), since it has a relatively large uncertainty and very nearly equals the Triebwasser, Dayhoff, and Lamb value, Eq. (140).] We find from Eqs. (140), (141), and (143) (in megahertz)

$$(\mathcal{S}_{D-H})_{\mathrm{TDL,TDL}} = 1.224 \pm 0.090,$$

$$(\mathcal{S}_{D-H})_{\mathrm{TDL,R}} = 1.141 \pm 0.090,$$

---

* It has been suggested by Fil'kov (1968) that the proton halo may be due to an "antibound" virtual p state of the $\pi\pi$ system.

$$(\mathcal{S}_{D-H})_{C,TDL} = 1.471 \pm 0.090,$$

$$(\mathcal{S}_{D-H})_{C,R} = 1.388 \pm 0.090. \qquad (145)$$

Defining $\Delta\mathcal{S}_{D-H} = \mathcal{S}_{D-H}(\text{exptl}) - \mathcal{S}_{D-H}(\text{theory})$, we obtain (in megahertz)

$$(\Delta\mathcal{S}_{D-H})_{TDL,TDL} = -0.039 \pm 0.093,$$

$$(\Delta\mathcal{S}_{D-H})_{TDL,R} = -0.122 \pm 0.093,$$

$$(\Delta\mathcal{S}_{D-H})_{C,TDL} = 0.209 \pm 0.093,$$

$$(\Delta\mathcal{S}_{D-H})_{C,R} = 0.126 \pm 0.093. \qquad (146)$$

Although the over-all agreement between theory and experiment is reasonably satisfactory, the uncertainties are too large to allow firm conclusions to be drawn. We do note that the good agreement for $(\Delta\mathcal{S}_{D-H})_{TDL,TDL}$ implies that if a systematic error is present, it is probably the same for both the $\mathcal{S}_H$ and $\mathcal{S}_D$ measurements. Alternatively, it may be concluded that since the only important difference between $\mathcal{S}_D(\text{theory})$ and $\mathcal{S}_H(\text{theory})$ is the nuclear structure correction [last term of Eq. (138a)], any error in it must be essentially the same for both H and D. [For further comments on the theory of the Lamb shift, see Erickson (1969).]

We conclude this section with some comments concerning the present status of the Lamb-shift measurements. First, we point out a possible weakness in the work of Triebwasser, Dayhoff, and Lamb. Robiscoe (1968) has noted that in the 1952 paper of Lamb an error appears in the analysis of the αe and αf hyperfine-splitting transition frequencies. If this error is corrected and Triebwasser, Dayhoff, and Lamb's values of $\mathcal{S}_H$ revised accordingly, Robiscoe finds

$$\mathcal{S}_H(\alpha e) = 1057.97 \pm 0.10 \text{ MHz},$$

$$\mathcal{S}_H(\alpha f) = 1057.45 \pm 0.10 \text{ MHz},$$

in significant disagreement with the original results, Eq. (139). Furthermore, the values from the two transitions are no longer consistent, thus implying the presence of additional systematic errors. One could perhaps argue that the good internal agreement of the values reported by Triebwasser, Dayhoff, and Lamb is evidence that the erroneous hfs correction was not in fact carried through to the final result. On the other hand, the good agreement could be fortuitous. (The effect of the "error" on the $\mathcal{S}_D$ measurements is negligible because of the relatively small size of the hyperfine splittings in D.) Robiscoe (private communication) has attempted to resolve this dilemma but it was not possible to find out if in fact the error did carry through. It must therefore be concluded that the Triebwasser, Dayhoff, and Lamb values of $\mathcal{S}$ should be viewed cautiously.

Our second comment concerns the uncertainties assigned the experimental values of the Lamb shift. The various measurements of $\mathcal{S}$ have usually been reported with a quoted error of 0.10 MHz. It was invariably stated by the authors that this 0.10 MHz represented a "limit of error" which is generally taken to mean two standard deviations. Thus, the final one-standard-deviation error would be 0.05 MHz. In our analysis, we have been somewhat more conservative and have assumed that the one-standard-deviation systematic error was 0.05 MHz and added to it RSS the one-standard-deviation statistical error. This procedure gave a 60-ppm or 0.064-MHz uncertainty in $\mathcal{S}$. But in view of the 0.25-MHz discrepancy between the Cosens and TDL value of $\mathcal{S}_D$, the 0.17-MHz correction recently discovered by Robiscoe, the uncertainty in the hfs correction used by Triebwasser, Dayhoff, and Lamb, and the many problems involved in splitting a line to one or two parts in two thousand, it may well be that even our estimates are too optimistic. We leave that judgement to the reader. (See also Notes Added in Proof.)

2. *Fine-Structure Splitting in H and D, $n=2$*

The theoretical equation for the fine-structure splitting ΔE may be obtained from the formulas given by Erickson and Yennie (1965a; 1965b) and Barker and Glover (1955), together with an expansion of the exact solution of the Dirac equation for the H atom (Bethe and Salpeter, 1957). For the nth level of a hydrogenic atom of charge Z and nuclear mass M_i, ΔE may be written as

$$\Delta E = \frac{Z^2 R_\infty (Z\alpha)^2 c}{2n^3} \left\{ [1+F_n(Z\alpha)^2] \left(1+\frac{m_e}{M_i}\right)^{-1} - \left(\frac{m_e}{M_i}\right)^2 \left(1+\frac{m_e}{M_i}\right)^{-3} + 2a_e \left(1+\frac{m_e}{M_i}\right)^{-2} - G_n \frac{4\alpha}{3\pi} (\alpha Z)^2 \ln (Z\alpha)^{-2} \right\}, \quad (147a)$$

$$F_n = (7n^2+18n-24)/16n^2; \qquad G_n = 1-(1/n^2). \quad (147b)$$

The first term in Eq. (147a) [brackets] comes from the Dirac solution (Bethe and Salpeter, 1957). The reduced-mass factor $(1+m_e/M_i)^{-1}$ has been obtained by Grotch and Yennie (1969) using an effective-potential model. The second term comes from the work of Barker and Glover (1955); it arises from the normal Dirac moments of the electron and nucleus. The third term is the contribution due to the anomalous moment of the electron and the last term is a radiative correction first calculated by Layzer (1960) and by Fried and Yennie (1960), and checked by Erickson and Yennie (1965a; 1965b). To this order, it is the only QED contribution other than the electron magnetic-moment

TABLE XXVII. Theoretical values of the fine-structure splitting ΔE and the $nS_{1/2}$–$nP_{3/2}$ splitting $\Delta E - \mathcal{S}$ in the $n=2$, 3, and 4 levels of H, D, and $^4He^+$ as calculated using the adjusted WQED value of α. The errors are standard deviations (see text).

n	Fine-structure splitting ΔE, and $\Delta E - \mathcal{S}$, (MHz)		
	H	D	$^4He^+$
$n=2$	10 969.026±0.042	10 972.020±0.042	175 593.12±0.71
	9 911.467±0.051	9 913.199±0.064	161 554.19±1.54
$n=3$	3 250.085±0.013	3 250.972±0.013	52 027.78±0.21
	2 935.293±0.015	2 935.806±0.019	47 845.13±0.46
$n=4$	1 371.128±0.005	1 371.502±0.005	21 949.117±0.089
	1 238.087±0.006	1 238.304±0.008	20 180.78±0.19

anomaly. For $n=2$ in H, it contributes only 1.2 ppm.* Brodsky and Parsons (1967) note a private communication from Erickson giving an estimated bound on the next term, $(\alpha/\pi)(Z\alpha)^2\Delta E \times (16/3)a$: $|a|<1$. Such a term would contribute less than 0.66 ppm for H and D, and 2.64 ppm for $^4He^+$. Since there are probably other uncalculated terms comparable in magnitude with $\alpha^2(m_e/M_i)$, $\alpha(m_e/M_i)^2$, etc. (Grotch and Yennie, 1969), it is probably more consistent to expand the reduced-mass factors in Eq. (147a) and to rewrite it as

$$\Delta E = \frac{Z^2 R_\infty (Z\alpha)^2 c}{2n^3} \left\{ [1+F_n(Z\alpha)^2]\left(1-\frac{m_e}{M_i}\right) + 2a_e\left(1-2\frac{m_e}{M_i}\right) - G_n \frac{4\alpha}{3\pi}(\alpha Z)^2 \ln (Z\alpha)^{-2} \right\}. \quad (148)$$

Note that the $(m_e/M_i)^2$ term cancels and that for H, $n=2$, Eq. (147) is equivalent to others which have appeared in the literature to within 0.02 ppm [see, for example, Brodsky and Parsons (1967)].

We have evaluated Eq. (147) for the $n=2$, 3, and 4 levels in H, D, and $^4He^+$ using α_{WQED}, the theoretical expression for a_e, Eq. (103), and the auxiliary constants of Table XI. The results are given in Table XXVII. The quoted uncertainties correspond to one

* We should emphasize that while this radiative correction term contributes only 1.2 ppm to ΔE for H, the electron moment anomaly contributes ~1000 ppm (0.1%). Thus, any value of α derived from ΔE would clearly not fit our definition of WQED—derived without essential use of QED theory. It may be argued that this criticism can be circumvented by using the experimental value for the anomaly, but it seems to us that this violates the spirit of the WQED concept; we believe if QED gives a theoretical expression for a quantity, then it should not be ignored in favor of an experimental value in any comparison of QED theory and experiment. Our set of WQED constants is derivable from quantities for which present QED theory gives no explicit equations and which do not require (at least to the negligible sub-part-per-million level) the use of QED theory for their analysis. This is not to say of course that reliable values of α cannot be obtained from measurements of ΔE. On the contrary, future high-accuracy measurements of ΔE in H and D may yet provide the best values of α. However, we do not feel these values can ever be considered WQED values.

standard deviation and are the RSS of 2×1.9 ppm $=$ 3.8 ppm for α^2 and 0.33 ppm (1.32 ppm for $^4He^+$) for possible contributions of uncalculated terms as indicated by the estimated bound given by Erickson. [Erickson (private communication) has suggested that his bound on uncalculated terms be interpreted as a limit of error.] Also given in Table XXVII are theoretical values for $\Delta E-\mathcal{S}$ based on the theoretical values of $\mathcal{S}$ given in Table XXVI. The quoted uncertainty here is also meant to be a standard deviation and thus the assigned errors for $\mathcal{S}$ were divided by 3 before being combined RSS with the ΔE uncertainties. (The uncertainties of ΔE and $\mathcal{S}$ are essentially independent since the error in $\mathcal{S}$ due to α_{WQED} is small compared with the total error in $\mathcal{S}$.)

Simultaneously with the publication of the Lamb-shift measurements in H and D, Dayhoff, Triebwasser, and Lamb (DTL) (1953) published results for the $2S_{1/2}$–$2P_{3/2}$ splitting in D, $\Delta E_{\text{D}}-\mathcal{S}_{\text{D}}$. Using the same general method as for the Lamb-shift work, these authors studied the αa transition at ≈ 0.0631 T ($\nu\approx$ 10.795 GHz) and the αc transition at ≈ 0.1189 T ($\nu\approx$ 7.195 GHz) (see Fig. 5). The final values for $\Delta E_{\text{D}}-\mathcal{S}_{\text{D}}$ obtained from the two transitions were (in megahertz)

$$\text{D}(\alpha a):\quad \Delta E_{\text{D}}-\mathcal{S}_{\text{D}}=9912.594\pm 0.056\ (5.6\ \text{ppm}),$$

$$\text{D}(\alpha c):\quad \Delta E_{\text{D}}-\mathcal{S}_{\text{D}}=9912.803\pm 0.094\ (9.5\ \text{ppm}). \tag{149}$$

The uncertainty originally quoted by Dayhoff, Triebwasser, and Lamb for the D(αa) transition was ± 0.10 MHz. This was a straight sum of a 0.05-MHz statistical limit of error and a 0.05-MHz possible systematic error. The latter was estimated as equal to the mean day-to-day scatter of the D(αa) results, with the idea that a nonrandom systematic change significantly larger than the day-to-day scatter could have been detected and eliminated. The standard-deviation uncertainty we give above is the RSS of this 0.05-MHz systematic error (which is identical to our adopted systematic error for this type of experiment) and the implied 0.025-MHz standard-deviation statistical error. The statistical standard deviation of the D(αc) results was 0.08 MHz, more than 3 times that of the D(αa) results. Adding the 0.05-MHz systematic error RSS gives 0.094 MHz. Thus, the D(αc) value exceeds the $D(\alpha a)$ value by 0.209 ± 0.110 MHz [(21 ± 11) ppm], or 1.9 times the standard deviation of the difference. The probability for this to occur by chance is about 6%. Dayhoff, Triebwasser, and Lamb chose to reject completely the D(αc) result for the following reason: The magnetic-field calibration in the D(αa) work was done using a proton resonance probe. In the D(αc) work it was done by observing the relatively sharp ($\sim$1 G wide) $\alpha\beta$ transition near the βe crossing point and using the Zeeman theory to calculate the field from this frequency splitting. Under the experimental conditions used, this is a second-order transition which goes via both e and f states. The theory of the position and shape of this resonance is thus a rather delicate matter, and the authors concluded that an error in field calibration by this method corresponding to 0.14 MHz in the final result could easily occur. This was large enough to account for the discrepancy between the αa and αc transitions, and it was therefore decided to retain only the D(αa) result.

Let us digress for a moment to learn an object lesson in the pitfalls of discarding data. The currently accepted value of α follows directly from the value of ΔE_{D} obtained by combining the Triebwasser, Dayhoff, and Lamb measurement of $\mathcal{S}_{\mathrm{D}}$ [Eq. (139), average value], and the value of $\Delta E_{\mathrm{D}}-\mathcal{S}_{\mathrm{D}}$ obtained by Dayhoff, Triebwasser, and Lamb from the D(αa) transition. The result usually quoted (Cohen and Dumond, 1965) is $\Delta E_{\mathrm{D}}=10971.59\pm0.10$ MHz, which gives $\alpha^{-1}=137.0388(6)$ (4.5 ppm) via Eq. (147) or a similar expression. As a consequence of neglecting the D(αc) transition result, this value of α rests on just six experimental runs on the D(αa) transition. But more important, the average experimental deviations in the six runs were such that the relative statistical weights of the runs varied from 3 to 48, with one particular run carrying a weight nearly equal to the remaining five together. The presently accepted value of α therefore rests to a considerable extent on a single experimental run! Furthermore, the rejected value of $\Delta E_{\mathrm{D}}-\mathcal{S}_{\mathrm{D}}$ obtained from the $\mathrm{D}(\alpha c)$ transition implies $\Delta E_{\mathrm{D}}=10971.80$, or $\alpha^{-1}=137.0375$, a value which differs significantly from the currently accepted $\mathrm{D}(\alpha a)$ value of α^{-1}. In view of these facts and the pivotal role which α plays in determining the fundamental physical constants and in comparing QED theory and experiment, it is difficult to understand why so much faith has been placed in the currently accepted α value and why some 15 years were allowed to pass before α was redetermined via fine-structure measurements. This situation provides further proof that for anyone who uses the present or any other set of fundamental constants, the guiding principle must be "*Caveat Emptor*!"

In obtaining the values for $\Delta E_{\mathrm{D}}-\mathcal{S}_{\mathrm{D}}$ given in Eq. (159), Dayhoff, Triebwasser, and Lamb used the constants of Bearden and Watts (1951) and the Thomas, Driscoll, and Hipple (1950a; 1950b) value for γ_p as did Triebwasser, Dayhoff, and Lamb in their analysis of their $\mathcal{S}_{\mathrm{D}}$ and $\mathcal{S}_{\mathrm{H}}$ data. We have therefore reevaluated the $\Delta E_{\mathrm{D}}-\mathcal{S}_{\mathrm{D}}$ results as we did the Lamb shift work. We find* (in megahertz)

$$\mathrm{D}(\alpha a):\quad \Delta E_{\mathrm{D}}-\mathcal{S}_{\mathrm{D}}=9912.607\pm0.056\ (5.6\ \mathrm{ppm}),$$

$$\mathrm{D}(\alpha c):\quad \Delta E_{\mathrm{D}}-\mathcal{S}_{\mathrm{D}}=9912.803\pm0.094\ (9.5\ \mathrm{ppm}). \tag{150}$$

[Our reevaluation procedure is not applicable to the $\mathrm{D}(\alpha c)$ transition and the original result has been retained. Recall, however, that this value is highly

* These values have been confirmed by Shawyer at NPL (private communication) using a matrix diagonalization method.

suspect because of the procedure used for magnetic-field calibration. We keep it as a matter of curiosity only.]

We now compare Eq. (150) with theory. Defining $\Delta E_D - \mathcal{S}_D \equiv \epsilon_D$, and $\Delta\epsilon_D$ as $\epsilon_D(\text{exptl}) - \epsilon_D(\text{theory})$, we find from Table XXVII (in megahertz)

$$\begin{aligned} \mathrm{D}(\alpha a): &\quad \Delta\epsilon_D = -0.592 \pm 0.085[(-60 \pm 8.6)\ \text{ppm}], \\ \mathrm{D}(\alpha c): &\quad \Delta\epsilon_D = -0.396 \pm 0.115[(-40 \pm 12)\ \text{ppm}]. \end{aligned} \tag{151}$$

Clearly, experiment and theory are in gross disagreement; the probability of the $\mathrm{D}(\alpha a)$ discrepancy occurring by chance is $\sim 3/10^{10}$ and for $\mathrm{D}(\alpha c)$, $\sim 0.05\%$. Because the theoretical expression for ΔE contains only a small QED contribution, it should be on somewhat firmer ground than the theoretical expression for $\mathcal{S}$ which owes its entire origin to QED. Thus, it is perhaps more instructive if instead of directly comparing the experimental values of $\Delta E_D - \mathcal{S}_D$ with theory, we combine them with the experimental measurements of $\mathcal{S}_D$ and compare the resulting values of ΔE_D with theory. The $\mathrm{D}(\alpha a)$ value of $\Delta E_D - \mathcal{S}_D$ and the values of $\mathcal{S}_D$ obtained by Triebwasser, Dayhoff, and Lamb and by Cosens give for ΔE_D (in megahertz)

$$\begin{aligned} \mathrm{D}(\alpha a): &\quad (\Delta E_D)_{\mathrm{TDL}} = 10971.603 \pm 0.085\ (7.7\ \text{ppm}), \\ &\quad (\Delta E_D)_{\mathrm{C}} \quad = 10971.851 \pm 0.085\ (7.7\ \text{ppm}). \end{aligned} \tag{152a}$$

Similarly, from the $\mathrm{D}(\alpha c)$ value we find

$$\begin{aligned} \mathrm{D}(\alpha c): &\quad (\Delta E_D)_{\mathrm{TDL}} = 10971.799 \pm 0.115\ (10\ \text{ppm}), \\ &\quad (\Delta E_D)_{\mathrm{C}} \quad = 10972.047 \pm 0.115\ (10\ \text{ppm}). \end{aligned} \tag{152b}$$

Defining $\delta\Delta E_D = \Delta E_D(\text{exptl}) - \Delta E_D(\text{theory})$, we finally obtain from Table XXVII (in megahertz)

D(αa):

$$(\delta\Delta E_{\mathrm{D}})_{\mathrm{TDL}} = -0.417 \pm 0.095[(-38 \pm 8.6)\ \mathrm{ppm}],$$
$$(\delta\Delta E_{\mathrm{D}})_{\mathrm{C}} = -0.170 \pm 0.095[(-15 \pm 8.6)\ \mathrm{ppm}],$$

D(αc):

$$(\delta\Delta E_{\mathrm{D}})_{\mathrm{TDL}} = -0.221 \pm 0.120[(-20 \pm 11)\ \mathrm{ppm}],$$
$$(\delta\Delta E_{\mathrm{D}})_{\mathrm{C}} = 0.026 \pm 0.120[(-2.4 \pm 11)\ \mathrm{ppm}]. \tag{153}$$

Since the experimental values of $\mathcal{S}_{\mathrm{D}}$ significantly exceed the theoretical value [see Eq. (144)], the discrepancies of Eq. (151) are somewhat reduced. Note that the least discrepant value of ΔE_{D} is that obtained from the measurements on the D(αc) transition originally discarded (justifiably so) by Dayhoff, Triebwasser, and Lamb and the recent measurement of $\mathcal{S}_{\mathrm{D}}$ by Cosens which itself disagrees with the earlier measurement of $\mathcal{S}_{\mathrm{D}}$ by Triebwasser, Dayhoff, and Lamb. We believe that the determination of an unambiguous value of the fine-structure constant from these data is precluded by their inconsistency, and that there is little justification for confidence in the presently accepted value of α based on these results. This situation lends further support to the idea that the uncertainties in this type of experiment may be underestimated.

A measurement of the $2S_{1/2}$–$2P_{3/2}$ splitting in hydrogen, $\Delta E_{\mathrm{H}} - \mathcal{S}_{\mathrm{H}}$, has been completed recently by Kaufman, Lamb, Lea, and Leventhal (KLLL) (1969a; 1969b) (Kaufman and Lea, private communication). These workers studied the αa transition at $\approx$0.1465 T ($\nu \approx$11.970 GHz—see Fig. 5), the αb transition at $\approx$0.1860 T ($\nu \approx$9.170 GHz), and the αc transition at $\approx$0.1090 T ($\nu \approx$7.430 GHz) using a technique similar to that used by Triebwasser, Dayhoff, and Lamb but with several important exceptions: (1) Dissociation of the molecular hydrogen and its excitation to the metastable $2S_{1/2}$ state are carried out in one step, and the microwave field applied in the same spatial region. (2) The number of metastables is determined by measuring the amount of Lyman-α radiation emitted at 1216 Å when the metastables are excited by the microwave field to the $2P_{3/2}$ state and subsequently decay to the ground state. As in the Dayhoff, Triebwasser, and Lamb experiments, the microwave frequency is held fixed and the magnetic field varied. The resonance curve is therefore obtained by plotting light intensity vs magnetic field. In order to locate accurately the center of the resonance curve, space-charge fields, stray fields of electron gun electrodes, and collisions with ions, electrons, and neutral particles must all be carefully considered. An important feature of this work is the procedure used to normalize the observed signal to the background which arises from the quenching of the metastables by mechanisms other than the applied microwave field (e.g., collisions with neutral molecules). The normalization method finally chosen by Kaufman, Lamb, Lea, and Leventhal [first introduced by Lipworth and Novick (1957)] was to use two microwave levels, one rather low, and the other sufficiently high to broaden the resonance without completely saturating it. Any asymmetry common to all the metastable states does not appear in the ratio of the two signals. In practice, the microwave power is sufficiently small so that there is no appreciable broadening of the observed resonance by overlapping resonances. To obtain the centers of the resonance lines, the working Hamiltonian, taken from the paper by Lamb (1952), is diagonalized, and the resulting frequencies, matrix elements, etc., are used in a partly theoretical, partly experimental quenching function which is then fitted to the resonance line. (The number of experimental points measured for each line is about 10.) The numerical value of $\Delta E_{\mathrm{H}} - \mathcal{S}_{\mathrm{H}}$ is a parameter in the line-shape equation and is varied to obtain the best fit, and therefore the best value for $\Delta E_{\mathrm{H}} - \mathcal{S}_{\mathrm{H}}$ implied by the data. Magnetic fields are measured in terms of the precession frequency of protons in water.

The final results for the αa, αb, and αc transitions are reported by Kaufman, Lamb, Lea, and Leventhal (1969a; 1969b) (Kaufman and Lea, private communication) to be (in megahertz)

$$\text{H}(\alpha a): \quad \Delta E_{\text{H}} - \mathcal{S}_{\text{H}} = 9911.363 \pm 0.031 \text{ (3.1 ppm)},$$

$$\text{H}(\alpha b): \quad \Delta E_{\text{H}} - \mathcal{S}_{\text{H}} = 9911.407 \pm 0.045 \text{ (4.5 ppm)},$$

$$\text{wt av}: \quad \Delta E_{\text{H}} - \mathcal{S}_{\text{H}} = 9911.377 \pm 0.026 \text{ (2.6 ppm)},$$

$$\text{H}(\alpha c): \quad \Delta E_{\text{H}} - \mathcal{S}_{\text{H}} = 9911.057 \pm 0.3 \text{ (30 ppm)}. \qquad (154)$$

The αc transition result is plagued by large uncertainties due to overlap of a nearby βd resonance and is not considered a high-precision determination. We shall therefore ignore it here. The αa result is the mean of 148 runs obtained by linearly extrapolating all of the runs to zero gas pressure and electron beam current. (The metastables are produced by bombarding molecular hydrogen with electrons at about 25 eV. The extrapolation presumably corrects for the effect of gas pressure and of space charge arising from the electron beam.) For the last 60 runs the total pressure–current correction amounts to less than 0.05 MHz, and its average value for all 148 runs is 0.125 MHz. The standard deviation of the 148 measurements is 0.22 MHz (22 ppm) and the standard deviation of the mean is 0.018 MHz (1.8 ppm). The uncertainty quoted for the αa result is the RSS of this statistical error of the mean and the following 70% confidence-level estimates of the systematic error: (1) uncertainty due to uncertainty in magnetic field, 0.012 MHz; (2) uncertainty in effect of overlapping β state resonances, 0.010 MHz, roughly $\frac{1}{3}$ the total overlap correction of 0.03 MHz; (3) uncertainty in Stark-shift correction, 0.020 MHz, about $\frac{1}{5}$ the total correction for this effect; (4) possible effects due to stray electric fields arising from charged insulating films in the apparatus, etc., 0.002 MHz; (5) computer roundoff in calculating matrix elements, 0.002 MHz. The αb result is the mean of 62 runs, again extrapolated to zero pressure and beam current. The standard deviation of the measurements is 0.082 MHz (8.3 ppm) and the standard deviation of the mean is 0.010 MHz (1 ppm). The quoted uncertainty is the RSS of this statistical error and the following estimates of systematic error: (1) uncertainty due to uncertainty in magnetic field, 0.003 MHz; (2) uncertainty in effect of overlapping β state resonances, 0.030 MHz, roughly $\frac{1}{3}$ the total 0.10 MHz overlap correction; (3) uncertainty in Stark-shift correction, 0.031 MHz, about $\frac{1}{5}$ the total correction for this effect; (4) possible stray electric-field effects, 0.006 MHz; (5) computer roundoff error, 0.002 MHz. The standard deviation of the αb data was reduced to $\frac{1}{3}$ that obtained for the αa data by using an improved preamplifier. However, this reduction in uncertainty was offset by the smaller number of measurements, the larger Stark-shift correction uncertainty, and the larger effect of overlapping resonances. On the other hand, the average correction for gas pressure and beam current was somewhat smaller for the αb measurements than for the αa measurements. We note that the results from the two transitions are in good agreement, their 0.044-MHz difference being 0.8 times the 0.055-MHz standard deviation of their difference. The probability for this to occur by chance is about 42%. We shall therefore use the weighted average of the two measurements as the final result of the experiment. Although the 0.026-MHz (2.6 ppm) uncertainty of this average value is considerably less than the uncertainty assigned other similar experiments, we shall provisionally retain it because of the rather careful analysis of the possible sources of systematic error carried out by Kaufman *et al.* (In the discussion of the measurement of ΔE_{H} by Metcalf *et al.*, the next experiment to be described, we consider the problem of error assignment in experiments in which the statistical scatter of the data is significantly larger than the quoted final total uncertainty. We shall later reconsider the assigned uncertainty in the Kaufman *et al.* experiments in light of this discussion.)

To compare the Kaufman *et al.* value of $\Delta E_{\text{H}} - \mathcal{S}_{\text{H}}$

with theory, we define $\Delta E_{\mathrm{H}} - \mathcal{S}_{\mathrm{H}} \equiv \epsilon_{\mathrm{H}}$ and $\Delta\epsilon_{\mathrm{H}} = \epsilon_{\mathrm{H}}(\mathrm{exptl}) - \epsilon_{\mathrm{H}}(\mathrm{theory})$. We find from Eq. (154), average value, and Table XXVII

$$\Delta\epsilon_{\mathrm{H}} = -0.090 \pm 0.057 \text{ MHz } [(-9.1 \pm 5.8) \text{ ppm}]. \tag{155}$$

The difference is 1.6 times the standard deviation of the difference and has a probability of occurring by chance of about 11%. This new experimental result is therefore reasonably compatible with theory, in marked contrast with the experimental values of $\mathcal{S}_{\mathrm{H}}$, $\mathcal{S}_{\mathrm{D}}$, and $\Delta E_{\mathrm{D}} - \mathcal{S}_{\mathrm{D}}$ we have thus far discussed. Equation (155) also implies near agreement with the theoretical value for $\mathcal{S}_{\mathrm{H}}$ as may be seen by computing the value of $\mathcal{S}_{\mathrm{H}}$ predicted by subtracting the Kaufman *et al.* result for $\Delta E_{\mathrm{H}} - \mathcal{S}_{\mathrm{H}}$ from the theoretical value of ΔE_{H} given in Table XXVII. (The theoretical expression for ΔE should be quite reliable in view of its weak dependence on QED.) We find

$$\mathcal{S}_{\mathrm{H}} = 1057.649 \pm 0.050 \text{ MHz}.$$

However, the agreement between the Kaufman *et al.* value of $\Delta E_{\mathrm{H}} - \mathcal{S}_{\mathrm{H}}$ and theory becomes somewhat worse when it is combined with the Triebwasser, Dayhoff, and Lamb and the Robiscoe values of $\mathcal{S}_{\mathrm{H}}$, and the resulting values of ΔE_{H} compared with theory. We find for ΔE_{H} (in megahertz)

$$\begin{aligned} (\Delta E_{\mathrm{H}})_{\mathrm{KLLL,TDL}} &= 10969.150 \pm 0.068 \text{ (6.2 ppm)}, \\ (\Delta E_{\mathrm{H}})_{\mathrm{KLLL,R}} &= 10969.233 \pm 0.068 \text{ (6.2 ppm)}, \\ (\Delta E_{\mathrm{H}})_{\mathrm{Av}} &= 10969.191 \pm 0.052 \text{ (4.7 ppm)}, \end{aligned} \tag{156}$$

where $(\Delta E_{\mathrm{H}})_{\mathrm{Av}}$ has been obtained from the average of the Robiscoe and TDL values of $\mathcal{S}_{\mathrm{H}}$, i.e., $(\mathcal{S}_{\mathrm{H}})_{\mathrm{Av}} = 1057.814 \pm 0.045$ MHz (42 ppm). Defining $\delta\Delta E_{\mathrm{H}} = \Delta E_{\mathrm{H}}(\mathrm{exptl}) - \Delta E_{\mathrm{H}}(\mathrm{theory})$ and using Table XXVII gives (in megahertz)

$$(\delta\Delta E_{\mathrm{H}})_{\mathrm{KLLL,TDL}} = 0.124 \pm 0.080 [(11 \pm 7.3) \text{ ppm}],$$

$$(\delta\Delta E_{\mathrm{H}})_{\mathrm{KLLL,R}}=0.206\pm0.080[(19\pm7.3)\text{ ppm}],$$
$$(\delta\Delta E_{\mathrm{H}})_{\mathrm{Av}}=0.165\pm0.067[(15\pm6.1)\text{ ppm}]. \tag{157}$$

The Triebwasser, Dayhoff, and Lamb value for $\mathcal{S}_{\mathrm{H}}$ gives reasonable agreement, the difference exceeding the standard deviation of the difference by a factor of 1.53. The probability for this to occur by chance is about 13%. On the other hand, the Robiscoe value of $\mathcal{S}_{\mathrm{H}}$ gives poorer agreement; the difference exceeds the standard deviation of the difference by a factor of 2.56; this has a probability of occurring by chance of only 1%. For $(\mathcal{S}_{\mathrm{H}})_{\mathrm{Av}}$, the difference is equal to about 2.47 standard deviations and has a probability of occurring by chance of ~1.4%. While these probabilities clearly indicate significant disagreement between the value of ΔE_{H} predicted by α_{WQED} and that implied by the Kaufman *et al.* result, they probably should not be taken too seriously at this time because of the uncertainties associated with the various $\mathcal{S}_{\mathrm{H}}$ measurements (see previous section) and the problem of realistically estimating the systematic errors present in the several experiments. We shall discuss this matter further in Sec. V.

A direct measurement of the fine-structure splitting ΔE_{H} in H, $n=2$, has been reported recently by Metcalf, Brandenberger, and Baird (MBB) (1968; private communication and to be published; Brandenberger, 1968). These workers used a level-crossing method to study the e–d crossing at ~0.3484 T. [This crossing has been discussed in detail by Himmell and Fontana (1967).] Lyman-α radiation is incident on a cloud of H atoms in a magnetic field, and the radiation scattered at right angles is detected using a nitrous oxide photoionization detector. Since no beam is used, certain corrections due to motional effects are avoided. The crossing occurs between two states of the same symmetry, and no external perturbation is necessary to mix the levels at the crossing. The observed level-crossing signal arises from the magnetic-field dependence of the distribution of the radiation scattered from atoms in the crossing states; as the magnetic field is swept through the crossing point, the amount of Lyman-α radiation reaching the detector changes. The magnetic field is measured in terms of the precession frequency f_0 of protons in oil and corrected to a value for protons in water. The observed crossing signal is approximately Lorentzian with full width at half-maximum of about 0.007 T. Thus, to measure ΔE_{H} to 10 ppm requires determining the crossing point to about 1/2000 of the linewidth or 0.003×10^{-3} T. A theoretical description of the line shape accurate to about 0.2% is required to attain this precision. In practice, the observed lines, consisting of about 25–30 discrete points, were fitted to an approximate line-shape equation, and the line center later corrected (by about 0.2 ppm) according to the theory of Brodsky and Parsons. The main problem in this work is the asymmetry in the line arising from nonorthogonal geometry, optical depth of the gas of scattering atoms, nonuniformity of the windows, and other causes. Since the magnitudes of these asymmetric contributions to the line shape can neither be accurately calculated or measured, they are taken into account by including a variable asymmetry parameter β in the line-shape equation which is fitted to the experimental points. This parameter was typically of the order of 0.02–0.03. However, because of the finite signal-to-noise ratio in the experiment and the several other floating parameters in the line-shape equation, the fitting process tends to scatter the adjusted values of β for consecutive data runs by as much as ± 0.01. Such changes are thought to be too large to correspond to actual changes in the experimental conditions and probably arise from the intimate relation which exists between the line-center frequency, f_0, and β; a fluctuation in β of 0.01 is nearly equivalent in terms of line shape to a 50-ppm fluctuation in f_0 and therefore ΔE_{H}. The net result is a large scatter in the experimental values of ΔE_{H}. Nevertheless, Metcalf, Brandenberger, and Baird see no reason why fluctuations should not be random, and tests of their line-shape-fitting program indicate that the scatter in f_0 due to the simultaneous variation of β is purely statistical to well below the 2-ppm level.

The unweighted average of 84 runs carried out by Metcalf *et al.* under a wide variety of experimental conditions is*

$$\Delta E_{\rm H} = 10969.127 \pm 0.095 \text{ MHz } (8.7 \text{ ppm}). \quad (158)$$

(This result includes a −1.5-ppm motional electric-field correction.) The quoted uncertainty is the RSS of the 0.074-MHz statistical error of the mean [as computed from Eq. (42)] and a 0.059-MHz RSS systematic error which arises from the following sources: (1) measurement of the magnetic field, 3 ppm (0.033 MHz); (2) magnetic-field dependence of lamp intensity, 2 ppm (0.022 MHz); (3) motional Stark effect, 0.2 ppm (0.002 MHz); (4) treatment of line-shape asymmetry, 0.2 ppm (0.002 MHz); (5) non-linearities in electronic instrumentation, 4 ppm (0.044 MHz). (Note that the statistical error is larger than the estimated systematic error.) We make the following additional comments concerning this experiment. First, while the statistical standard deviation of the 84 measurements is 0.68 MHz or 62 ppm, the runs are quite normally distributed and thus the statistical reduction of this uncertainty by the factor $(84)^{1/2}$ would appear to be justified, i.e., the scatter in the data seems to be purely random in nature. [We have computed Fisher's (1925) measures of normality, γ_1 and γ_3, and find them to be −0.080 and −0.086, respectively. These are well within the expected range.] Second, Metcalf, Brandenberger, and Baird claim to have found no correlation between the observed values of $\Delta E_{\rm H}$ and the gas pressure, no measurable Stark shift due to charges on the walls or to ions within the scat-

* The 84 runs do not include 10 exploratory (low signal to noise) runs using argon as a buffer gas. The result of these runs is $\Delta E_{\rm H} = 10969.18(27)$ (25 ppm), statistical error only, in agreement with Eq. (158).

tering region, and no effect of magnetic-field modulation amplitude. Nevertheless, the rather large scatter of the measurements makes any investigation of possible systematic errors a rather difficult matter. In most of the experiments we have discussed, the estimated systematic error is large compared with the final statistical error (standard deviation of the mean of the set of measurements) and is comparable with or larger than the standard deviation of the measurements. The presence or absence of a particular source of systematic error can therefore be established with some confidence using just a few control experiments. In the work of Metcalf *et al.*, this is not so. For example, these workers carried out a series of experiments in an attempt to understand the origin of the asymmetry parameter β. In a group of 14 runs with greater than usual amounts of water vapor in the system, they found an increase in the amount of scattered light and in the lifetime (caused by coherence narrowing) accompanied by an increase in the average value of β (for one or two runs, β exceeded 0.05). They concluded from this and other evidence that the optical depth effect was responsible for β. The value of ΔE_{H} obtained in these experiments was 10969.101 ± 0.180 MHz (16 ppm), where the error is the statistical standard deviation of the mean computed via Eq. (42). It differs from the mean of all the runs, 10969.127 ± 0.074 MHz (6.8 ppm) (statistical error only), by 2.4 ± 18 ppm. Although the agreement is good, the uncertainty is sufficiently large that on *experimental grounds alone*, one cannot rule out the possibility that there is a systematic effect of order, say, 10 ppm arising from variations in β. Since the randomness of the fluctuations of β and ΔE_{H} are of prime concern in this experiment, the lack of a more stringent experimental test is cause for concern.

Metcalf *et al.* also made five runs using a magnetic-field modulation amplitude different from their customary one in order to investigate the effect of this parameter on ΔE_{H}. The mean value of ΔE_{H} implied by the five runs is $\Delta E_{\mathrm{H}} = 10968.769 \pm 0.339$ MHz (31 ppm) (statistical error only), which differs from the mean of all the data by 33±32 ppm. Now this difference is not really statistically significant, but the reason it is not is the relatively large statistical error of the five control runs. In this particular case, there are no theoretical reasons to suspect a significant dependence of ΔE_{H} on modulation amplitude, but it is again clear that on experimental grounds alone, it is not possible to rule out the presence of a systematic effect which might be several times the quoted total error of the experiment! Similar considerations also apply to other possible systematic errors in the experiment of Metcalf, Brandenberger, and Baird, for example, the effect of buffer gas pressure. We could surmount this difficulty by simply expanding the estimated systematic error using, say, the criterion employed by Dayhoff, Triebwasser, and Lamb (1953) (see Sec. IV.C.2). The latter workers estimated their possible systematic error as equal to the mean day-to-day scatter of their measurements, with the idea that a nonrandom systematic change significantly larger than this scatter could have been detected and eliminated. Application of this criterion to the result of Metcalf *et al.* would require expanding the assigned error by almost a factor of 10!

It should be noted that the same considerations apply (but to a lesser extent) to the Kaufman, Lamb, Lea, and Leventhal measurement of $\Delta E_{\mathrm{H}} - \mathcal{S}_{\mathrm{H}}$. In these experiments, the standard deviations of the data for the αa and αb transitions were 22 ppm and 8.3 ppm, respectively. The final uncertainties (including systematic error) assigned in the two cases were 3.1 and 4.5 ppm, respectively. These may be compared with the corresponding quantities in the experiment of Metcalf *et al.*, 62 and 8.7 ppm. For the most recent measurements of Kaufman *et al.*, those on the αb transition, the standard deviation of the data is less than twice the final assigned uncertainty, but for the earlier αa transition result, the ratio is greater than 7, just as for the result of Metcalf *et al.* Further discussion of the assigned errors in the experiments of Metcalf *et al.* and Kaufman *et al.* will be deferred to Sec. V. For purposes of comparison with theory, we shall retain the uncertainty given by Metcalf *et al.* as we did for the measurement of Kaufman *et al.*

Defining $\delta\Delta E_H = \Delta E_H(\text{exptl}) - \Delta E_H(\text{theory})$ and using Table XXVII yields for the result of Metcalf *et al.*, Eq. (158),

$$\delta\Delta E_H = 0.101 \pm 0.105 \text{ MHz } [(9.2 \pm 9.5) \text{ ppm}]. \quad (159)$$

The agreement is quite reasonable, the probability for the difference to occur by chance being 33%. We also note that the result of Metcalf, Brandenberger, and Baird is in rather good agreement with the values of ΔE_H implied by the measurement by Kaufman *et al.* of $\Delta E - \mathcal{S}_H$ and the Robiscoe and Triebwasser, Dayhoff, and Lamb measurements of $\mathcal{S}_H$, Eq. (156). It also agrees with a much less accurate value of ΔE_H obtained by Wing (1968; private communication) using the same method as Metcalf *et al.*, but in a somewhat less refined form. Wing obtained $\Delta E_H = 10969.6 \pm 0.7$ MHz (64 ppm). (See also Notes Added in Proof.)

3. *Other Fine-Structure Measurements*

There have been several other fine-structure measurements in hydrogenic atoms in addition to those we have so far discussed. However, most are of relatively large uncertainty and therefore of limited utility. For completeness, we include some of them in the next section, where we compare, via the value of α^{-1} they imply, all of the experimental results given in Secs. IV.A–IV.C. However, the preliminary measurements of Mader and Leventhal (1968a; 1968b)* on the fine structure of the $n=3$ level of ionized helium and the measurements of the Lamb shift in the $n=2$ level of ionized helium by Narasimham (1968; private communication) and also by Lipworth and Novick (1957)† are suffi-

* These measurements supersede the earlier ones by Leventhal, Lea and Lamb (1965).

† This measurement is in agreement with, but is significantly more accurate than, the earlier result of Novick, Lipworth, and Yergin (1955).

ciently accurate to require special mention here. Using methods very similar to those used by Kaufman, Lamb, Lea, and Leventhal, Mader and Leventhal have determined $\mathcal{S}_{\mathrm{He}^+}$, $n=3$, from measurements on the αe and αf transitions, and $\Delta E_{\mathrm{He}^+} - \mathcal{S}_{\mathrm{He}^+}$, $n=3$, from measurements on the αc and βd transitions. Their preliminary results (Mader, private communication) are (in megahertz)

$$n=3,\ {}^4\mathrm{He}^+(\alpha e, \alpha f): \quad \mathcal{S}_{\mathrm{He}^+} = 4182.4 \pm 1.0,$$

$$n=3,\ {}^4\mathrm{He}^+(\alpha c, \beta d): \quad \Delta E_{\mathrm{He}^+} - \mathcal{S}_{\mathrm{He}^+} = 47843.8 \pm 0.5,$$

$$n=3,\ {}^4\mathrm{He}^+: \quad \Delta E_{\mathrm{He}^+} = 52026.2 \pm 1.1.$$

The quoted value $\mathcal{S}_{\mathrm{He}^+}$ is an average value obtained from measurements on both the αe and αf transitions. Similarly, the value given for $\Delta E_{\mathrm{He}^+} - \mathcal{S}_{\mathrm{He}^+}$ was obtained from measurements on both the αc and βd transitions. The fine-structure splitting ΔE_{He^+} is simply the sum of $\mathcal{S}_{\mathrm{He}^+}$ and $\Delta E_{\mathrm{He}^+} - \mathcal{S}_{\mathrm{He}^+}$. The error quoted for $\mathcal{S}_{\mathrm{He}^+}$ is primarily due to small systematic effects still under investigation. When these are understood, the final error may be 0.5 MHz or less. Defining

$$\Delta\mathcal{S}_{\mathrm{He}^+} = \mathcal{S}_{\mathrm{He}^+}(\mathrm{exptl}) - \mathcal{S}_{\mathrm{He}^+}(\mathrm{theory}),$$

$$\epsilon_{\mathrm{He}^+} = \Delta E_{\mathrm{He}^+} - \mathcal{S}_{\mathrm{He}^+},$$

$$\delta\epsilon_{\mathrm{He}^+} = \epsilon_{\mathrm{He}^+}(\mathrm{exptl}) - \epsilon_{\mathrm{He}^+}(\mathrm{theory}),$$

and

$$\delta\Delta E_{\mathrm{He}^+} = \Delta E_{\mathrm{He}^+}(\mathrm{exptl}) - \Delta E_{\mathrm{He}^+}(\mathrm{theory}),$$

we compare experiment with theory and find from Tables XXVI and XXVII (in megahertz),

$$\Delta\mathcal{S}_{\mathrm{He}^+} = -0.26 \pm 1.10[(-61 \pm 260)\ \mathrm{ppm}],$$

$$\Delta\epsilon_{\mathrm{He}^+} = -1.33 \pm 0.68[(-28 \pm 14)\ \mathrm{ppm}],$$

$$\delta\Delta E_{\mathrm{He}^+} = -1.58 \pm 1.30[(-30 \pm 25)\ \mathrm{ppm}].$$

The agreement is not unreasonable.

Using a microwave–optical technique very similar to that used by Kaufman, Lamb, Lea, and Leventhal and by Mader and Leventhal, Narasimham (1968) has recently made a measurement of the Lamb shift in the $n=2$ level of ionized helium. From measurements of the αe transition at about 1.6 T ($\nu \approx 29.3$ GHz), he obtained*

$$n=2,\ {}^4\mathrm{He}^+(\alpha e):\quad \mathcal{S}_{\mathrm{He}^+}=14045.4\pm1.2\ \mathrm{MHz}\ (89\ \mathrm{ppm}),$$

where the quoted error is our own estimate and is meant to be approximately a standard deviation. It is the RSS of the 0.411-MHz statistical error and the 1.177-MHz uncertainty in various corrections required in the experiment, interpreted as a 70% confidence-level estimate. (Narasimham gives 1.7 MHz as an experimental uncertainty obtained by taking the RSS of 3 times the statistical error, and the 1.177-MHz correction uncertainty.) Defining $\Delta\mathcal{S}_{\mathrm{He}^+}=\mathcal{S}_{\mathrm{He}^+}(\mathrm{exptl})-\mathcal{S}_{\mathrm{He}^+}(\mathrm{theory})$, we compare this experimental result with the theoretical value given in Table XXVII:

$$\Delta\mathcal{S}_{\mathrm{He}^+}=6.5\pm1.8\ \mathrm{MHz}\ [(460\pm130)\ \mathrm{ppm}].$$

Clearly, theory and experiment are in disagreement, the difference exceeding the standard deviation of the difference by a factor of 3.5. The probability for this to occur by chance is only 0.05%. We are thus faced with still another apparent discrepancy between a theoretical prediction from QED and an experimental result. It is interesting to note that the proton halo hypothesis of Barrett, Brodsky, Erickson, and Goldhaber (1968) discussed previously predicts that the theoretical value for $\mathcal{S}_{\mathrm{He}^+}$, $n=2$, should be increased by some 4.1 MHz. Such an increase in $\mathcal{S}_{\mathrm{He}^+}(\mathrm{theory})$ would imply $\Delta\mathcal{S}_{\mathrm{He}^+}=2.4\pm1.8$ MHz, significantly reducing the disagreement between the Narasimham result and theory, thus giving further support to the idea that the proton indeed has a "tail" if the Narasimham result is correct. On the other hand, Narasimham's result is in disagreement with a similar measurement reported in 1957 by Lipworth and

---

* This value should perhaps be regarded as preliminary since Narasimham (private communication) is presently carrying out more precise calculations regarding certain possible systematic effects.

Novick which is in excellent agreement with theory. Using what has become the standard microwave–optical technique, these workers also carried out measurements on the αe transition at ≈ 1.6 T and obtained

$$n=2,\ {}^4\mathrm{He}^+(\alpha e):\quad \mathcal{S}_{\mathrm{He}^+}=14040.2\pm 1.8\ \mathrm{MHz}\ (130\ \mathrm{ppm}).$$

Again, the error is our own estimate and is meant to be a standard deviation. It was obtained by combining RSS Lipworth and Novick's 1.0 MHz statistical standard deviation with their 1.5-MHz estimated uncertainty in the corrections required in the experiment. (Lipworth and Novick originally gave 4.5 MHz as a limit of error, obtained by adding to the 1.5-MHz uncertainty in the corrections 3 times the 1.0-MHz statistical standard deviation.) In comparing the Lipworth and Novick result with that obtained by Narasimham, we find the latter exceeds the former by 5.2 ± 2.2 MHz or by 2.4 standard deviations. The probability for this to occur by chance is less than 2%. However, in comparing the Lipworth–Novick result with theory, we find

$$\Delta\mathcal{S}_{\mathrm{He}^+}=1.3\pm 2.2\ \mathrm{MHz}\ [(91\pm 160\ \mathrm{ppm})],$$

where, as before, $\Delta\mathcal{S}_{\mathrm{He}^+}=\mathcal{S}_{\mathrm{He}^+}(\mathrm{exptl})-\mathcal{S}_{\mathrm{He}^+}(\mathrm{theory})$. Clearly, the agreement is excellent, in marked contrast with the majority of the Lamb-shift measurements we have discussed thus far.

Highly accurate fine-structure measurements in atomic He have been carried out recently by Pichanick, Swift, Johnson, and Hughes (1968), but the present state of the theory does not warrant a detailed discussion of this work here. For a summary of the present situation, see Hughes (1969).

D. Comparison of Experimental Data via the Fine Structure Constant

In the preceding sections we have compared experimental values of several quantities of interest with the corresponding values calculated from theoretical equations using our WQED adjusted value of α. In this section, we reverse the procedure and calculate values of α from the experimental measurements and the appropriate theoretical equations. These values may then be compared with one another and with α_{WQED} in order to check the over-all consistency of the data. This procedure affords us the opportunity of comparing, for example, measurements of ν_{Hhfs} and g_μ, and provides us with potential candidates for inclusion in our final least-squares adjustment to obtain a best or recommended set of physical constants.

Table XXVIII summarizes the more important QED experimental measurements we have discussed and the implied values of α^{-1}. The theoretical equations used are those presented in the last several sections. An iterative procedure was used to solve for α and where applicable, the α dependence of each term was written explicitly, e.g., the theoretical expression for a_e, Eq. (103), was used in the equations for ν_{Hhfs}, ν_{Mhfs}, and ΔE rather than an experimentally determined value. All uncertainties in Table XXVIII are meant to be one standard deviation. The uncertainty in α^{-1} includes the uncertainty in the theoretical equation from which it was derived due to (1) uncertain constants, (2) approximately calculated terms, and (3) estimates of uncalculated terms. (As usual, the auxiliary constants were assumed to be exact.) Combining these three uncertainties RSS, the errors in the theoretical equations turned out to be as follows: for a_e [Eq. (103)], 2.2 ppm assuming the uncertainty in the coefficient of the sixth-order term to be ± 0.20; for a_μ [Eq. (118)], $+120$ or -6 ppm due to the sixth-order term; for ν_{Hhfs} [Eqs. (120) and (123)], 5.1 ppm due mainly to the uncertainty in the polarizability contribution $\delta_N^{(2)}$ (to be discussed in detail in the next paragraph); for ν_{Mhfs} [Eqs. (120) and (126)], 13 ppm due to the uncertainty in μ_μ'/μ_p' (we assign no error to the chemical shift for muons in water but calculate α^{-1} for both the Ruderman and standard proton correction); for ν_{Phfs} [Eq. (136)], 29 ppm due to theoretical

Table XXVIII. A comparison of QED experimental data via the implied value of the fine structure constant. The deviations are $[\alpha^{-1}-\alpha_{WQED}{}^{-1}]/\alpha_{WQED}{}^{-1}$, and the uncertainties in the deviations are the RSS of the errors in $\alpha_{WQED}{}^{-1}$ and α^{-1}.

Quantity	Source	Experimental value	α^{-1}	Error in α^{-1} (ppm)	Deviation (ppm)
	WQED least-squares adjusted value, $\alpha_{WQED}{}^{-1}$		137.03608(26)	1.9	0
	Anomalous moment of the electron, positron, and muon				
1. a_e^-	Wilkinson and Crane (revised by Rich)	0.001159549(30)	137.0467(36)	26	77±26
2. $g_s=2\mu_e/\mu_B$	μ_p'/μ_B by Klein plus μ_e/μ_p' by Lambe	1.00115980(49)	137.018(58)	420	−135±420
3. a_e^+	Rich and Crane	0.0011680(55)	136.05(65)	4710	−7170±4710
4. a_μ	CERN (muon storage ring)	0.00116616(31)	$136.976\binom{+0.040}{-0.037}$	+290 −270	$-440\binom{+290}{-270}$

TABLE XXVIII (*Continued*)

Quantity	Source	Experimental value	α^{-1}	Error in α^{-1} (ppm)	Deviation (ppm)
Ground-state hyperfine splitting in hydrogen, muonium, and positronium (MHz)					
5. ν_{Hhfs}	Vessot *et al.*, hydrogen maser	1420.4057517864(17)	137.03591(35)	2.6	-1.2 ± 3.2
6. ν_{Mhfs}	a. Yale group; α^{-1} calc. with Ruderman correction	4463.255(40)	137.0363(11)	7.9	1.9 ± 8.1
	b. α^{-1} calc. with standard proton correction	4463.255(40)	137.0374(11)	7.9	9.8 ± 8.1
7. ν_{Phfs}	Theriot, Beers, and Hughes	$2.03403(12)\times10^5$	137.0349(45)	33	-8.3 ± 33
Fine structure of hydrogen and deuterium, $n=2$, directly measured quantities (MHz)					
8. $\mathcal{S}_{\mathrm{H}}$	Triebwasser, Dayhoff, and Lamb (TDL)	1 057.772(63)	137.0260(32)	23	-73 ± 23
9. $\mathcal{S}_{\mathrm{H}}$	Robiscoe	1 057.855(63)	137.0221(32)	23	-100 ± 23
10. $\mathcal{S}_{\mathrm{D}}$	TDL	1 058.996(64)	137.0279(36)	26	-60 ± 26
11. $\mathcal{S}_{\mathrm{D}}$	Cosens	1 059.244(64)	137.0162(36)	26	-145 ± 26
12. $\Delta E_{\mathrm{H}}-\mathcal{S}_{\mathrm{H}}$	Kaufman, Lamb, Lea, and Leventhal (KLLL)	9 911.377(26)	137.03673(25)	1.8	4.7 ± 2.6

13. $\Delta E_D - \mathcal{S}_D$	a. Dayhoff, Triebwasser, and Lamb (DTL), $D(\alpha a)$	9 912.607(56)	137.04034(51)	3.7	31±4.2
	b. DTL, $D(\alpha c)$	9 912.803(94)	137.03893(74)	5.4	21±5.8
14. ΔE_H	Metcalf, Brandenberger, and Baird (MBB)	10 969.127(95)	137.03545(59)	4.3	−4.6±4.7
	Fine structure of hydrogen and deuterium, $n=2$, combinations of directly measured quantities (MHz)				
15. $\mathcal{S}_H$	MBB minus KLLL	1 057.750(99)	137.0271(48)	35	−66±35
16. ΔE_H	a. KLLL plus TDL $\mathcal{S}_H$	10 969.150(68)	137.03531(43)	3.1	−4.9±3.7
	b. KLLL plus Robiscoe $\mathcal{S}_H$	10 969.233(68)	137.03479(43)	3.1	−9.4±3.7
	c. KLLL plus $(\mathcal{S}_H)_{Av}$	10 969.191(52)	137.03505(32)	2.4	−7.5±3.0
17. ΔE_D	a. DTL $D(\alpha a)$ plus TDL $\mathcal{S}_D$	10 971.603(85)	137.03868(53)	3.9	19±4.3
	b. DTL $D(\alpha a)$ plus Cosens $\mathcal{S}_D$	10 971.851(85)	137.03714(53)	3.9	7.7±4.3
	c. DTL $D(\alpha c)$ plus TDL $\mathcal{S}_D$	10 971.799(115)	137.03746(71)	5.2	10±5.5
	d. DTL $D(\alpha c)$ plus Cosens $\mathcal{S}_D$	10 972.047(115)	137.03592(71)	5.2	−1.2±5.5

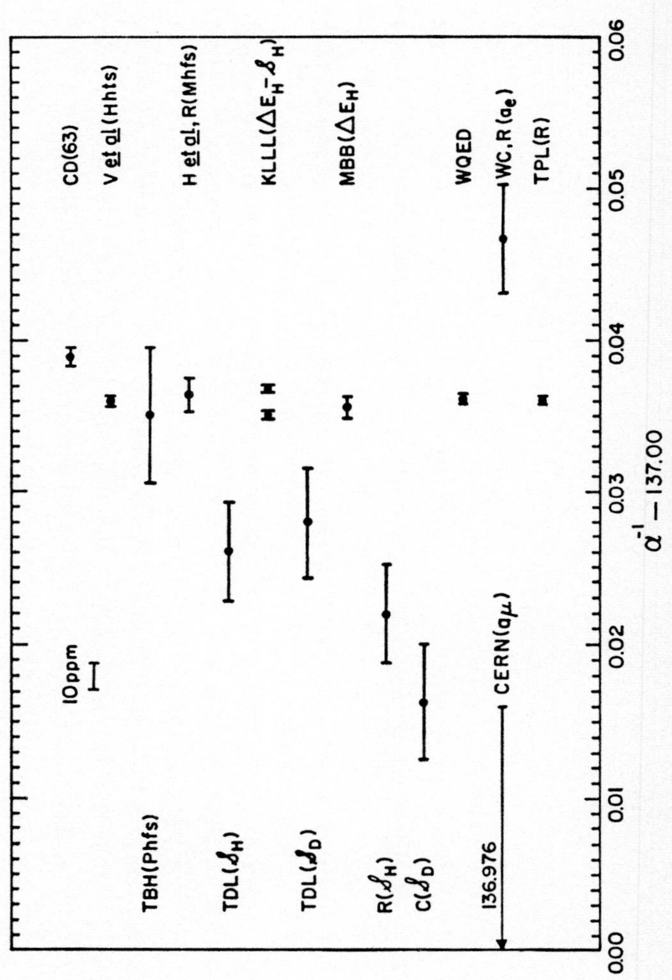

10ppm
CD(63)
V et al (Hhfs)
TBH(Phfs)
H et al, R(Mhfs)
TDL($\mathcal{S}_H$)
KLLL($\Delta E_H - \mathcal{S}_H$)
TDL($\mathcal{S}_D$)
MBB(ΔE_H)
R($\mathcal{S}_H$)
C($\mathcal{S}_D$)
WQED
136.976
CERN(a_μ)
WC, R(a_e)
TPL(R)
0.00
0.01
0.02
0.03
0.04
0.05
0.06
$\alpha^{-1} - 137.00$

FIG. 6. Graphical comparison of some of the data discussed in Sec. IV via the implied value of the inverse fine structure constant. (All numerical values are taken from Table XXVIII.) Also included are $\alpha_{\mathrm{WQED}}^{-1}$, our final recommended value [labeled TPL(R)], and the 1963 adjusted value of Cohen and DuMond (labeled CD) which is essentially the same as that obtained by combining the Dayhoff, Triebwasser, and Lamb and Triebwasser, Dayhoff, and Lamb measurements of $\Delta E_{\mathrm{D}} - \mathcal{S}_{\mathrm{D}}$ and $\mathcal{S}_{\mathrm{D}}$. Note that two different values are given for the Kaufman, Lamb, Lea, and Leventhal measurement of $\Delta E_{\mathrm{H}} - \mathcal{S}_{\mathrm{H}}$. The lower value was obtained by combining the result of Kaufman *et al.* with the average of the Triebwasser, Dayhoff, and Lamb, and Robiscoe measurements of $\mathcal{S}_{\mathrm{H}}$ and then calculating α^{-1} from the theoretical equation for ΔE_{H}. The higher value was calculated directly from the result of Kaufman *et al.* using the theoretical equation for $\Delta E_{\mathrm{H}} - \mathcal{S}_{\mathrm{H}}$. The errors for the Kaufman, Lamb, Lea, and Leventhal and the Metcalf, Brandenberger, and Baird values are based on the uncertainties assigned by the experimenters. (For a similar but magnified comparison of the more precise data contained in this figure, see Fig. 7.)

uncertainties; for $\mathcal{S}_H$ and $\mathcal{S}_D$ [Eq. (138), $n=2$], 22 and 40 ppm, respectively (the contribution due to the uncertainty in α has been subtracted out of the uncertainties given in Table XXVI and the resulting values divided by 3 to convert to a standard deviation); for ΔE_H and ΔE_D [Eq. (147), $n=2$], 0.33 ppm due to uncalculated terms; for $\Delta E_H-\mathcal{S}_H$ and $\Delta E_D-\mathcal{S}_D(n=2)$, 2.3 and 4.3 ppm, respectively, as implied by the individual errors in ΔE and $\mathcal{S}$. The final error quoted for a particular value of α^{-1} in Table XXVIII was obtained by combining RSS the error in the theoretical equation with the error in the experimental value used in the equation, and taking into account how α appears in the equation. Note that for the Lamb shift, a relative change in α gives rise to a relative change in $\mathcal{S}$ about 2.7 times as large (depending on n and Z) rather than 3 times because of the ln $(Z\alpha)^{-2}$ term and the constant terms [see Eq. (138)]. Similarly, a relative change in α causes a relative change in $\Delta E-\mathcal{S}$ about 1.9 times as large.

A discussion of the 5.1-ppm uncertainty assigned the theoretical equation for ν_{Hhfs} is in order. We saw in Sec. IV.B.1 that there was some uncertainty concerning the proton polarizability contribution to ν_{Hhfs}, $\delta_N^{(2)}$. Briefly, Drell and Sullivan (1967) find no candidate for contributing as much as 10 ppm but cannot unequivocally rule out such a possibility, while several calculations by other authors (mainly of the 33 resonance) never give more than 1 or 2 ppm. We believe this situation is not very different from that which exists in an experimental determination of a quantity, i.e., the experimenter is never sure there are no systematic errors present, but he does his best to find them and to allow for them in his final error assignment. In the case of ν_{Hhfs}, the proton polarizability contribution is analogous to an uncertain experimental systematic error. Thus, on the basis of the theoretical investigations of $\delta_N^{(2)}$, we believe a reasonable 70% confidence-level estimate of this quantity is 0 ± 5 ppm. It therefore follows that the total error in

ν_{Hhfs}(theory) is the RSS of this 5 ppm with 0.9 ppm due to $\delta_N^{(1)}$ and 0.6 ppm due to uncalculated terms [primarily, the uncertainty in the $\alpha(Z\alpha)^2$ term calculated by Brodsky and Erickson—see Sec. IV.B.1]. The result is 5.1 ppm.

On examining Table XXVIII for trends, discrepancies, etc, it is immediately evident that the derived values of α^{-1} are highly variable. (For a graphical comparison of some of these data, see Fig. 6.) The two most accurate anomalous moment experiments (items 1 and 4) give values of α^{-1} which disagree significantly with each other and with $\alpha_{\mathrm{WQED}}^{-1}$. The other two experiments (items 2 and 3) are more consistent, but the uncertainties are rather large. The hyperfine splittings in hydrogen, muonium, and positronium appear to be in better shape. The value of α^{-1} predicted by ν_{Hhfs} is in excellent agreement with the adjusted WQED value of α^{-1} and is well supported by the ν_{MHfs} results using the corrected Ruderman diamagnetic shielding correction. Turning to the fine structure of the $n=2$ state of H and D, we see that the values of α^{-1} obtained from the direct Lamb-shift measurements (items 8–11) and the indirect value (item 15) are reasonably consistent among themselves but are quite inconsistent with $\alpha_{\mathrm{WQED}}^{-1}$. They are also inconsistent with the values predicted by the remaining fine-structure measurements (and for that matter, all of the other experiments). The most consistent fine-structure measurements appear to be those recently carried out by Metcalf, Brandenberger, and Baird (1968) and Kaufman, Lamb, Lea, and Leventhal (1969) (items 14 and 16) when the latter is combined with the various $\mathbb{S}_{\mathrm{H}}$ measurements. The α^{-1} values they predict are in good agreement with one another, but in only fair agreement with $\alpha_{\mathrm{WQED}}^{-1}$ if the assigned uncertainties in these experiments are taken at face value. In contrast, the values of α^{-1} derived from ΔE_{D} (item 17) vary over rather wide limits. Further discussion of Table XXVIII will be given in the next section where we select the most reliable values of α for incorporation in our final

TABLE XXIX. A comparison of some relatively low-accuracy fine-structure measurements via the implied value of the fine structure constant. The deviations are $(\alpha^{-1}-\alpha_{\mathrm{WQED}}^{-1})/\alpha_{\mathrm{WQED}}^{-1}$, and the uncertainties in the deviations are the RSS of the errors in $\alpha_{\mathrm{WQED}}^{-1}$ and α^{-1}. (All helium measurements are for $^4He^+$.)

Quantity	Source	Experimental value (MHz)	α^{-1}	Error in α^{-1} (ppm)	Deviation (ppm)
WQED least-squares adjusted value, $\alpha_{\mathrm{WQED}}^{-1}$			137.03608(26)	1.9	0
1. $\mathcal{S}_{\mathrm{He}^+}$, $n=2$	Lipworth and Novick (1957)	14 040.2(1.8)	137.0315(80)	58	−34±58
2. $\mathcal{S}_{\mathrm{He}^+}$, $n=2$	Narasimham (1968)	14 045.4(1.2)	137.0127(65)	47	−170±47
3. $\mathcal{S}_{\mathrm{H}}$, $n=3$	Kleinpoppen[a]	313.6(2.9)	137.22(46)	3 370	1380±3370
4. $\mathcal{S}_{\mathrm{D}}$, $n=3$	Wilcox and Lamb[b] (WL)	315.30(40)	137.015(63)	465	−185±465
5. $\mathcal{S}_{\mathrm{D}}$, $n=4$	WL	133.0(5.0)	137.1(1.9)	13 700	620±13 700
6. $\mathcal{S}_{\mathrm{He}^+}$, $n=3$	Mader and Leventhal (ML) (preliminary; private communication)	4 182.4(1.0)	137.039(13)	94	23±94

7. $\mathcal{S}_{He^+}$, $n=4$	Jacobs, Lea, and Lamb[c] (JLL)	1 768(5)	137.05(14)	1 040	71±1040
8. $\mathcal{S}_{He^+}$, $n=4$	Hatfield and Hughes[d]	1 766.0(7.5)	137.10(21)	1 560	490±1560
9. $\Delta E_{He^+}-\mathcal{S}_{He^+}$, $n=3$	ML (prelim.)	47 843.8(0.5)	137.03804(94)	6.8	14±7.1
10. $\Delta E_{He^+}-\mathcal{S}_{He^+}$, $n=4$	JLL[c]	20 179.7(1.2)	137.0398(42)	31	27±31
11. ΔE_H, $n=2$	Wing (1968)	10 969.6(7)	137.0325(44)	32	−26±32
12. ΔE_D, $n=3$	WL	3 250.7(1.0)	137.042(21)	155	42±155
13. ΔE_{He^+}, $n=3$	ML plus ML (prelim.)	52 026.2(1.1)	137.0382(15)	11	19±11
14. ΔE_{He^+}, $n=4$	JLL plus JLL	21 947.7(5.1)	137.040(16)	115	32±115

[a] H. Kleinpoppen, Z. Physik **164,** 174 (1961).

[b] L. R. Wilcox and W. E. Lamb, Jr., Phys. Rev. **119,** 1915 (1960).

[c] R. R. Jacobs, K. R. Lea, and W. E. Lamb, Jr., Bull. Am. Phys. Soc. **14,** 525 (1969), and private communication. These measurements are in agreement with but are more accurate than the earlier measurements of K. R. Lea, M. Leventhal, and W. E. Lamb, Jr., Phys. Rev. Letters **16,** 163 (1966).

[d] L. L. Hatfield and R. H. Hughes, Phys. Rev. **156,** 102 (1967). H. J. Byer and H. Kleinpoppen, [Z. Physik **206,** 177 (1967)] report a value of 1751(13) MHz which gives $\alpha^{-1}=137.54(38)$.

adjustment to obtain a best or recommended set of constants. (See also Notes Added in Proof.)

For completeness, we give in Table XXIX the results of other fine-structure measurements reported in the literature, including those discussed in Sec. IV.C.3. Unfortunately, much of this work is of relatively low accuracy and is therefore of limited usefulness in the present work.

V. Final Recommended Set of Fundamental Constants

A. Selection of Input Data

We now turn our attention to obtaining a final best or recommended set of fundamental physical constants. In principle, such a set may readily be found by combining the most reliable data contained in Table XXVIII with all of the input data used to obtain our WQED values of the constants. However, the general disagreement of the α values in this table makes it rather difficult to decide objectively just which of the measurements should be retained and which should be discarded. Indeed, it may be argued that because they are so discrepant, it is best to eliminate them entirely as was done for the x-ray data. We believe this would amount to "throwing out the baby with the bathwater" and will therefore attempt to identify the most reliable values of the fine-structure constant in Table XXVIII in as logical a way as possible. We are of course aware that our decisions will not be unique and that there are other possible choices which would lead to slightly different sets of constants. This ambiguity simply underscores our previous discussion in Sec. I.C to the effect that *no* set of recommended constants should be used uncritically.

We shall decide which values of α to include in the final adjustment with the aid of the following three criteria: (1) Any input datum must be reasonably consistent with the value of α^{-1} implied by the WQED data, i.e., $\alpha_{\mathrm{WQED}}^{-1}=137.03608(26)$ (1.9 ppm). (By reasonably consistent we mean within two or perhaps three standard deviations.) This requirement follows from our general principle (see Sec. II.B.2) that it is incorrect to average together data which are in gross disagreement. It is in accord with the principal aim of this paper, namely, to investigate the implications of the Josephson-effect measurement of $2e/h$ and other WQED experiments for both QED and the fundamental constants. It is also consistent with our belief that the state of agreement between quantum electrodynamic theory and the relevant experiments is sufficiently unsatisfactory that these experiments cannot be viewed as sources of information on the fundamental constants with quite the same degree of confidence as other types of experiments. (2) Any input datum must have a sufficiently small uncertainty so that it will carry a meaningful weight in the adjustment. Since α_{WQED} has an uncertainty of 1.9 ppm, the factor-of-3 rule of thumb (see Secs. II.A.3 and II.A.4) indicates that any value of α with an uncertainty exceeding about 6 ppm should be excluded. (3) Any input datum must not be in gross disagreement with other experimental values of the same quantity which are of comparable reliability.

To illustrate the application of this last requirement, let us consider the values of α^{-1} derived by combining the experimental value of $\Delta E_{\mathrm{D}}-\mathcal{S}_{\mathrm{D}}$ with experimental values of $\mathcal{S}_{\mathrm{D}}$, and using the theoretical equation for ΔE_{D}. [See items 17a and 17b, Table XXVIII. We ignore items 17c and 17d, even though item 17d is in excellent agreement with $\alpha_{\mathrm{WQED}}^{-1}$, because of the experimental uncertainties associated with the D(αc) measurements; see Sec. IV.C.2.] Clearly item 17b is reasonably consistent with $\alpha_{\mathrm{WQED}}^{-1}$ and is sufficiently accurate to warrant inclusion in our adjustment, i.e., it satisfies requirement (2). However, it is not consistent with the similar value, item 17a. (This is due of course to the discrepancy between the Triebwasser, Dayhoff, and Lamb and the Cosens values of $\mathcal{S}_{\mathrm{D}}$.) Since there is no experimental justification for using only item 17b and discarding item 17a, both of them must be discarded. Stated another way, the disagreement between the two values implies the presence of systematic error, and thus precludes the use of either one of them. Furthermore, it seems entirely unjustifiable to use the value of Dayhoff *et al.* for $\Delta E_{\mathrm{D}} - \mathcal{S}_{\mathrm{D}}$ but not their value of $\mathcal{S}_{\mathrm{D}}$ since the two values were measured under essentially identical conditions; it is quite likely that if $\mathcal{S}_{\mathrm{D}}$ is in error, $\Delta E_{\mathrm{D}} - \mathcal{S}_{\mathrm{D}}$ is also in error. We also note that the experimental result for $\Delta E_{\mathrm{D}} - \mathcal{S}_{\mathrm{D}}$ (item 13a) yields a value of α^{-1} via the theoretical expression for $\Delta E_{\mathrm{D}} - \mathcal{S}_{\mathrm{D}}$ which is in gross disagreement with $\alpha_{\mathrm{WQED}}^{-1}$. Hence, it does not meet requirement (1) and must be discarded. (See also our comments below about the general reliability of the theoretical equation for $\Delta E - \mathcal{S}$.)

Turning now to the beginning of Table XXVIII, we see that our three criteria immediately eliminate all of the anomalous moment measurements (items 1–4) since these are of comparatively low accuracy and are inconsistent with α_{WQED} and with each other. Of the hyperfine-splitting measurements (items 5–7), only the hydrogen result (item 5) merits retention. The muonium hfs value of α^{-1} (item 6) is too uncertain even assuming, as we have, that the Ruderman correction is exact. (If it is assumed to be uncertain by some reasonable amount, say 10 ppm, then the uncertainty in $\alpha_{\mathrm{Mhfs}}^{-1}$ would be 9.4 ppm.) An additional problem with the muonium work is the still somewhat questionable pressure correction. However, we do note that $\alpha_{\mathrm{Mhfs}}^{-1}$ tends to support $\alpha_{\mathrm{WQED}}^{-1}$. The positronium result (item 7) is clearly much too uncertain for use in any adjustment.

The Lamb-shift results (items 8–11) are of low relative accuracy and are highly inconsistent with α_{WQED} as well as with almost all other values of α. Consequently, they are not directly useable. Moreover, the general disagreement of these Lamb-shift measurements with theoretical values calculated using any reasonable value of α plays an important role in the over-all problem of obtaining α values from the fine-structure measurements. This is because the disagreement creates a suspicion that the experiments may be in error due to some undetected common systematic effect or that the theory of the Lamb shift is incorrect. Both possibilities are of importance since a value of α can be obtained from a measurement of $\Delta E - \mathcal{S}$ in one of two ways: α may be calculated directly from the theoretical formula for $\Delta E - \mathcal{S}$ or by combining the measured value of $\Delta E - \mathcal{S}$ with an experimental value of $\mathcal{S}$ and using the theoretical formula for ΔE. In the first case, the result is suspect because the theoretical formula for $\Delta E - \mathcal{S}$ includes the possibly inaccurate Lamb-shift theory. In the second case, the result is suspect because the experimental value of $\mathcal{S}$ used may be in error, even though the theoretical formula for ΔE is almost free of questionable QED contributions. The discrepancies between theory and experiment for $\mathcal{S}$ are of order 100 ppm and imply discrepancies between values of α obtained from ΔE and $\Delta E - \mathcal{S}$ of order 10 ppm. For example, we observe from Table XXVIII that the value of α^{-1} implied by the Kaufman, Lamb, Lea, and Leventhal measurement of $\Delta E_{\mathrm{H}} - \mathcal{S}_{\mathrm{H}}$ and the theoretical expression for $\Delta E_{\mathrm{H}} - \mathcal{S}_{\mathrm{H}}$ (item 12) is $\alpha^{-1} = 137.03673(25)$ (1.8 ppm). If the measurement by Kaufman *et al.* is combined with the weighted average of the Triebwasser, Dayhoff, and Lamb and Robsicoe measurements of $\mathcal{S}_{\mathrm{H}}$, and if α^{-1} is calculated from the theoretical formula for ΔE_{H} (item 16c), the result is $\alpha^{-1} = 137.03505(32)$ (2.9 ppm). The two values differ by (12 ± 3) ppm. One might argue that because the theoretical expression for $\mathcal{S}$ is based on complex QED calculations, it is more suspect than $\mathcal{S}$(exptl), and therefore calculating α^{-1} from experimental values of $\Delta E - \mathcal{S}$, $\mathcal{S}$, and the theoretical equation for ΔE should give a more reliable result. The fact that the more accurate Lamb-shift measurements all exceed their corresponding theoretical values tends to support this viewpoint, and we ourselves lean in that direction. However, the history of fine-structure measurements clearly shows that the discrepancy could easily have an experimental origin. This implies that the disagreement between theory and experiment should perhaps be regarded as casting suspicion equally on both.

For hydrogen, there is the direct experimental measurement of ΔE_{H} by Metcalf, Brandenberger, and

Baird which might be expected to shed some light on the situation. However, for the reasons discussed in Sec. IV.C.2, we do not believe that the uncertainty of this measurement can be established with sufficient confidence to permit any definitive conclusions to be drawn from its agreement or disagreement with the other hydrogen data, i.e., the Triebwasser *et al.* and Robiscoe measurements of $\mathcal{S}_H$ and the Kaufman *et al.* measurement of $\Delta E_H - \mathcal{S}_H$. Equally important, the errors to be assigned these experiments are also open to question, and as discussed in Sec. IV.C.1, it is quite possible that the result of Triebwasser *et al.* contains a large systematic error due to an incorrect hfs correction. Nevertheless, we shall investigate the implications of the hydrogen fine-structure measurements for the fundamental constants along with the hydrogen hyperfine-splitting result. For the purposes of this study, which will be in the form of an analysis of variance, we shall make the same assumptions as were made in Table XXVIII, viz. (1) The uncertainties assigned by Kaufman *et al.* to their experimental value of $\Delta E_H - \mathcal{S}_H$ and by Metcalf, Brandenberger, and Baird to their similar value of ΔE_H are correct. (We shall, however, investigate the effect of expanding these errors.) (2) The uncertainties we have adopted for the Triebwasser *et al.* and Robiscoe measurements are correct and the two measurements are of equal quality. (But we shall also note the result of using either one of the measurements separately.) (3) The error in the theoretical equation for $\mathcal{S}_H$ estimated by Erickson and Yennie is correct and corresponds to three standard deviations as intended. Note that this implies a one-standard-deviation uncertainty in the theoretical equation for $\mathcal{S}_H$ of 0.069 MHz/3 = 0.023 MHz or 22 ppm. It therefore contributes only about 2.2 ppm to the uncertainty of the theoretical expression for $\Delta E_H - \mathcal{S}_H$ and about 1.2 ppm to the uncertainty of any value of α derived from this theoretical expression.

The four new pieces of stochastic input data we wish to investigate are thus as follows (these have already been included in Table XVI and are numbered accordingly):

I20, Hhfs: $\alpha^{-1}=137.03591(35)$ (2.6 ppm),

I21, ΔE_{H}, MBB: $\alpha^{-1}=137.03545(59)$ (4.3 ppm),

I22a, $\Delta E_{\mathrm{H}}-\mathcal{S}_{\mathrm{H}}$, KLLL, plus $\mathcal{S}_{\mathrm{H}}$, TDL, R:

$$\alpha^{-1}=137.03505(32) \quad (2.4\text{ ppm}),$$

I22b, $\Delta E_{\mathrm{H}}-\mathcal{S}_{\mathrm{H}}$, KLLL:

$$\alpha^{-1}=137.03673(25) \quad (1.8\text{ ppm}).$$

(These and other values are graphically compared with $\alpha_{\mathrm{WQED}}{}^{-1}$ in Fig. 7.) To summarize, I20 is the value of the inverse fine structure constant derived from the hydrogen-maser measurements of, and the theoretical equations for, ν_{Hhfs} [Eqs. (120) and (123)], assuming $\delta_{\mathrm{N}}{}^{(2)}=0\pm5$ ppm (this is item 5, Table XXVIII). Item 21 is the value of α^{-1} derived from the Metcalf, Brandenberger, and Baird measurement of ΔE_{H} and Eq. (147) (this is item 14, Table XXVIII). Item 22a is the value of α^{-1} derived by combining the Kaufman, Lamb, Lea, and Leventhal measurement of $\Delta E_{\mathrm{H}}-\mathcal{S}_{\mathrm{H}}$ with the weighted average of the Triebwasser, Dayhoff, and Lamb and Robiscoe measurements of $\mathcal{S}_{\mathrm{H}}$ and using Eq. (147) (this is item 16c, Table XXVIII). Item 22b is the value of α^{-1} which follows directly from the Kaufman, Lamb, Lea, and Leventhal measurement of $\Delta E_{\mathrm{H}}-\mathcal{S}_{\mathrm{H}}$ and the theoretical expression for $\Delta E_{\mathrm{H}}-\mathcal{S}_{\mathrm{H}}$, Eqs. (138) and (147) (this is item 12, Table XXVIII). Investigating the implications of the two separate values of α^{-1} derivable from the work of Kaufman *et al.* is in keeping with the idea that the disagreement between $\mathcal{S}_{\mathrm{H}}$(theory) and $\mathcal{S}_{\mathrm{H}}$(exptl) casts doubt equally on both theory and experiment. However, since I22a and I22b both involve the Kaufman *et al.* measurement, they are not independent values. Thus, we first present the results of least-squares adjustments in which I22b is excluded as a stochastic input datum, and then the results of similar adjustments in which I22a is excluded.

Table XXX gives the results of six adjustments involving the WQED data listed in Table XVI as well as

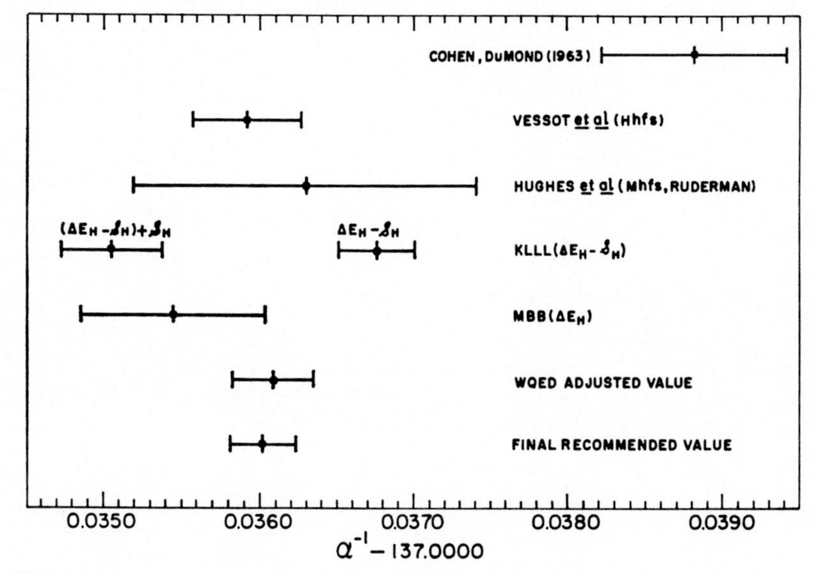

FIG. 7. Graphical comparison of the more precise values of the inverse fine structure constant with some least-squares adjusted values. (All numerical values are taken from Table XXVIII. For an explanation of the two KLLL values of α^{-1}, see the caption of Fig. 6 and the text.)

Table XXX. Results of six least-squares adjustments involving both WQED and QED data. The number of unknowns or adjustable constants is four (α^{-1}, e, K, N), and the initial set of data consists of items I1–I10, I13, I20, I21, and I22a of Table XVI. The number of degrees of freedom, $N-J$, is therefore $14-4=10$, minus the number of items deleted. The quantity e is in units of 10^{-19} C, and N is in units of 10^{26} kmole^{-1}. Also given are the Birge ratio R, χ^2, and the residual of each stochastic input datum, r.

	Adjustment number and items deleted					None
	11. None	12. I22a	13. I21, I22a	14. I1	15. I1, I20	16. (expanded error)
α^{-1}	137.03571(17)	137.03596(20)	137.03602(21)	137.03545(22)	137.03514(28)	137.03590(19)
e	1.6021996(62)	1.6021933(68)	1.6021917(70)	1.6022123(93)	1.602223(11)	1.6021948(67)
K	1.0000100(26)	1.0000093(26)	1.0000091(26)	1.0000090(27)	1.0000090(27)	1.0000095(26)
N	6.022147(39)	6.022165(39)	6.022169(40)	6.022090(49)	6.022050(55)	6.022161(39)
R	0.89	0.49	0.41	0.70	0.44	0.60
χ^2	7.87	2.18	1.36	4.47	1.56	3.65
$N-J$	10	9	8	9	8	10
$r1$	−0.89	−0.30	−0.14	Deleted	Deleted	−0.44

*r*2	−0.01	−0.09	−0.11	−0.12	−0.12	−0.07
*r*3	0.10	0.01	−0.01	−0.04	−0.04	0.03
*r*4	0.34	0.22	0.19	0.16	0.16	0.25
*r*5	0.22	0.18	0.17	0.16	0.16	0.19
*r*6	0.63	0.04	−0.11	−0.26	−0.26	0.18
*r*7	0.95	0.57	0.47	0.38	0.38	0.66
*r*8	0.01	−0.09	−0.11	−0.14	−0.14	−0.06
*r*9	0.69	0.64	0.63	0.62	0.62	0.65
*r*10	0.13	0.11	0.11	0.11	0.11	0.12
*r*13	−0.68	−0.71	−0.71	−0.72	−0.72	−0.70
*r*20	−0.58	0.13	0.31	−1.32	Deleted	−0.04
*r*21	0.44	0.85	Deleted	0.00	−0.52	0.66
*r*22a	2.04	Deleted	Deleted	1.23	0.28	1.24

I20, I21, and I22a. The two high values of μ_p/μ_n, I11 and I12, and the x-ray data have been deleted because the conclusions reached in Sec. III.B.2 concerning these items remain essentially unchanged by the inclusion of the new data. In adjustment No. 11, all of the data have been included and we see that its over-all consistency is quite good; for 10 degrees of freedom, the probability that χ^2 will equal or exceed 7.87 is about 65%. However, examination of the residuals shows that the Kaufman *et al.*–Triebwasser *et al.*–Robiscoe value of α^{-1} (I22a) contributes over half the strain in the system. We also note that: (1) The value of α^{-1} resulting from this adjustment is 2.7 ppm less than $\alpha_{\mathrm{WQED}}{}^{-1}$, 1.4 times the latter's standard deviation. For e_{WQED}, K_{WQED}, and N_{WQED}, the changes are $+1.2$, $+0.4$, and -0.7 standard deviations, respectively. (2) If Kaufman *et al.*'s measurement of $\Delta E_{\mathrm{H}} - \mathcal{S}_{\mathrm{H}}$ is combined with only Triebwasser *et al.*'s measurement of $\mathcal{S}_{\mathrm{H}}$ and the resulting value of α^{-1} used in place of I22a, then we find $\alpha^{-1} = 137.03584(18)$, $\chi^2 = 4.06$, and a residual for this datum of 1.24. If instead the α^{-1} value implied by combining the Robiscoe measurement of $\mathcal{S}_{\mathrm{H}}$ with the Kaufman, Lamb, Lea, and Leventhal result is used in place of I22a, then we find $\alpha^{-1} = 137.03575(18)$, $\chi^2 = 8.26$, and a residual for this datum of 2.24. Clearly, the Triebwasser, Dayhoff, and Lamb value of $\mathcal{S}_{\mathrm{H}}$ is in somewhat better agreement with the other data then the Robiscoe value.

In adjustment No.12, I22a has been deleted and the strain in the system decreases significantly as expected. Furthermore, the resulting value of α^{-1} is only 0.9 ppm less than $\alpha_{\mathrm{WQED}}{}^{-1}$ or about half the latter's standard deviation. We conclude that the Metcalf *et al.* and Hhfs values of α^{-1} are quite compatible with the WQED data. In adjustment No. 13, both the Metcalf, Brandenberger, and Baird and the Kaufman *et al.* results have been deleted and the over-all compatability of the data improves still more. The adjusted value of α^{-1} is only 0.4 ppm less than $\alpha_{\mathrm{WQED}}{}^{-1}$, about $\frac{1}{5}$ the latter's standard deviation. This is of course a result of the excellent agreement between $\alpha_{\mathrm{WQED}}^{-1}$ and $\alpha_{\mathrm{Hhfs}}^{-1}$. In adjustment No. 14, the critical role played by the Josephson-effect value of $2e/h$ (I1) is demonstrated by deleting this datum alone. The adjusted value of α^{-1} is seen to decrease by 1.9 ppm compared with that obtained from adjustment No. 11 in which I1 is included. Deleting *both* I1 and I20 (adjustment No. 15) results in a highly consistent set of data as evidenced by the small values of R and χ^2, but the adjusted values of the constants differ considerably from their corresponding WQED values; the resulting value of α^{-1} is 6.9 ppm less than $\alpha_{\mathrm{WQED}}^{-1}$ or 3.6 times the latter's standard deviation. In adjustment No. 16, we have increased the errors assigned I21 and I22a to 5 ppm in order to show what happens if the errors assigned these quantities more nearly reflect what we believe their true uncertainty might be. Clearly, even with this relatively small error expansion, the influence of these items is greatly reduced as may be seen by comparing adjustment No. 16 with adjustment No. 11.

In Table XXXI, we present the results of a similar series of adjustments in which I22a has been replaced by I22b, the value of α^{-1} obtained directly from the Kaufman, Lamb, Lea, and Leventhal measurement of $\Delta E_{\mathrm{H}} - \mathcal{S}_{\mathrm{H}}$ and the theoretical expression for $\Delta E_{\mathrm{H}} - \mathcal{S}_{\mathrm{H}}$. In adjustment No. 17, all of the data have been included and we see that the over-all agreement is again quite reasonable; for 10 degrees of freedom, the probability that χ^2 will equal or exceed 8.02 is about 65%. However, the result of Kaufman *et al.* still contributes about half the total strain of the system. We also note that the value of α^{-1} resulting from this adjustment exceeds $\alpha_{\mathrm{WQED}}^{-1}$ by 1.3 ppm, 0.7 times its standard deviation. For e_{WQED}, K_{WQED}, and N_{WQED}, the changes are -0.6, -0.2, and $+0.3$ standard deviations, respectively. In adjustment No. 18, the Josephson-effect value of $2e/h$ has been deleted but the over-all compatability of the data is little affected. The changes in the adjusted constants as compared with their WQED values are, however, somewhat larger than for adjustment No.17.

TABLE XXXI. Results of four least-squares adjustments involving the same data used in the adjustments of Table XXX but with I22a replaced by I22b.

	Adjustment number and items deleted			
	17. None	18. I1	19. I1, I20	20. None (expanded error)
α^{-1}	137.03625(16)	137.03635(19)	137.03653(23)	137.03603(19)
e	1.6021857(60)	1.6021807(85)	1.6021742(96)	1.6021914(67)
K	1.0000085(26)	1.0000090(27)	1.0000090(27)	1.0000091(26)
N	6.022186(38)	6.022209(47)	6.022234(50)	6.022170(39)
R	0.90	0.90	0.80	0.56
χ^2	8.02	7.34	5.14	3.11
$N-J$	10	9	8	10
$r1$	0.42	Deleted	Deleted	−0.12

*r*2	−0.18	−0.12	−0.12	−0.11
*r*3	−0.10	−0.04	−0.04	−0.02
*r*4	0.08	0.16	0.16	0.18
*r*5	0.13	0.16	0.16	0.17
*r*6	−0.68	−0.26	−0.26	−0.14
*r*7	0.12	0.38	0.38	0.46
*r*8	−0.21	−0.14	−0.14	−0.12
*r*9	0.59	0.62	0.62	0.63
*r*10	0.09	0.11	0.11	0.11
*r*13	−0.74	−0.72	−0.72	−0.72
*r*20	0.98	1.24	Deleted	0.34
*r*21	1.35	1.51	1.83	0.85
*r*22b	−1.89	−1.52	−0.77	−1.02

(Note that the next two adjustments would logically have been those in which I22b and then I21 and I22b were deleted. However, these are the same as adjustments No.12 and 13, Table XXX.) In adjustment No. 19, *both* the Josephson-effect and hyperfine-splitting values of α^{-1} have been deleted and the adjusted constants change still more. We note that the main factor contributing to the strain of the system is the Metcalf, Brandenberger, and Baird result, I21. Thus, the value of α^{-1} calculated from the Kaufman, Lamb, Lea and Leventhal measurement of $\Delta E_{\mathrm{H}} - \mathcal{S}_{\mathrm{H}}$ and the theoretical expression for $\Delta E_{\mathrm{H}} - \mathcal{S}_{\mathrm{H}}$ is somewhat inconsistent with the value of Metcalf *et al.* On the other hand, Metcalf *et al.*'s value of α^{-1} and that obtained from Kaufman *et al*,'s measurement in combination with experimental values of $\mathcal{S}_{\mathrm{H}}$ are in good agreement (see adjustment No. 15). In the last adjustment, No. 20, we have increased the errors assigned I21 and I22b to 5 ppm as in adjustment No. 16. Again this error expansion significantly reduces the influence of these items as may be seen by comparing adjustment No. 20 with No. 17.

It is perhaps worthwhile to give also the results of least-squares adjustments involving only the different values of α^{-1}, i.e., $\alpha_{\mathrm{WQED}}{}^{-1}$, I20, I21, and either I22a or I22b. Although the adjusted values of α^{-1} will be the same as those given in Tables XXX and XXXI, the resulting values of R and χ^2 will be rather different and will give a better indication of the relative compatability of the various fine structure constants. Using I22a, we find $\alpha^{-1} = 137.03571(17)$ (1.2 ppm), $R = 1.49$, and $\chi^2 = 6.66$. For three degrees of freedom, the probability for χ^2 to equal or exceed this value by chance is on the order of 9%. [If we include the muonium hfs value of α^{-1} calculated using the Ruderman correction with the latter assumed to be exact (item 6a, Table XXVIII), the adjusted value of α^{-1} increases by less than 0.2 ppm, $R = 1.33$, and $\chi^2 = 7.22$.] With the uncertainties of I21 and I22a increased to 5 ppm, we find $\alpha^{-1} = 137.03590(19)$ (1.4 ppm), $R = 0.90$ and $\chi^2 = 2.44$. For three degrees of freedom, the chance probability for χ^2 to equal or exceed this value is about 45%. Replacing I22a by I22b yields $\alpha^{-1}=137.03625(16)$ (1.1 ppm), $R=1.51$, and $\chi^2=6.81$. Increasing the error of I21 and I22b to 5 ppm yields $\alpha^{-1}=137.03603(19)$ (1.4 ppm), $R=0.80$, and $\chi^2=1.90$. For three degrees of freedom, the chance probabilities for these last two χ^2 values are on the order of 8% and 60%, respectively.

We may summarize the main conclusions to be drawn from these α adjustments and the adjustments given in Tables **XXX** and **XXXI** as follows: (1) Even if the errors assigned by the experimenters are used, the over-all compatability of the hydrogen fine-structure data with the hyperfine-splitting measurement and the WQED data is reasonably satisfactory. (2) With all the data included, changes in the WQED values of the constants are not excessive, i.e., typically less than 1 or 2 times the standard deviations of the latter. This conclusion is independent of which value of α^{-1} implied by the Kaufman, Lamb, Lea, and Leventhal measurement is used. (3) If the errors assigned the hydrogen fine-structure data are expanded to values which may more closely reflect their proper weights, the adjusted constants differ only by small fractions of a standard deviation from the WQED values or the values obtained by including $\alpha_{\mathrm{Hhfs}}{}^{-1}$ with the WQED data.

The question which must now be answered is: Should the hydrogen hyperfine-splitting and fine-structure data be included in our final least-squares adjustment to obtain a recommended set of fundamental constants? After considerable thought, we have decided that it is best at this time to exclude from that adjustment all of the hydrogen fine-structure values of α, but to include the hydrogen hyperfine-splitting value. We base this decision on facts which have been discussed at some length throughout this section as well as in Sec. IV and which we now briefly summarize. (1) The present theory of the Hhfs, in particular the proton polarizability contribution, seems to us to have been studied sufficently carefully to permit a reliable value of α to be derived from the very accurate experimental measurements of ν_{Hhfs} (see Sec. IV.D).* (2) The scatter in the experimental data of Metcalf *et al.* prevented them from carefully investigating possible sources of systematic error at the level of their final assigned uncertainty. This leads us to believe that the accuracy of the experiment may well have been overestimated. Although it is extremely difficult to estimate what the true 70% confidence-level uncertainty might be, any appreciable increase in the 4.3-ppm error assigned $\alpha_{\mathrm{MBB}}^{-1}$ would make the result of Metcalf, Brandenberger, and Baird sufficiently uncertain that it would carry negligible weight in our adjustment. (3) The scatter in the experimental data of Kaufman *et al.*, while not as large as in the experiment of Metcalf *et al.*, also implies that the assigned error may be overly optimistic. Furthermore, there remains the problem of the unreliability of either the theoretical formula for, or the experimental values of, $\mathcal{S}_{\mathrm{H}}$. The net result is that if the uncertainties of either of Kaufman *et al.*'s values of α^{-1} are expanded to reflect this situation, they carry negligible weight in our adjustment.

In the next section, we shall see that the recommended values of the constants resulting from our final adjustment differ by several tens of parts per million from the currently accepted values resulting from the 1963 adjustment of Cohen and DuMond (1965). These differences are generally more than an order of magnitude greater than any changes in the values of our recommended constants which would result from including the hydrogen fine-structure measurements as input data (with or without an expanded error). We therefore believe the following viewpoint should be adopted at this time: The necessary major changes in the currently accepted values of the constants are adequately given by including in our final adjustment only the Josephson-effect value of $2e/h$ and $\alpha_{\mathrm{Hhfs}}^{-1}$. Any further changes due to the hydrogen fine-structure

* Iddings (1969) also agrees that a value of α reliable to 2–3 ppm may be obtained from the hydrogen hyperfine splitting.

data will be relatively so small and unimportant that they are unwarranted by the present level of confidence which can be placed in these data. In essence, we feel that since there is no compelling need to include these data at this time, it is best to wait until the several questions we have raised about them are satisfactorily resolved.*

B. Final Adjustment and Recommended Constants

We present here a recommended set of fundamental constants obtained from our final least-squares adjustment. This adjustment (which is in fact adjustment No. 13, Table XXX) is based on the stochastic input data used to obtain the WQED values of the constants plus the value of the fine structure constant obtained from the hydrogen hyperfine-structure measurement. The set of stochastic input data used is therefore that listed in Table XVI with the following deletions: I11 and I12, the two high values of μ_p/μ_n as measured by Boyne and Franken and by Mamyrin and Frantsuzov, all of the x-ray data, I14–I19, and the hydrogen fine-structure data, I21 and I22a and I22b. The adjusted constants α^{-1}, e, N and K and a fairly complete set of constants derived from these quantities and the necessary auxiliary constants are given in Tables

* Brodsky (private communication) has raised the point that since QED may be in trouble in the Lamb shift and magnetic moment anomaly, it is questionable to use $\alpha_{\mathrm{Hhfs}}^{-1}$ in our final adjustment. He notes that if for some reason the $\alpha(Z\alpha)^5$ terms in the Lamb shift require a 10% revision (experiment and theory for $\mathcal{S}_H$ and $\mathcal{S}_D$ would then agree), then the value of $\alpha_{\mathrm{Hhfs}}^{-1}$ would change by ~10 ppm since the same operator structure yields $\alpha(Z\alpha)$ terms in the hfs. The point of view to be adopted depends of course on how much faith one wants to place in the Lamb-shift discrepancies and whether one believes they are due to some fundamental cause like the one suggested. An equally plausible guess as to the source of the discrepancies is an unknown experimental effect, an uncalculated term, or a miscalculated term. We do note that: (1) if we were to expurgate $\alpha_{\mathrm{Hhfs}}^{-1}$, then our final recommended set of constants would be identical to our WQED set since we believe there would then be no QED value of α sufficiently reliable for inclusion in our final adjustment, and (2) including $\alpha_{\mathrm{Hhfs}}^{-1}$ makes little practical difference.

TABLE XXXII.[a] Final list of recommended physical constants. The stochastic input data include I1-I10, I13, and I20 of Table XVI. $R=0.412$, $\chi^2=1.360$, and $N-J=8$. The numbers in parentheses are the standard-deviation uncertainties in the last digits of the quoted value, computed on the basis of internal consistency.

Quantity	Symbol	Value	Error (ppm)	Units: SI	Units: cgs
Velocity of light	c	2.9979250(10)	0.33	10^8 m sec^{-1}	10^{10} cm sec^{-1}
Fine-structure constant, $[\mu_0 c^2/4\pi](e^2/\hbar c)$	α	7.297351(11)	1.5	10^{-3}	10^{-3}
	α^{-1}	137.03602(21)	1.5		
Electron charge	e	1.6021917(70)	4.4	10^{-19} C	10^{-20} emu
		4.803250(21)	4.4		10^{-10} esu
Planck's constant	h	6.626196(50)	7.6	10^{-34} J·sec	10^{-27} erg·sec
	$\hbar=h/2\pi$	1.0545919(80)	7.6	10^{-34} J·sec	10^{-27} erg·sec
Avogadro's number	N	6.022169(40)	6.6	10^{26} kmole^{-1}	10^{23} mole^{-1}
Atomic mass unit	amu	1.660531(11)	6.6	10^{-27} kg	10^{-24} g
Electron rest mass	m_e	9.109558(54)	6.0	10^{-31} kg	10^{-28} g
	m_e^*	5.485930(34)	6.2	10^{-4} amu	10^{-4} amu
Proton rest mass	M_p	1.672614(11)	6.6	10^{-27} kg	10^{-24} g
	M_p^*	1.00727661(8)	0.08	amu	amu

Neutron rest mass	M_n	1.674920(11)	6.6	10^{-27} kg	10^{-24} g
	M_n^*	1.00866520(10)	0.10	amu	amu
Ratio of proton mass to electron mass	M_p/m_e	1836.109(11)	6.2		
Electron charge to mass ratio	e/m_e	1.7588028(54)	3.1	10^{11} C kg^{-1}	10^7 emu g^{-1}
		5.272759(16)	3.1		10^{17} esu g^{-1}
Magnetic flux quantum, $[c]^{-1}(hc/2e)$	Φ_0	2.0678538(69)	3.3	10^{-15} T·m^2	10^{-7} G·cm^2
	h/e	4.135708(14)	3.3	10^{-15} J·sec C^{-1}	10^{-7} erg·sec emu^{-1}
		1.3795234(46)	3.3		10^{-17} erg·sec esu^{-1}
Quantum of circulation	$h/2m_e$	3.636947(11)	3.1	10^{-4} J·sec kg^{-1}	erg·sec g^{-1}
	h/m_e	7.273894(22)	3.1	10^{-4} J·sec kg^{-1}	erg·sec g^{-1}
Faraday constant, Ne	F	9.648670(54)	5.5	10^7 C kmole^{-1}	10^3 emu mole^{-1}
		2.892599(16)	5.5		10^{14} esu mole^{-1}
Rydberg constant, $[\mu_0 c^2/4\pi]^2(m_e e^4/4\pi\hbar^3 c)$	R_∞	1.09737312(11)	0.10	10^7 m^{-1}	10^5 cm^{-1}
Bohr radius, $[\mu_0 c^2/4\pi]^{-1}(\hbar^2/m_e e^2) = \alpha/4\pi R_\infty$	a_0	5.2917715(81)	1.5	10^{-11} m	10^{-9} cm
Classical electron radius $[\mu_0 c^2/4\pi](e^2/m_e c^2) = \alpha^3/4\pi R_\infty$	r_0	2.817939(13)	4.6	10^{-15} m	10^{-13} cm

TABLE XXXII. (*Continued*)

Quantity	Symbol	Value	Error (ppm)	Units	
				SI	cgs
Electron magnetic moment in Bohr magnetons	μ_e/μ_B	1.0011596389(31)	0.0031		
Bohr magneton, $[c](e\hbar/2m_e c)$	μ_B	9.274096(65)	7.0	10^{-24} J T^{-1}	10^{-21} erg G^{-1}
Electron magnetic moment	μ_e	9.284851(65)	7.0	10^{-24} J T^{-1}	10^{-21} erg G^{-1}
Gyromagnetic ratio of protons in H_2O	γ_p' $\gamma_p'/2\pi$	2.6751270(82) 4.257597(13)	3.1 3.1	10^8 rad $sec^{-1} \cdot T^{-1}$ 10^7 Hz T^{-1}	10^4 rad $sec^{-1} \cdot G^{-1}$ 10^3 Hz G^{-1}
γ_p' corrected for diamagnetism of H_2O	γ_p $\gamma_p/2\pi$	2.6751965(82) 4.257707(13)	3.1 3.1	10^8 rad $sec^{-1} \cdot T^{-1}$ 10^7 Hz T^{-1}	10^4 rad $sec^{-1} \cdot G^{-1}$ 10^3 Hz G^{-1}
Magnetic moment of protons in H_2O in Bohr magnetons	μ_p'/μ_B	1.52099312(10)	0.066	10^{-3}	10^{-3}
Proton magnetic moment in Bohr magnetons	μ_p/μ_B	1.52103264(46)	0.30	10^{-3}	10^{-3}
Proton magnetic moment	μ_p	1.4106203(99)	7.0	10^{-26} J T^{-1}	10^{-23} erg G^{-1}

Magnetic moment of protons in H_2O in nuclear magnetons	μ_p'/μ_n	2.792709(17)	6.2		
μ_p'/μ_n corrected for diamagnetism of H_2O	μ_p/μ_n	2.792782(17)	6.2		
Nuclear magneton, $[c](e\hbar/2M_pc)$	μ_n	5.050951(50)	10	10^{-27} J T^{-1}	10^{-24} erg G^{-1}
Compton wavelength of the electron, h/m_ec	λ_C	2.4263096(74)	3.1	10^{-12} m	10^{-10} cm
	$\lambda_C/2\pi$	3.861592(12)	3.1	10^{-13} m	10^{-11} cm
Compton wavelength of the proton, h/M_pc	$\lambda_{C,p}$	1.3214409(90)	6.8	10^{-15} m	10^{-13} cm
	$\lambda_{C,p}/2\pi$	2.103139(14)	6.8	10^{-16} m	10^{-14} cm
Compton wavelength of the neutron, h/M_nc	$\lambda_{C,n}$	1.3196217(90)	6.8	10^{-15} m	10^{-13} cm
	$\lambda_{C,n}/2\pi$	2.100243(14)	6.8	10^{-16} m	10^{-14} cm
Gas constant	R_0	8.31434(35)	42	10^3 J $kmole^{-1} \cdot K^{-1}$	10^7 erg $mole^{-1} \cdot K^{-1}$
Boltzman's constant, R_0/N	k	1.380622(59)	43	10^{-23} J K^{-1}	10^{-16} erg K^{-1}
Stefan–Boltzman constant, $\pi^2k^4/60\hbar^3c^2$	σ	5.66961(96)	170	10^{-8} W m^{-2} K^4	10^{-5} erg $sec^{-1} \cdot cm^{-2} \cdot K^{-4}$

TABLE XXXII. (*Continued*)

Quantity	Symbol	Value	Error (ppm)	Units: SI	Units: cgs
First radiation constant, $8\pi hc$	c_1	4.992579(38)	7.6	10^{-24} J·m	10^{-15} erg·cm
Second radiation constant, hc/k	c_2	1.438833(61)	43	10^{-2} m·K	cm·K
Gravitational constant	G	6.6732(31)	460	10^{-11} N·m^2 kg^{-2}	10^{-8} dyn·cm^2 g^{-2}

[a] Note that the unified atomic mass scale $^{12}C \equiv 12$ has been used throughout, that amu = atomic mass unit, C = coulomb, G = gauss, Hz = hertz = cycles/sec, J = joule, K = kelvin (degrees kelvin), T = tesla (10^4 G), V = volt, and W = watt. In cases where formulas for constants are given (e.g., R_∞), the relations are written as the product of two factors. The second factor, in parentheses, is the expression to be used when all quantities are expressed in cgs units, with the electron charge in electrostatic units. The first factor, in brackets, is to be included only if all quantities are expressed in SI units. We remind the reader that with the exception of the auxiliary constants which have been taken to be exact, the uncertainties of these constants are correlated, and therefore the general law of error propagation must be used in calculating additional quantities requiring two or more of these constants. (See Appendix A; for further comments on the table, see Sec. III.C.)

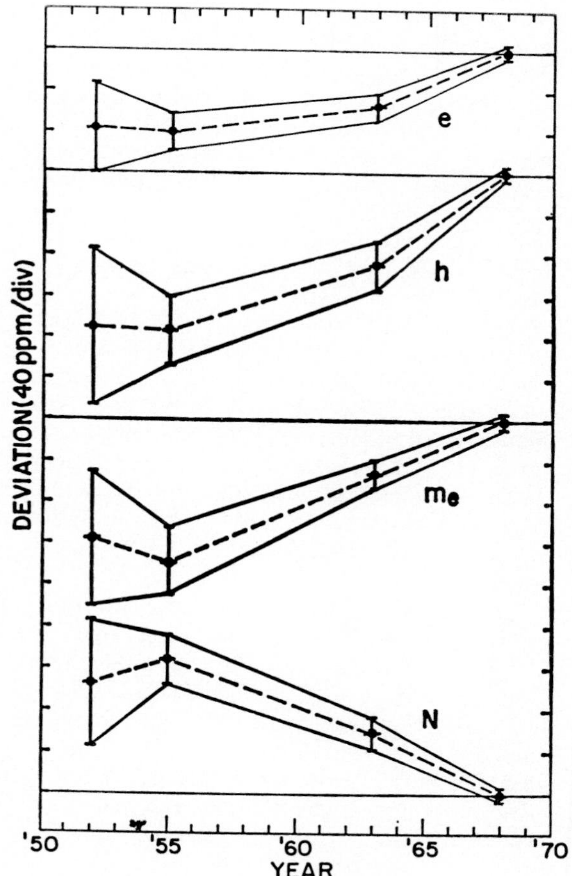

FIG. 8. Plot showing how the recommended numerical values and standard deviations of the electron charge, Planck's constant, electron mass, and Avogadro's number have changed since 1952. The final recommended values of the present adjustment are used as references. [Values for 1963, Cohen and DuMond (1965); values for 1955, Cohen, DuMond, Layton, and Rollett (1955); values for 1952, DuMond and Cohen (1953).] For a similar plot of α^{-1}, see Fig. 1.

XXXII–XXXV. The standard-deviation errors quoted in these tables have been computed from the appropriate variance–covariance matrix which is presented and discussed in Appendix A. All of the comments made in Sec. III.C concerning the similar tables of WQED constants (Tables XX–XXIII) apply here as well.

Since the numerical values of these constants differ considerably from the currently accepted values, i.e., those given by Cohen and DuMond (1965) in their 1963 adjustment, it is perhaps of interest to compare the values of some of the more important constants resulting from the two adjustments. This is done in Table XXXVI. We note that the values have changed by several standard deviations and that the uncertainties resulting from our adjustment are significantly less than the corresponding uncertainties given by the 1963 adjustment. These changes are primarily due to the values of α^{-1} implied by the Josephson-effect measurement of $2e/h$ and the hydrogen hyperfine splitting. These are significantly smaller and have smaller uncertainties than the value of α^{-1} derived by Cohen and DuMond from the deuterium fine-structure measurements.

It is also of interest to see how our knowledge of the fundamental constants has changed over the years. This is shown in Fig. 8 for the constants e, h, m_e, and N (see Fig. 1 for α^{-1}). This figure graphically illustrates our contention that no set of fundamental constants should be taken as Gospel truth. While we may hope that the present adjustment brings us closer to that truth, realism compels us to recognize that further significant changes in our knowledge of the constants may well take place. Goethe might well have been speaking of this when he wrote "Es irrt der Mensch, solang' er strebt."

VI. Summary and Conclusions

A. Conclusions Concerning Quantum Electrodynamics

We believe there is some cause for concern in the present unsatisfactory state of agreement between quantum electrodynamic theory and experiment. Comparisons of experimental data with theoretical values calculated using α_{WQED} have been given throughout Sec. IV. Since our final adjusted value of α differs very little from α_{WQED}, these comparisons remain essentially unchanged. The general picture which emerges from a study of these comparisons can be summarized as follows: For quantities which involve relatively small QED corrections and for which the experimental situation is reasonably satisfactory, the agreement between theory and experiment is adequate. For quantities which are totally quantum electrodynamic in origin, the agreement between theory and experiment is considerably worse.

The weakly QED-dependent quantities which are essentially in agreement with theory include the ground-state hyperfine splitting in atomic hydrogen and muonium and the fine-structure splitting in the $n=2$ state of atomic hydrogen. The accuracy of the hydrogen-maser determinations of the hydrogen hyperfine splitting is unmatched anywhere in physics; the uncertainty in any comparison between theory and experiment is due entirely to the theory and revolves mainly around the proton polarization correction. However, the best theoretical estimates of this correction yield excellent agreement with experiment. The experimental status of the muonium hyperfine splitting is less satisfying than for the hydrogen hyperfine

TABLE XXXIII. Final recommended energy conversion factors. (See Sec. III.C for discussion.)

Quantity	Value	Unit	Error (ppm)
1 kg	5.609538(24)	10^{29} MeV	4.4
1 amu	931.4812(52)	MeV	5.5
Electron mass	0.5110041(16)	MeV	3.1
Proton mass	938.2592(52)	MeV	5.5
Neutron mass	939.5527(52)	MeV	5.5
1 electron volt	1.6021917(70)	10^{-19} J	4.4
		10^{-12} erg	
	2.4179659(81)	10^{14} Hz	3.3
	8.065465(27)	10^{5} m^{-1}	3.3
		10^{3} cm^{-1}	
	1.160485(49)	10^{4} K	42
Energy–wavelength conversion	1.2398541(41)	10^{-6} eV·m	3.3
		10^{-4} eV·cm	

Rydberg constant, R_∞	2.179914(17)	10^{-18} J	7.6
		10^{-11} erg	
	13.605826(45)	eV	3.3
	3.2898423(11)	10^{15} Hz	0.35
	1.578936(67)	10^{5} K	43
Bohr magneton, μ_B	5.788381(18)	10^{-5} eV T^{-1}	3.1
	1.3996108(43)	10^{10} Hz T^{-1}	3.1
	46.68598(14)	$m^{-1} \cdot T^{-1}$	3.1
		10^{-2} $cm^{-1} \cdot T^{-1}$	
	0.671733(29)	K T^{-1}	43
Nuclear magneton, μ_n	3.152526(21)	10^{-8} eV T^{-1}	6.8
	7.622700(42)	10^{6} Hz T^{-1}	5.5
	2.542659(14)	10^{-2} $m^{-1} \cdot T^{-1}$	5.5
		10^{-4} $cm^{-1} \cdot T^{-1}$	
	3.65846(16)	10^{-4} K T^{-1}	44
Gas constant, R_0	8.20562(35)	10^{-2} $m^3 \cdot$atm $kmole^{-1} \cdot K^{-1}$	42
Standard volume of ideal gas, V_0	22.4136	m^3 $kmole^{-1}$	

TABLE XXXIV. Final recommended values of various quantities involving as-maintained electrical units. (See Sec. III.C for further explanation.)

Quantity	Symbol	Value (Prior to 1 Jan. 1969)	Value (After 1 Jan. 1969)	Error (ppm)	Units
Ratio of NBS ampere to absolute ampere	$K \equiv A_{NBS}/A_{ABS}$	1.0000091(26)	1.0000007(26)	2.6	
Ratio of NBS volt to absolute volt	V_{NBS}/V_{ABS}	1.0000088(26)	1.0000004(26)	2.6	
Ratio of BIPM ampere to absolute ampere	A_{BIPM}/A_{ABS}	1.0000115(26)	1.0000005(26)	2.6	
Ratio of BIPM volt to absolute volt	V_{BIPM}/V_{ABS}	1.0000114(26)	1.0000004(26)	2.6	

Ratio absolute ohm to NBS ohm	$\Omega_{ABS}/\Omega_{NBS}$	1.00000036(70)	1.00000036(70)	0.7	
Ratio absolute ohm to BIPM ohm	$\Omega_{ABS}/\Omega_{BIPM}$	1.00000017(70)	1.00000017(70)	0.7	
Josephson frequency–voltage ratio	$2e/h$	4.835974(11)	4.835934(11)	2.2	10^{14} Hz V_{NBS}^{-1}
Faraday constant	F	9.648581(55)	9.648662(55)	5.7	10^{7} $A_{NBS}\cdot$sec kmole^{-1}
Gyromagnetic ratio (low field) of protons in H_2O	γ_p' $\gamma_p'/2\pi$	2.6751514(71) 4.257636(11)	2.6751289(71) 4.257600(11)	2.6 2.6	10^{8} rad sec$^{-1}\cdot T_{NBS}^{-1}$ 10^{7} Hz T_{NBS}^{-1}
γ_p' (low field) corrected for diamagnetism of H_2O	γ_p $\gamma_p/2\pi$	2.6752210(71) 4.257746(11)	2.6751985(71) 4.257711(11)	2.6 2.6	10^{8} rad sec$^{-1}\cdot T_{NBS}^{-1}$ 10^{7} Hz T_{NBS}^{-1}
Voltage–wavelength conversion, $V\lambda$	hc/e	1.2398433(28)	1.2398537(28)	2.2	10^{-6} $V_{NBS}\cdot$m

TABLE XXXV. Final recommended values for various quantities involving the kilo-x-unit and angstrom-star x-ray scales. (For a detailed discussion, see Sec. III.C.)

Quantity	Symbol	Value	Error (ppm)	Units
kx-unit-to-angstrom conversion factor, $\Lambda = \lambda(\text{Å})/\lambda(\text{kxu})$; $\lambda(\text{Cu}K\alpha_1) \equiv 1.537400$ kxu	Λ	1.0020764(53)	5.3	
Å*-to-angstrom conversion factor, $\Lambda = \lambda(\text{Å})/\lambda(\text{Å}^*)$; $\lambda(\text{W}K\alpha_1) \equiv 0.2090100$ Å*	Λ^*	1.0000197(56)	5.6	
Voltage-wavelength conversion product, $V\lambda = hc/e$	$V\lambda(\text{kxu})$	1.2372855(69)	5.6	10^4 V·kxu
	$V\lambda(\text{Å}^*)$	1.2398301(73)	5.9	10^4 V·Å*
$V\lambda$ in NBS units prior to 1 Jan. 1969	$V_{\text{NBS}}\lambda(\text{kxu})$	1.2372742(66)	5.3	10^4 V_{NBS}·kxu
	$V_{\text{NBS}}\lambda(\text{Å}^*)$	1.2398188(70)	5.6	10^4 V_{NBS}·Å*
$V\lambda$ in NBS units after 1 Jan. 1969	$V_{\text{NBS}}\lambda(\text{kxu})$	1.2372845(66)	5.3	10^4 V_{NBS}·kxu
	$V_{\text{NBS}}\lambda(\text{Å}^*)$	1.2398292(70)	5.6	10^4 V_{NBS}·Å*
Compton wavelength of the electron	$\lambda_C(\text{kxu})$	2.421282(14)	5.7	10^{-2} kxu
	$\lambda_C(\text{Å}^*)$	2.426262(14)	6.0	10^{-2} Å*

splitting, but the theory should be somewhat more trustworthy because no polarizability correction is expected. Although earlier experimental measurements of the muonium hyperfine splitting were in good agreement with the theoretical value calculated from the 1963 adjusted value of α, the results of the most recent measurements combined with the revised (Ruderman) diamagnetic shielding correction are now in excellent agreement with the theoretical value calculated from our recommended value of α. The experimental data relating to the hydrogen fine structure are quite consistent and yield values of the fine-structure splitting which differ from the theoretical value calculated using our recommended α value by one to two and a half standard deviations, even if the rather small uncertainties assigned by the experimenters are accepted at face value. These discrepancies (if one may call them that) would of course be smaller if the hydrogen fine-structure measurements had been included as input data in our final adjustment. For this reason and because of the error assignment problem in these experiments, we do not believe the discrepancies should be taken too seriously at this time. The two available measurements of the Lamb shift in deuterium are not very consistent so it is difficult to draw any conclusions about agreement between theory and experiment for the deuterium fine structure. However, it may be noted that the most recent measurement of $\mathcal{S}_D$, when combined with the single available measurement of $\Delta E_D - \mathcal{S}_D$, gives a value of ΔE_D which differs from the theoretical value by less than two standard deviations (although in the opposite direction from the hydrogen fine-structure-splitting discrepancies).

In contrast with the fairly satisfactory situation which exists for these weakly QED-dependent quantities, the situation is markedly worse for the g-factor anomalies and the Lamb shift, which are totally quantum electrodynamic quantities. For the electron moment anomaly, theory exceeds experiment by (77 ± 26) ppm; the probability for this to occur by chance is 0.3%. For the muon moment anomaly, the discrepancy is in the opposite direction, with experiment exceeding theory by [440(+270 or −290] ppm; the chance probability for this is about 11%. For the Lamb shift in the $n=2$ state of H, experiment exceeds theory by between (180±97) ppm and (280±66) ppm. For D, experiment exceeds theory by between (165±75) ppm and (400±75) ppm. The chance probabilities for these discrepancies are between 6% and $<10^{-3}$% for H and between 3% and $<10^{-5}$% for D. The discrepancy in the muon moment anomaly is not insufferable. The discrepancy in the electron moment anomaly looks more serious. It seems reasonable to suppose that if QED is capable of supplying an accurate description of anything, it should be the properties of a single free electron. Since there is only one really accurate experimental measurement of the electron moment anomaly, it would appear that the principal burden of resolving this problem falls on the experimentalists. (Fortunately, redeterminations of the electron moment anomaly are currently in progress in several laboratories.) With regard to the Lamb-shift discrepancies, it could be argued that the errors assigned either or both the experimental and theoretical values of $\mathcal{S}$ are too small, and that the disagreements are simply artifacts. However, we believe that the number and sizes of the discrepancies and the care and effort which have been lavished on both the experiments and the theory require that serious attention be given this problem. With four different Lamb-shift measurements presently available, all in significant disagreement with theory, in the same direction, and by roughly comparable amounts, the finger of suspicion here would seem to point in the direction of the theory. (Further measurements of the Lamb shift would of course still be useful.) We would not venture to guess where in the limitless forests of diagrams salvation may be found, but a search seems advisable. In any case, it appears that all is not well with quantum electrodynamics.

B. Conclusions Concerning Superconductivity

In Sec. II.C.1 of this paper and in Parker *et al.* (1969) we have presented the theoretical and experimental reasons for believing in the general validity of the Josephson frequency–voltage relation on which our measurement of $2e/h$ is based. The arguments presented there were drawn from within the field of superconductivity. Here we would like to ask what may be concluded about the exactness of the frequency–voltage relation within the over-all framework of the present adjustment, i.e., how well does the Josephson-effect value of $2e/h$ agree with the best independent, non-Josephson-effect value of the same quantity. To answer this question, we carry out a least-squares adjustment identical to the adjustment on which our recommended set of constants is based, but with the Josephson-effect value of $2e/h$ deleted. From the resulting adjusted values of α^{-1}, e, and K we compute $2e/h$ to be

$$2e/h = 4.835964(29) \times 10^{14}\ \mathrm{Hz}/\mathrm{V_{NBS}}\ (5.9\ \mathrm{ppm}).$$

This value is to be compared with the experimental value, Eq. (31):

$$(2e/h)_{\mathrm{exptl}} = 4.835976(12) \times 10^{14}\ \mathrm{Hz}/\mathrm{V_{NBS}}\ (2.4\ \mathrm{ppm}).$$

The two values differ by (2.5 ± 6.4) ppm. We may therefore conclude that, on the basis of consistency with the most reliable data available, the Josephson frequency–voltage ratio is in fact equal to $2e/h$ to within approximately 6 ppm.

It may be argued of course that this is a somewhat biased result since the only direct value of the fine-structure constant retained in our final adjustment was that obtained from the hydrogen hyperfine splitting. However, our conclusion is not significantly altered if we compare the experimental value of $2e/h$ with values resulting from adjustments in which it is deleted, but

TABLE XXXVI. A comparison of the final recommended values of α^{-1}, e, h, m_e, and N resulting from the present adjustment with those resulting from the 1963 adjustment of Cohen and DuMond (1965).

Quantity	Units	Value, this adjustment	Error (ppm)	Value, 1963 adjustment	Error (ppm)	Change (ppm)
α^{-1}		137.03602(21)	1.5	137.0388(6)	4.4	−20
e	10^{-19} C	1.6021917(70)	4.4	1.60210(2)	12	+57
h	10^{-34} J·sec	6.626196(50)	7.6	6.62559(16)	24	+91
m_e	10^{-31} kg	9.109558(54)	6.0	9.10908(13)	14	+52
N	10^{26} kmole^{-1}	6.022169(40)	6.6	6.02252(9)	15	−58

the hydrogen fine-structure values of α are included along with the hyperfine splitting value. If the experimental $2e/h$ is deleted from the two adjustments in which the uncertainties of the fine-structure data have been expanded (adjustment No. 16, Table XXX and adjustment No. 20, Table XXXI), we find $2e/h(\text{exptl}) - 2e/h(\text{adjust}) \equiv \Delta(2e/h) = (5.8 \pm 5.6)$ ppm and (1.5 ± 5.6) ppm, respectively. The two adjustments in which the fine-structure uncertainties have not been expanded and in which the experimental $2e/h$ has already been deleted (adjustment No. 14, Table XXX and adjustment No. 18, Table XXXI) yield $\Delta(2e/h) = (9.2 \pm 5.0)$ ppm and $\Delta(2e/h) = (-4.7 \pm 4.8)$ ppm, respectively. Clearly, even the latter two cases, which we believe are rather unrealistic treatments of the data, are reasonably consistent with the conclusion reached above.

C. Weaknesses in the Present Adjustment, and Future Work

We would like to discuss here what we consider to be the major weaknesses in the present set of recommended constants and to indicate what experiments might be carried out in the near future to remedy these defects.

Perhaps the least aesthetically satisfying feature of the present adjustment is the rejection of all data from QED experiments except the hydrogen hyperfine-splitting value of α. Although we feel that the case for rejecting most of the QED data is clear, the decision to expurgate the hydrogen fine-structure data was made with some trepidation. The reason is that in the absence of these data, the adjusted value of α is determined by the indirect value of α obtained primarily from γ_p and the Josephson-effect measurement of $2e/h$ (uncertainty about 1.9 ppm) and the hydrogen hfs value of α (uncertainty 2.6 ppm). This means that the $2e/h$ measurement carries considerable weight in the adjustment, more than one would perhaps like for a single measurement which is the first of its kind. However, as noted in Sec. V.A, the effect of the deletion of the hydrogen fine-structure data on the final recommended constants is small compared with the differences between our present set of constants and the constants of the 1963 adjustment. Thus, whatever the aesthetic deficiencies of this deletion may be, its practical consequences are negligible. Still, it is to be hoped that further development of the hydrogen fine-structure experiments and redeterminations of $2e/h$ using the Josephson effect (see Sec. VI.D) will clarify this situation. We might also note that other potentially useful sources of information on the fine-structure constant include further measurements of the fine structure of hydrogenic atoms, D and He^+ in particular, and measurement of the Compton wavelength of the electron via pair annihilation. (Measurement of the Compton wavelength using flux quantization in superconductors will be discussed in the next section.) Improvements in the theory of the polarizability of the proton and the diamagnetic shielding correction for muons in water would make the present measurements of the hyperfine splitting in hydrogen and muonium more useful (as would an improved measurement of μ_μ'/μ_p'). To this we may add that at present, theoretical uncertainties prevent utilization of the very accurate measurements of the hfs in D and T (Mathur, Crampton, Kleppner, and Ramsey, 1967), as well as the accurate fine-structure measurements in atomic helium (Pichanick, Swift, Johnson, and Hughes, 1968).

The major problem of the present work is undoubtedly the discrepancies among the μ_p/μ_n data. While all evidence tends to support our decision to discard the two high values of Boyne and Franken and Mamyrin and Frantzusov, the very fact that they exist casts some doubt on the other three measurements which we retained. (This suspicion also extends to the NBS measurement of the Faraday because of its intimate relationship with μ_p/μ_n via γ_p.) While this discrepancy is relatively unimportant in determining a best value of α, it is of major importance in determining Avogadro's number N and the several additional constants which require N for their evaluation (e.g., the mass of the electron and proton in absolute units). We therefore strongly urge remeasurement of μ_p/μ_n by any and all means available. This includes methods such as the nuclear reaction-energy technique of Marion and Winkler, redetermination of the Faraday, measurements of N via the x-ray crystal-density method, and spectroscopic determination of the ratio $M_p/m_e = (\mu_p/\mu_n)/(\mu_p/\mu_B)$.

Compared with the μ_p/μ_n problem, all other uncertainties seem minor, and we shall therefore only touch upon them briefly:

(1) There is only one high-accuracy determination of the Faraday. In view of its close connection with the μ_p/μ_n discrepancy, new measurements are certainly called for. (Fortunately, NBS is currently undertaking a redetermination of F.)

(2) The low-field measurements of γ_p are not as well in hand as one might like due to the somewhat discrepant values of Hara *et al.* and Studentsov *et al.* The gyromagnetic ratio of the proton is of course quite important because of the critical role it plays in determining α from $2e/h$. Additional measurements are therefore desirable. (Such experiments are underway at NPL and several other national laboratories.) Note also that a significant reduction in the error assigned γ_p would lead to a reduction in the uncertainty in α. This may be even more important in the future since it appears possible to determine $2e/h$ via the ac Josephson effect to within a few tenths of a part per million (see Sec. VI.D).

(3) There is only one reliable high-accuracy, high-field measurement of γ_p. In view of its major role in determining the constant $K \equiv A_{NBS}/A_{ABS}$, additional measurements are clearly in order. (We might add that unless a significant improvement in accuracy can be achieved, the general agreement of the three different direct current-balance measurements of K would seem to place this quantity rather low on a remeasurement priority list.)

(4) As discussed in Sec. II.C.6, the currently available x-ray data can contribute very little to our knowledge of the constants. However, we believe the information potentially obtainable from sufficiently high-precision x-ray experiments warrants further effort in this direction. Since one of the major stumbling blocks is the x-unit-to-milliangstrom conversion factor, continued work on its direct determination to 1 or 2 ppm should be given high priority.

(5) Our present knowledge of the various auxiliary constants listed in Table XI is quite adequate, with no one item being glaringly suspect. However, the general age and quality of the spectroscopic data used to obtain the Rydberg constant (as well as the required assumptions concerning line intensities) are such that new and improved measurements would be most desirable, the recent work of Csillag notwithstanding. Improved measurements of the velocity of light are also desirable since the uncertainty assigned to it is among the largest of the group. If the error could be reduced to, say 0.1 ppm from its present 0.3 ppm, then the uncertainty in $c\Omega_{ABS}/\Omega_{NBS}$ would be reduced to 0.2 ppm. This reduction may be of some importance in the near future since, as just noted, it is quite likely that $2e/h$ can be measured via the Josephson effect to 0.1 to 0.2 ppm (see Sec. VI.D), and the quantity $c\Omega_{ABS}/\Omega_{NBS}$ is required in order to derive α from $2e/h$. (Such improved measurements of c are in fact well under way at NBS, Boulder, Colorado.)

D. Further Fundamental-Constant Experiments Utilizing Macroscopic Quantum Phase Coherence in Superfluids

The Josephson-effect e/h measurement reported by Parker *et al.* (1969) represents the first application of macroscopic quantum phase coherence in superconductors to the accurate determination of a fundamental physical constant. It is therefore highly desirable that this type of experiment be repeated in order to check the result of the first measurement.

Such a redetermination of e/h is being carried out at NBS in a joint effort with a University of Pennsylvania group. Preliminary work indicates that it should be possible to reduce the uncertainty in the value of $2e/h$ measured in terms of V_{NBS} to several tenths of a part per million. Similar experiments are planned or under way in at least three other national standards laboratories. [Part of the motivation for all of these efforts is the potential utility of Josephson junctions in maintaining and comparing primary voltage standards (Taylor, Parker, Langenberg, and Denenstein, 1967).] It should be noted that if an accuracy of several tenths of a part per million can actually be realized, the Josephson-effect value of $2e/h$ would become an auxiliary constant and the indirect value of α determined by $2e/h$ and γ_p would have an uncertainty of about 1.5 ppm. If the accuracy of γ_p determinations were to be increased to 2 ppm, a value which should be realizable in practice, the indirect value of α would have an uncertainty of 1 ppm or less. If the accuracy of the QED experiments could not also be significantly increased, they would then cease to contribute effectively to our knowledge of the constants in much the same way as the x-ray measurements have ceased to do so. It is worth commenting that, since a large part of the uncertainty in the γ_p experiments is associated with the coils which produce the magnetic field, some serious thought ought to be given to the possible advantages of the unique properties of superconducting coils and their cryogenic environment [see Meservey (1968)].

There are several other types of experiments in which it should be possible to take advantage of macroscopic quantum phase coherence in superconductors (and also in superfluid helium) to determine fundamental constants. We briefly indicate here what some of these experiments are, what quantities they measure, and what accuracies are required to be useful.

1. *Measurement of the Magnetic Flux Quantum*, $\Phi_0 = h/2e$

The quantization of flux in superconductors was predicted by London (1935; 1950) and first observed experimentally by Deaver and Fairbank (1961) and Doll and Näbauer (1961). The basic result is that the magnetic flux Φ linking a multiply connected superconductor (e.g., a ring) is quantized in units of the flux quantum, $\Phi_0 = h/2e$. This phenomenon, like the Josephson effect, is a consequence of quantum phase coherence in superconductors. There are several ways of detecting and measuring the number of flux quanta (or the change in flux) in a superconducting ring. One of the most useful methods utilizes a superconducting de Broglie-wave interferometer (Jaklevic, Lambe, Mercereau, and Silver, 1965). [For a recent review, see Mercereau (1969).] The procedure for determining Φ_0 is, in principle, quite straightforward. A magnetic field B is applied to a superconducting loop of area A and the number of flux quanta contained in the loop is determined. Hence, $BA = \Phi = n\Phi_0 = nh/2e$. (In practice, it will probably be easier to measure the change in the number of flux quanta contained in the loop for a given change in applied field.) The main experimental difficulty lies in determining the area of the loop with sufficient accuracy. The diameter of a loop 1 cm in diameter would have to be measured to 5 ppm or 500 Å in order to achieve an accuracy of 10 ppm in $h/2e$. Actually, there are two distinct methods for carrying out this experiment. In the first, the magnetic field is provided by a precision solenoid and is determined in terms of the as-maintained ampere in the same manner as in a low-field γ_p measurement (Meservey, 1968). The observational equation for $\Phi_0 = h/2e$ measured in this way is identical to that for $2e/h$ measured via the ac Josephson effect [Eq. (7)], except that the factor $\Omega_{\mathrm{ABS}}/\Omega_{\mathrm{NBS}}$ is absent since Φ_0 is measured in terms of the NBS ampere rather than the NBS volt:

$$(\alpha^{-1})^{-1} e^{-1} K^1 N^0 = (\mu_0 c/4)(\Phi_0^{-1})_{\mathrm{NBS}}.$$

Here Φ_0 is to be expressed in the units in terms of which it is measured, $\mathrm{tesla}_{\mathrm{NBS}}\cdot$meters squared. In view of the present 2.4-ppm error in $2e/h$ as obtained via the ac Josephson effect, Φ_0 would have to be determined to an accuracy of something like 7 ppm to be of value.

In the second method, the magnetic field is measured in terms of the angular precession frequency of protons in water, ω_p'. Defining ω_{p0}' as the proton precession frequency corresponding to the magnetic field increment B_0 which causes the flux in the loop to change by exactly one flux quantum, we may write $\hbar\omega_{p0}'=2\mu_p'B_0$ so that $B_0=\hbar\omega_{p0}'/2\mu_p'=m_e\omega_{p0}'(\mu_p'/\mu_B)^{-1}e$. Thus, since $B_0A=\Phi_0=h/2e$, we obtain

$$\frac{h}{m_ec}=\lambda_C=\frac{2\omega_{p0}'A}{c(\mu_p'/\mu_B)}\,.$$

This experiment therefore measures $h/2m_e$ (often called the quantum of circulation) or, equivalently, the electron Compton wavelength. The observational equation for λ_C was discussed in Sec. II.C.6 and is

$$(\alpha^{-1})^{-2}e^0K^0N^0=2R_\infty\lambda_C.$$

We note that this experiment would give a direct WQED value of the fine-structure constant. Since α^{-1} is known from the ac Josephson effect to about 2 ppm, λ_C would have to be measured to 4 ppm in order to carry equal weight. However, even a 10-ppm measurement would be useful. Although this implies a 10-ppm measurement of the area of the interferometer, it appears to be well within the realm of possibility.

It is worth noting that the requirement for an accurate measurement of the interferometer area can be eliminated simply by combining the results of these two types of flux-quantum experiments in such a way as to eliminate the interferometer area. However, the result is not particularly interesting since it amounts to using a superconducting de Broglie-wave interferometer as a transfer device in a regular low-field γ_p experiment, i.e., it simply adds extra complexity to what is otherwise a conventional measurement of γ_p.

2. *Measurement of the Electron Compton Wavelength*, λ_C

This experiment measures the quantum of circulation $h/2m_e$ (or λ_C) in a somewhat different manner than does the second flux-quantization experiment just discussed.

It depends on the fact that the maximum supercurrent flow through a superconducting de Broglie-wave interferometer can be modulated by the mechanical momentum of the electrons. This momentum may be provided indirectly by current flow or by direct mechanical rotation of the interferometer. For the latter case it can be shown that the maximum supercurrent through the interferometer for an angular rotation rate Ω varies as $|\cos\pi[(2m_e/h)(2\pi r^2)\Omega]|$ assuming the interferometer is circular with a radius r and that any external magnetic field remains constant (Zimmerman and Mercereau, 1965). [For a recent review, see Mercereau (1969).] A measurement of the maximum supercurrent as a function of rotation rate thus yields a direct measurement of $h/2m_e$ or λ_C. Here again the principal accuracy-determining factor is the measurement of the interferometer dimensions. Zimmerman and Mercereau (1965) have demonstrated the effect and, more recently, a 0.05% measurement has been reported by Simmonds, Parker, and Nisenoff (1968). A 10-ppm measurement appears quite feasible, and, as noted above, would bring the accuracy into the useful range.

Here again, the problem of measuring interferometer dimensions to high accuracy can be eliminated by combining the flux-quantization experiments with the rotating-superconductor experiment in such a way as to eliminate the interferometer area. If the rotating-superconductor experiment is combined with the second flux-quantum experiment, i.e., the one in which the field is measured in terms of the proton precession frequency, we find

$$2\mu_p'/\mu_B = \omega_{p0}'/\Omega_0,$$

where Ω_0 is the rotation frequency increment which causes the circulation in the loop to change by exactly one quantum, i.e., by $h/2m_e$. However, since μ_p'/μ_B is already known with an accuracy of 0.07 ppm from the microwave absorption experiment of Lambe, it seems unlikely that the superconductivity experiments will be able to contribute significantly to our knowledge of this quantity.

If instead we combine the rotating-superconductor experiment with the first flux-quantum experiment, i.e., the one in which the field is calculated from the dimensions of a precision solenoid, we obtain

$$(\alpha^{-1})^{-3}e^{-1}K^1N^0 = 2\mu_0 R_\infty[\Omega_0/(B_0)_{\mathrm{NBS}}],$$

where the field increment B_0 is to be expressed in the units in which it is measured, $\mathrm{tesla}_{\mathrm{NBS}}$. This observational equation is essentially identical to that for γ_p as measured by the low-field method [Eq. (69)], and thus these two superconducting experiments together are equivalent to measuring γ_p. While this method of determining γ_p does not bypass the problem of accurately measuring the dimensions of the precision solenoid required in the usual low-field γ_p experiments, it does have the advantage of not requiring the measurement of decaying proton precession signals. Furthermore, it may have certain advantages when used in conjunction with precision superconducting solenoids because all of the required measurements can be carried out in the same cryogenic environment.

3. *Measurement of h/M_{He} using the Analog of the ac-Josephson Effect in Superfluid Helium*

Superfluid helium or He II is characterized by a highly correlated phase coherent quantum state of macroscopic scale which is analogous to the superconducting state (except that it is uncharged). Under suitable conditions, one would expect to be able to observe analogs of the Josephson effects in He II. Indeed, the analog of the ac Josephson effect has been observed by Richards and Anderson (1965) and by Khorana and Chandrasekhar (1967). In these experiments, two containers of superfluid helium are coupled together via a small orifice ($\sim$15 μm diameter). An ultrasonic transducer operating at a frequency ν($\sim$100 kHz) synchronizes the motion of quantized vortices across the orifice, resulting in a height difference Z between the two helium baths given by

$$M_{\mathrm{He}}gZ = nh\nu/n', \tag{160}$$

where M_{He} is the mass of the helium atom, g is the acceleration due to gravity, and n and n' are integers. The difference in bath height is therefore quantized and changes in a steplike manner. The left side of Eq. (160) will be recognized as the difference in chemical potential between the two baths and is exactly analogous to the term $2eV$ in the Josephson frequency–voltage relation. An experiment to determine h/M_{He} by this method would consist of measuring the heights of the steps, the transducer frequency ν, and the local value of g. The latter two quantities can be determined to $\sim$0.1 ppm. Typical step heights in present experiments are $\sim$1 mm, so that a 10-ppm measurement would require determining Z to about 100 Å. This is clearly a rather formidable task. However, it might be possible to take advantage of higher-order steps as in the Josephson-effect measurement of e/h in order to increase the effective step height to be measured. Increasing the synchronization frequency would also be helpful.

The observational equation for h/M_{He} follows from the fact that $M_{\mathrm{He}}^* = M_{\mathrm{He}}N$ and $h = (\mu_0 c/2)\alpha^{-1}e^2$. The result is

$$(\alpha^{-1})^1 e^2 K^0 N^1 = (2M_{\mathrm{He}}^* g/\mu_0 c)(h/gM_{\mathrm{He}}),$$

where h/gM_{He} is to be expressed in terms of the units in which it is measured, i.e., meters per hertz. (Note that $h/gM_{\mathrm{He}} \sim 10$ mm/MHz.) Clearly, such a measurement would be of great importance since it represents a new source of information about N. Using the variance–covariance matrix given in Table AII, we find that the uncertainty in h/M_{He} calculated from our recommended set of adjusted constants is 6.8 ppm. Consequently, for any measurement of this quantity to be useful, it should have an error no larger than 20 ppm. Unfortunately, many theoretical and experimental questions will have to be answered before such accuracy can be achieved. [A 1% measurement has recently been reported by Khorana and Douglass (1969).]

E. Recommendations for Reporting Experimental Results

During the preparation of this review, it became apparent that our task would have been much easier if certain procedures had been generally followed in the reporting of experimental results. More important, we would have been able to compare theory with experiment and to incorporate data into our least-squares adjustment with a great deal more confidence. We would therefore like to make several suggestions for presenting experimental results which, if followed, will facilitate the work of future reviewers and materially increase the reliability of direct comparisons between theory and experiment and the output values of any least-squares adjustment.

(1) It should be made quite clear exactly what steps were taken to arrive at the final result from the actual experimental observations. If at all possible, the data analysis should be presented in such a way that the reader can follow each important step of the analysis and duplicate the final reported results. This means that particular attention must be paid to the way in which presumably well-known constants and corrections of various kinds enter the calculations. The experimenter should ask himself, *"Have I provided enough information in a sufficiently clear manner so that 10 years hence my result can be updated in light of any new information or data which may become available?"*

(2) Every effort should be made to discuss in some detail each important systematic error believed to be present in the experiment and to list estimates of these errors in tabular form, preferably as 70% confidence-level estimates (i.e., standard deviations). We strongly deplore statements such as "the systematic error is not believed to exceed 0.5" without any further discussion of the individual systematic errors contributing to this value or whether the quoted value is meant to be a 50%, 70%, or 95% confidence-level estimate. We also urge that the statistical uncertainty be computed using

standard methods and presented in the form of a standard deviation. Since the uncertainty assigned a particular experimental datum plays a major role in determining whether it is in agreement or disagreement with theory and how much weight it will carry in a least-squares adjustment, correct error estimates are of the utmost importance.

(3) The numerical value of a datum and its uncertainty should be presented with a sufficient number of significant figures to prevent rounding errors when it is used. As a rule of thumb, the error should usually be stated to two-digit accuracy, and the datum itself should contain enough digits so that the quoted uncertainty corresponds to the last two figures.

(4) When the experiment is completed and the results published, the original laboratory notebooks and other pertinent information should be deposited in duplicate in secure places; use of a departmental safe or bank vault is not presumptuous, but prudent. No matter how detailed the published report may be, these items constitute the *only* really complete record of the results of the expenditure of man-years of effort and many thousands of dollars of someone's money. Their preservation may prevent much uncertainty and unnecessary future expenditures.

We recognize that these recommendations will be considered a nuisance by some workers, but it seems to us that anyone who can spend two or three (or perhaps five) years measuring a particular quantity can certainly spend an extra month or two writing up his results in a manner which will maximize their future usefulness. We also believe rather strongly that the responsibilities of an experimenter include assuring complete and open access to his results and procedures for a considerable time after the completion of his experiments.

Lest our theoretical colleagues feel slighted, may we suggest that when they write up the results of their calculations, they do the following: (1) Clearly summarize the final results in one place so that the reader doesn't have to search through innumerable pages. (2) Discuss the effects of the important approximations and/or those of uncalculated terms. (3) Present the results in such a way that corrections can easily be made at a later date for changes in any of the numerical constants used in the calculations. (4) Discuss any differences, even if they are apparently insignificant, between their results and those of other workers. (5) Keep in mind that experimentalists may wish to read and understand the paper and use its results.

VII. Notes Added in Proof

We note here several recent developments that have occurred or have come to our attention after this paper was accepted for publication.

1. μ_p/μ_B

Winkler and Kleppner (private communication) have noted that a slight error was made in converting from the UT2 to the A1 time standard in the work of Myint *et al.* (1966) (Sec. II.B.6) and that the value of $g_j(\mathrm{H})/g_p(\mathrm{H})$ determined by these workers should be increased by 0.06 ppm to

$$g_j(\mathrm{H})/g_p(\mathrm{H}) = 658.21053(20) \quad (0.30 \text{ ppm}).$$

This result should replace Eq. (21).

The measurements of Myint *et al.* have been continued at M.I.T. by Winkler, Walther, Myint, and Kleppner with several improvements in apparatus and technique [private communication, and to be published; see also *Physics of the One- and Two-Electron Atoms*, F. Bopp and H. Kleinpoppen, Eds., (North-Holland Publishing Co., Amsterdam, 1969)]. Their result is

$$g_j(\mathrm{H})/g_p(\mathrm{H}) = 658.21073(10) \quad (0.15 \text{ ppm}).$$

This value is to be regarded as preliminary, but it is not expected to change by more than one or two digits in the last place. However, the uncertainty may be reduced by a factor of 2 or more when certain small systematic effects due to imperfect averaging of the magnetic field by the H atoms are understood. This new preliminary value is 0.30 ppm larger than Myint *et al.*'s revised value and 0.36 ppm larger than their original value. The changes are not inconsistent with the assigned uncertainties.

To obtain a value of μ_p/μ_B from $g_j(\mathrm{H})/g_p(\mathrm{H})$, one must convert the latter quantity to its free-space value (see Sec. II.B.6). Recall that, to order α^2, the bound-state correction to the electron g factor just cancels the Lamb dimagnetic shielding correction to the proton g factor and $g_s/g_p = g_j(\mathrm{H})/g_p(\mathrm{H})$. Recently, Hegstrom [private communication, and to be published in Phys. Rev. **184,** 5 August (1969)] has extended the bound-state corrections to order α^3. He reports

$$g_j(\mathrm{H}) = g_s\left[1-\tfrac{1}{3}\alpha^2\left(\frac{M_p}{M_p+m_e}\right)^2 + \frac{\alpha^3}{12\pi}\left(\frac{M_p}{M_p+m_e}\right)^2 - \frac{26\alpha^3}{15\pi}\right]$$

$$= g_s(1-17.935 \text{ ppm}),$$

$$g_p(\mathrm{H}) = g_p\{1-\tfrac{1}{3}\alpha^2[M_p/(M_p+m_e)]^2\}$$

$$= g_p[1-17.731 \text{ ppm}],$$

$$g_j(\mathrm{H})/g_p(\mathrm{H}) = (g_s/g_p)[1-(33\alpha^3/20\pi)]$$

$$= (g_s/g_p)(1-0.204 \text{ ppm}).$$

The last term in the expression for $g_j(\mathrm{H})$ was actually calculated in 1955 by E. H. Lieb [Phil. Mag. **46,** 311 (1955)]. It is by far the dominant α^3 correction, being equal to 0.214 ppm. However, it has not yet been verified by Hegstrom, and additional terms of order $\alpha^2 m_e/M_p$ may also exist. [These corrections to $g_j(\mathrm{H})$ and $g_p(\mathrm{H})$ should be compared with the correction $(1-\alpha^2/3) = (1-17.750$ ppm$)$ usually assumed.] Applying the new corrections to Winkler *et al.*'s measurement of $g_j(\mathrm{H})/g_p(\mathrm{H})$ and proceeding as in Sec. II.B.6, we find

$$\mu_p/\mu_B = 0.00152103178(23) \quad (0.15 \text{ ppm}).$$

This value is 0.57 ppm smaller than the value derived

from the earlier Myint *et al.* measurement and used in the present work, Eq. (22).

The only calculation in which μ_p/μ_B entered here was that for the hfs in the ground state of atomic hydrogen, Eq. (120). The uncertainty in various theoretical terms in this equation is 5.1 ppm, and therefore the 0.57-ppm change in μ_p/μ_B has negligible effects. For example, it implies a decrease in $\alpha_{\mathrm{Hhfs}}^{-1}$ of 0.28 ppm to 137.03587(35) (2.6 ppm). The value used in the text (Table XVI, item 20 and Table XXVIII, line 5) is 137.03591(35) (2.6 ppm). Clearly our final set of recommended constants would remain essentially unchanged if the newer value were used instead.

The Lieb–Hegstrom corrections also affect the value of μ_p'/μ_B derived from the Lambe measurement of $g_j(\mathrm{H})/g_p(\mathrm{H_2O})$. Again following the procedure of Sec. II.B.6, we find

$$\mu_p'/\mu_B = 0.00152099284(10) \quad (0.066 \text{ ppm}).$$

This is 0.184 ppm less than the value used in the text, Eq. (26). The principal places in the present work where μ_p'/μ_B appeared were in the observational equations for γ_p', Eqs. (69) and (70), and for μ_p'/μ_n, Eq. (82). However, the change is so small compared with the uncertainties of the stochastic input data which enter these and the other observational equations that it can be ignored. The effect on the adjusted values of the constants is entirely negligible; the change in $\alpha_{\mathrm{WQED}}^{-1}$ would be about -0.09 ppm.

We note that the value of the diamagnetic shielding correction for protons in H_2O (spherical sample) obtained by combining the new Winkler *et al.* measurement of $g_j(\mathrm{H})/g_p(\mathrm{H})$ with the Lambe value of $g_j(\mathrm{H})/g_p(\mathrm{H_2O})$ is $\sigma(\mathrm{H_2O}) = 25.60 \pm 0.17$ ppm. This should be compared with the value $\sigma(\mathrm{H_2O}) = 26.0 \pm 0.3$ ppm calculated in the text from the Myint *et al.* result (Sec. II.B.6).

2. 2e/h

Petley and Morris at NPL [private communication, and Phys. Letters **29A,** 289 (1969)] have completed a measurement of $2e/h$ using the ac Josephson effect in solder–drop junctions. These authors used the method of microwave-induced dc current steps and worked at a frequency of about 36.8 GHz. Their result in terms of the NPL volt as maintained after 1 January 1969 is

$$2e/h = 4.8359393(100) \times 10^{14}\ \mathrm{Hz}/\mathrm{V}_{\mathrm{NPL}}\ (2.2\ \mathrm{ppm}).$$

In order to compare this value with that obtained by the present authors and A. Denenstein, Eq. (31), (Parker *et al.*, 1969), we note that the post 1 January 1969 NPL as-maintained volt has been obtained from the pre 1 January 1969 NPL volt by a 13-ppm downwards adjustment of the latter (P. Vigoureux, private communication), and that the 1967 comparisons of as-maintained volts (Table I) implies that $\mathrm{V}_{\mathrm{NPL}}(67)/\mathrm{V}_{\mathrm{NBS}}(67) = 1.0000052$. If we assume that this relation still holds at the present time, the Petley and Morris result in terms of the NBS volt as maintained prior to 1 January 1969 becomes

$$2e/h = 4.835977(10) \times 10^{14}\ \mathrm{Hz}/\mathrm{V}_{\mathrm{NBS}}\ (2.2\ \mathrm{ppm}).$$

This value exceeds Eq. (31) by only 0.2 ± 3.3 ppm. The excellent agreement is reassuring in view of the major role the ac Josephson effect value of $2e/h$ plays in the present work. Note that if this new value of $2e/h$ were in fact included in the present adjustments along with our own, the adjusted values of the constants would remain essentially unchanged but would have somewhat smaller uncertainties.

Finnegan, Denenstein, Langenberg, McMenamin, Novoseller, and Cheng (private communication, and to be published) have shown with the aid of a HCN gas laser that $2e/h$ is independent of the frequency of the incident radiation used to induce current steps for frequencies up to 891 GHz. Parker *et al.* (1969) had previously demonstrated frequency independence up to 70 GHz.

McCumber (private communication, and to be published) has reconsidered the frequency pulling effects calculated by Stephen (1968) and by Scully and Lee (1969) (Sec. II.C.1). He finds their conclusions concerning the existence of such effects to be incorrect due to improper use of the electrostatic potential for the electrochemical potential in the Josephson frequency–voltage relation.

3. μ_e/μ_B, a_e, *and* a_μ

Mignaco and Remiddi [Nuovo Cimento **60A,** 519 (1969)] have calculated the contribution of fourth-order vacuum polarization to the sixth-order radiative correction to the magnetic moment of the electron. They find an additional contribution of $0.055(\alpha/\pi)^3$, so that A_3 (see Secs. II.B.5 and IV.A.1) becomes 0.185. However, this increase is well within the ± 0.2 error we have assigned the Parsons value ($A_3=0.13$) used in the text. The change in μ_e/μ_B [Eq. (20)] implied by the new term is only $+0.0007$ ppm, an increase of less than one digit in the last place. The change in a_e [Eq. (104)] is $+0.6$ ppm or $+7$ in the last place. This is about 1/40 of the uncertainty in the experimental value of a_e and therefore does not significantly affect our comparison of a_e(exptl) with a_e(theory) (Sec. II.B.5). A similar situation obtains for a_μ; Eq. (114) becomes $a_\mu^{(6)}=(3.00\pm 0.35)(\alpha/\pi)^3$ and the theoretical value of a_μ, Eq. (118), increases by 0.6 ppm, less than one digit in the last place. We also note that Mignaco and Remiddi did not include the fourth-order vacuum polarization correction to $a_e^{(6)}$ from light by light scattering. This contribution is presently being calculated by S. J. Brodsky (private communication).

Bailey and Picasso (private communication, to be published in *Progress in Nuclear Physics*) note a private communication from Bailey, Petermann, and de Rafael pointing out that by using experimental knowledge of the $\pi^+\pi^-$ system near the ρ mass, Terazawa's upper limit to the contribution of all hadrons to a_μ [see Eq. (116)] can be reduced by roughly a factor of 5. Thus, Eq. (117) should perhaps read something like $[8.1(+3 \text{ or } -0.5)](\alpha/\pi^3)$, and Eq. (118), $a_\mu=$ 0.001165643(+38 or −7)(+32 or −6 ppm). However, since the uncertainty in the experimental value of a_μ is so much larger than the uncertainty in the theoretical value, our comparisons between a_μ(exptl) and a_μ(theory) remain essentially unchanged (see Sec. IV.A.2).

Gräff, Klempt, and Werth [Z. Physik **222,** 201 (1969)] have continued the measurements of Gräff *et al.* (1968) (Sec. IV.A.1) but have measured a_e directly. This was done by inducing electron spin-flips with an rf field of frequency $a_e\omega_e$ and separately measuring the electron cyclotron frequency ω_e. Their result is $a_e=0.00115966(30)$ (260 ppm). Although this result exceeds the theoretical value [Eq. (104)] by only 19 ppm, its 260-ppm uncertainty also makes it consistent with Wilkinson and Crane's value as revised by Rich, Eq. (105). Thus it contributes little to the resolution of the present discrepancy between a_e(theory) and a_e(exptl). However, this was a preliminary measurement and is likely to improve with time. If the uncertainty can be reduced by a factor of 10 or 20, it clearly will be of great importance.

4. μ_p/μ_n

Correspondence and discussions with P. L. Bender, D. Fystrom, B. W. Petley, and H. Sommer have led us to examine the original laboratory data books pertaining to the 1950 μ_p/μ_n experiment of Sommer, Thomas, and Hipple. On the basis of this examination, we have tentatively concluded that the uncertainty assigned by Sommer *et al.* to their result and described by them as "several times the estimated probable error" may more nearly represent the standard-deviation error. This implies that the uncertainty we have assigned their result in this paper may be too small by a factor of 2 or more (see Sec. II.C.5). This further beclouds the already confused state of the μ_p/μ_n data and reinforces our statement in Sec. VI.C that the major problem of the present work is the discrepancies among these data. New μ_p/μ_n experiments are currently under way in several laboratories and hopefully the results of these experiments will resolve this question. We believe that detailed reconsideration of the μ_p/μ_n situation should await these results. We do note that if we were simply to expand the uncertainty we have assigned to the Sommer *et al.* value of μ_p/μ_n, the resulting changes in our final output values for the constants would be small fractions of their standard deviations. (See also the discussion in Sec. III.B.2, and Table XVIII.)

5. *Fine Structure*

(a) $\mathcal{S}$: Robiscoe and Shyn (private communication; to be published) have found theoretical and experimental evidence which indicates that the effective velocity distribution for metastable hydrogen atoms produced in a typical atomic-beam apparatus (including types used for many fine-structure measurements) may be significantly altered from a Maxwellian distribution due to recoil and beam collimation effects. They find the velocity distribution may be more closely characterized by a function of the form $U^n \exp(-U^2)$ with $n \simeq 4$ rather than by this same function with $n \simeq 2$ as is usually assumed. (U is the atom velocity normalized to its thermal velocity.) Robiscoe and Shyn have considered the effects of this new velocity distribution on the Robiscoe and the Triebwasser, Dayhoff, and Lamb measurements of the Lamb shift in the $n=2$ level of atomic hydrogen. This was done by recalculating with the new distribution the velocity-dependent corrections applied in these experiments. For the Robiscoe measurements, the only velocity-dependent correction is that for motional field asymmetry as discussed by Robiscoe (1968). The result is that both the H(538) and H(605) values of $\mathcal{S}_H$ [see Eq. (141)] are increased by 0.040 MHz (38 ppm), about $\frac{2}{3}$ of the 0.063-MHz (60-ppm) standard-deviation uncertainty we have assigned. Robiscoe and Shyn believe the H(605) corrections are not as well specified as the H(538) corrections and therefore recommend a relative weighting of 2:1 for the two values. Thus, the average value given in Eq. (141) becomes

$$\mathcal{S}_{\rm H} = 1057.896 \pm 0.063 \text{ MHz } (60 \text{ ppm}).$$

This revised value exceeds theory by 4.8 standard deviations; the former value exceeded theory by 4.3 standard deviations [see Eq. (144)].

For the Triebwasser *et al.* measurements of $\mathcal{S}_{\rm H}$ the major correction affected is that for the Stark effect. Robiscoe and Shyn find these corrections to be increased by a factor of about 5/3. This results in a surprisingly large increase in $\mathcal{S}_{\rm H}(\alpha e)$ of 0.222 MHz, and a decrease in $\mathcal{S}_{\rm H}(\alpha f)$ of 0.049 MHz. Thus Eq. (140) becomes (in megahertz)

$$\text{H}(\alpha e): \quad \mathcal{S}_{\rm H} = 1057.960 \pm 0.095,$$

$$\text{H}(\alpha f): \quad \mathcal{S}_{\rm H} = 1057.758 \pm 0.089,$$

$$\text{Average:} \quad \mathcal{S}_{\rm H} = 1057.859 \pm 0.063 \ (60 \text{ ppm}).$$

The important thing to note is that the values obtained from the two transitions now differ by 0.202 MHz. This is rather larger than one would expect from the scatter in the data. (Recall that the quoted error for each transition is the average deviation of the several separate runs of which it is the mean.) Indeed, the difference is roughly three times the standard deviation of the difference. It must be concluded that the new velocity-distribution correction decreases the confidence which can be placed in the Triebwasser *et al.* results. On the other hand, it is interesting to note that the new Triebwasser *et al.* average value is only 0.037 MHz less than the new Robiscoe average value, whereas previously the difference between the two measurements was 0.083 MHz.

Cosens (private communication) has also corrected his measurement of $\mathcal{S}$ in the $n=2$ level of atomic deuterium for the new velocity distribution. He finds that both the D(574) and D(584) values of $\mathcal{S}_{\rm D}$ should be increased by 0.038 MHz (36 ppm), about $\frac{2}{3}$ of the 0.064-MHz (60-ppm) assigned error [see Eq. (143)]. Thus the average value given in Eq. (143) becomes

$$\mathcal{S}_{\rm D} = 1059.282 \pm 0.064 \text{ MHz } (60 \text{ ppm}).$$

This result now exceeds that predicted by theory by 5.8 standard deviations; formerly it exceeded theory by 5.3 standard deviations [see Eq. (144).]

In summary, the new velocity distribution correction, if valid, slightly worsens the already poor agreement between theory and experiment for the Lamb shift, and casts further doubts on the existing Lamb-shift measurements themselves.

(b) $\Delta E - \mathcal{S}$, ΔE, *and* α: If we assume in the light of the above discussion of the Lamb shift that the most reliable experimental value of $\mathcal{S}_H$ now presently available is the newly revised value of Robiscoe, then the most reliable value of ΔE_H which can be derived from the Kaufman, Lamb, Lea, and Leventhal value of $\Delta E_H - \mathcal{S}_H$ [see Eqs. (154) and (156)] is

$$(\Delta E_H)_{KLLL,R} = 10969.274 \pm 0.068 \text{ MHz } (6.2 \text{ ppm}).$$

(Note that the new velocity-distribution correction does not apply to the Kaufman *et al.* experiment.) This result now exceeds the theoretical value calculated using α_{WQED} by 0.247 ± 0.080 MHz (23 ± 7.3 ppm), 3.1 times the standard deviation of the difference. Previously, when the average of the Triebwasser *et al.* and Robiscoe results was used as the best experimental value for $\mathcal{S}_H$, the discrepancy was 2.5 standard deviations [see Eq. (157)]. The value of α^{-1} implied by the above value of ΔE_H is 137.03454(43) (3.1 ppm), 3.8 ppm less than $\alpha^{-1} = 137.03505(32)$ (2.4 ppm) as obtained in the text using the average value of $\mathcal{S}_H$ (see Table XVI, item 22a, and Table XXVIII, line 16c). It is also 11 ppm less than α_{WQED}^{-1}, 3.1 times the 3.6-ppm standard deviation of the difference.

There are two newly completed measurements of $\Delta E_H - \mathcal{S}_H$ which disagree with that of Kaufman *et al.* but which give values of ΔE_H in better agreement with theory. Shyn, Williams, Robiscoe, and Rebane (SWRR) [private communication; Phys. Rev. Letters **22,** 1273 (1969); erratum **23,** 62 (1969)] have measured $\Delta E_H - \mathcal{S}_H$ in a manner similar to that used by Triebwasser *et al.* (1953), but with two important differences.

First, the beam of metastables follows a trajectory parallel to the dc magnetic field, thereby reducing Stark quenching of the atoms due to motional electrical fields. Such fields cause asymmetries in the resonance lines and, if large, require accurate knowledge of the velocity distribution in the beam for correction. Second, the observed transitions are between individual hyperfine levels as in the Robiscoe (1965) and Robiscoe and Cosens (1966a) experiments. The new measurements were made using both the β^+b^+ and β^+d^+ transitions at about 0.079 T and 11.8 GHz, and 0.0825 T and 8.8 GHz, respectively (see Fig. 5). No data were obtained from the β^-b^- and β^-d^- transitions because the method employed to produce these states (Majorana transitions) gives them a complex oscillatory velocity distribution. Using a perturbation calculation to extrapolate along the Zeeman lines to zero field, Shyn *et al.* find (in megahertz)

$$\beta^+b^+: \quad \Delta E_{\mathrm{H}} - \mathcal{S}_{\mathrm{H}} = 9911.256 \pm 0.079 \ (8.0\ \mathrm{ppm}),$$

$$\beta^+d^+: \quad \Delta E_{\mathrm{H}} - \mathcal{S}_{\mathrm{H}} = 9911.166 \pm 0.073 \ (7.4\ \mathrm{ppm}),$$

$$\text{Average:} \quad \Delta E_{\mathrm{H}} - \mathcal{S}_{\mathrm{H}} = 9911.213 \pm 0.058 \ (5.9\ \mathrm{ppm}).$$

The quoted errors are the standard deviations of the means and are statistical only. Systematic uncertainties were estimated at less than 1 ppm. The average value was obtained by combining all of the data from the two transitions and then calculating the mean and standard deviation of the mean. A total of 60 line centers was obtained in four runs for the β^+b^+ transition and 55 line centers in five runs for the β^+d^+ transition. Corrections applied to the data were (β^+b^+ and β^+d^+ transitions, respectively): Stark matrix element variation, 4 ppm and 0 ppm; overlapping transitions, 1.7 ppm and 0.1 ppm; motional and stray fields, 1 ppm and 1 ppm. A $U^4 \exp(-U^2)$ velocity distribution was assumed. We note that this new measurement is 0.164 ± 0.063 MHz (17 ± 6.4 ppm) less than that of Kaufman *et al.*

The difference is 2.6 times the standard deviation of the difference and has a 1% probability of occurring by chance. The two measurements are therefore somewhat inconsistent.

The theoretical value of $\Delta E_{\mathrm{H}} - \mathcal{S}_{\mathrm{H}}$ calculated using $\alpha_{\mathrm{WQED}}^{-1}$ (see Table XXVII) exceeds the result of Shyn *et al.* by 0.254 ± 0.077 MHz (26 ± 7.8 ppm), 3.3 times the standard deviation of the difference. Clearly, the agreement is poor. This may also be seen by comparing $\alpha_{\mathrm{WQED}}^{-1}$ and the value $\alpha^{-1} = 137.03791(44)$ (3.3 ppm), derived directly from the Shyn *et al.* value of $\Delta E_{\mathrm{H}} - \mathcal{S}_{\mathrm{H}}$. On the other hand, if we combine their result with the newly revised Robiscoe value of $\mathcal{S}_{\mathrm{H}}$, we find

$$(\Delta E_{\mathrm{H}})_{\mathrm{SWRR,R}} = 10969.109 \pm 0.086 \text{ MHz } (7.8 \text{ ppm}).$$

This value of ΔE_{H} exceeds the theoretical value by 0.083 ± 0.096 MHz (7.6 ± 8.7 ppm) and implies a value for α^{-1} of 137.03556(53) (3.9 ppm). Thus the agreement with theory is reasonable, in marked contrast to the situation which exists when a similar comparison is made using the Kaufman *et al.* result. However, the Kaufman *et al.* value of $\Delta E_{\mathrm{H}} - \mathcal{S}_{\mathrm{H}}$ is in reasonable agreement with the theoretical value of $\Delta E_{\mathrm{H}} - \mathcal{S}_{\mathrm{H}}$ [see Eq. (155)] while the Shyn *et al.* value is not. The reason for this, of course, is the discrepancy which exists between the Robiscoe value of $\mathcal{S}_{\mathrm{H}}$ and the theoretical value.

Vorburger and Cosens (VC) [private communication; Bull. Am. Phys. Soc. **14,** 525 (1969); to be published] have measured $\Delta E_{\mathrm{H}} - \mathcal{S}_{\mathrm{H}}$ in the same way as did Shyn *et al.* except that they worked at magnetic fields only half as large, i.e., about 0.04 T. At these fields, the β–b transition occurs at about 10.8 GHz and the β–d transition at about 9.4 GHz. Because they worked at lower fields, Vorburger and Cosens were troubled with overlapping αa and αc resonances for the β–b and β–d transitions, respectively. To alleviate this problem, they quenched the β-state atoms in the beam, thereby determining the contribution of the α transitions to the resonance signal. This was then subtracted from the resonance signal obtained with the unquenched beam. Another difference between their experiment and that of Shyn *et al.* is that they carried out measurements on the β^-b^- and β^-d^- transitions as well as on the β^+b^+ and β^+d^+ transitions. Vorburger and Cosen's results are (in megahertz)

$$\beta^-b^-:\quad \Delta E_{\mathrm{H}} - \mathcal{S}_{\mathrm{H}} = 9911.272 \pm 0.092 \text{ (9.3 ppm)},$$

$$\beta^+b^+:\quad \Delta E_{\mathrm{H}} - \mathcal{S}_{\mathrm{H}} = 9911.117 \pm 0.085 \text{ (8.6 ppm)},$$

$$\beta^-d^-:\quad \Delta E_{\mathrm{H}} - \mathcal{S}_{\mathrm{H}} = 9911.196 \pm 0.076 \text{ (7.7 ppm)},$$

$$\beta^+d^+:\quad \Delta E_{\mathrm{H}} - \mathcal{S}_{\mathrm{H}} = 9911.084 \pm 0.084 \text{ (8.5 ppm)},$$

$$\text{wt av}:\quad \Delta E_{\mathrm{H}} - \mathcal{S}_{\mathrm{H}} = 9911.165 \pm 0.042 \text{ (4.3 ppm)}.$$

About 50 measurements of $\Delta E_{\mathrm{H}} - \mathcal{S}_{\mathrm{H}}$ were made for each transition. The quoted errors are standard deviations of the means; about 75% of each error is statistical. The remainder is due to the uncertainties in the applied corrections, which ranged from about 0 to 0.05 MHz. A $U^4 \exp(-U^2)$ velocity distribution was assumed and a Brodsky and Parsons (1967) type calculation was used to obtain $\Delta E_{\mathrm{H}} - \mathcal{S}_{\mathrm{H}}$ from the observed resonance centers.

In comparing the Vorburger and Cosens result with that of Kaufman *et al.*, we find the latter exceeds the former by 0.212±0.049 MHz (21±5.0 ppm). The difference is 4.3 times the standard deviation of the difference and has a probability of occurring by chance of less than 0.01%. The two measurements are clearly in significant disagreement. By contrast, the Vorburger and Cosens result is only 0.048±0.072 MHz (4.8±7.2 ppm) less than that of Shyn *et al.* The Kaufman *et al.* value would thus seem to be suspect.

The theoretical value of $\Delta E_{\mathrm{H}} - \mathcal{S}_{\mathrm{H}}$ calculated using $\alpha_{\mathrm{WQED}}^{-1}$ (see Table XXVII) exceeds the result of Vorburger and Cosens by 0.302±0.066 MHz (31±6.7 ppm), 4.6 times the standard deviation of the difference. Clearly, the agreement is poor. This may also be seen by comparing $\alpha_{\mathrm{WQED}}^{-1}$ and the value $\alpha^{-1}=137.03825(35)$ (2.5 ppm) derived directly from the Vorburger and Cosens value of $\Delta E_{\mathrm{H}}-\mathcal{S}_{\mathrm{H}}$. On the other hand if we combine their result with the newly revised Robiscoe value of $\mathcal{S}_{\mathrm{H}}$, we find

$$(\Delta E_{\mathrm{H}})_{\mathrm{VC,R}}=10969.061\pm0.076 \text{ MHz } (6.9 \text{ ppm}).$$

This value of ΔE_{H} exceeds the theoretical value by only 0.035 ± 0.087 MHz (3.2 ± 7.9 ppm) and implies a value for α^{-1} of 137.03586(48) (3.5 ppm). The agreement with theory is rather good.

We shall leave further analysis and discussion of the $\Delta E_{\mathrm{H}}-\mathcal{S}_{\mathrm{H}}$ and Lamb shift data for the future. However, we would like to conclude with several general remarks. First, the scatter in the measurements of $\Delta E_{\mathrm{H}}-\mathcal{S}_{\mathrm{H}}$ made by different workers and on different transitions is, we believe, strong evidence for the existence in these experiments of systematic effects which are not really well understood. For example, the difference between the Kaufman *et al.* αb measurement [Eq. (154)] and the Vorburger and Cosens $\beta^{+}d^{+}$ measurement is, astonishingly, 0.323 MHz or 33 ppm. Second, we believe the present experimental values of the Lamb shift are sufficiently uncertain that a reliable value of ΔE_{H} and hence α^{-1} cannot now be derived from the $\Delta E_{\mathrm{H}}-\mathcal{S}_{\mathrm{H}}$ measurements *even* if the latter are considered to be trustworthy. Our conclusion is that the new work strongly supports our decision not to use the fine-structure data in the final adjustment to obtain our best or recommended set of constants. Furthermore, the large discrepancies between the two new measurements of $\Delta E_{\mathrm{H}}-\mathcal{S}_{\mathrm{H}}$ and the theoretical value provide additional support for our conclusion in Sec. IV.A that the disagreement between $\mathcal{S}$(exptl) and $\mathcal{S}$(theory) is probably real and that the theory of the Lamb shift needs revision.

Appendix: VARIANCE–COVARIANCE MATRICES

As noted in Sec. III.B.1, the output values of a least-squares adjustment have correlated errors. As a result, when numerical values of other quantities are calculated from formulas containing two or more adjusted constants, the general law of error propagation must be used [Bearden and Thomsen (1957); see also Cohen and DuMond (1965); Cohen, Crowe and DuMond (1957)]. This requires not only the variances or squares of the standard deviations of the adjusted constants, but also the covariances associated with pairs of constants. The variances are given by the diagonal elements of the error matrix $\mathbf{G}^{-1}$ (see Sec. III.B.1) and the covariances by the off-diagonal elements. For example, the variance of the ith adjusted constant, $v_{ii}=\epsilon_i^2$, is equal to $(\mathbf{G}^{-1})_{ii}$, while the covariance of the ith and jth adjusted constants, v_{ij}, is equal to $(\mathbf{G}^{-1})_{ij}$. Note that the error matrix is symmetrical so that $v_{ij}=v_{ji}$.

If a quantity Q depends on N statistically correlated quantities x_i according to the equation

$$Q=Q(x_1, x_2\cdots x_N), \qquad \text{(A1)}$$

then the variance in Q, ϵ_Q^2, is given by

$$\epsilon_Q^2=\sum_{i=1}^{N}\sum_{j=1}^{N}\frac{\partial Q}{\partial x_i}\frac{\partial Q}{\partial x_j}v_{ij}, \qquad \text{(A2)}$$

where v_{ij} is the covariance of x_i and x_j. For most cases of interest involving the fundamental constants, Q will depend on a number of constants Z_j in the following way:

$$Q=q\prod_{j=1}^{J}Z_j^{Y_{Qj}} \qquad \text{(A3)}$$

TABLE AI. Expanded variance–covariance matrix for the adjustment used to obtain the WQED values of the constants. The variances and covariances, which are on and above the main diagonal, are in (parts per million)^2. The correlation coefficients are in boldface type.

	α^{-1}	e	K	N	Λ	F	h	m_e
α^{-1}	3.670	−8.022	−1.468	5.983	−1.994	−2.039	−12.375	−5.035
e	**−0.832**	25.347	8.670	−20.419	6.806	4.928	42.671	26.627
K	**−0.285**	**0.642**	7.203	−5.781	1.927	2.889	15.873	12.938
N	**0.457**	**−0.594**	**−0.315**	46.676	−15.559	26.257	−34.854	−22.888
Λ	**−0.195**	**0.253**	**0.135**	**−0.427**	28.503	−8.752	11.618	7.629
F	**−0.191**	**0.175**	**0.193**	**0.688**	**−0.294**	31.185	7.817	3.739
h	**−0.756**	**0.992**	**0.692**	**−0.597**	**0.255**	**0.164**	72.968	48.219
m_e	**−0.426**	**0.856**	**0.781**	**−0.542**	**0.231**	**0.108**	**0.914**	38.150

(q is just a numerical factor). Thus, Eq. (A2) becomes

$$\epsilon_Q^2 = \sum_{i=1}^{J} \sum_{j=1}^{J} Y_{Qi} Y_{Qj} v_{ij}, \tag{A4}$$

where the v_{ij} are to be expressed in (parts per million)^2. Equation (A2) may also be written in terms of correlation coefficients defined by $r_{ij} \equiv v_{ij}/(v_{ii}v_{jj})^{1/2} \equiv v_{ij}/\epsilon_i\epsilon_j$ (note that $r_{ii}=1$):

$$\epsilon_Q^2 = \sum_{i=1}^{N} \left(\frac{\partial Q}{\partial x_i}\right)^2 \epsilon_i^2 + \sum_{i\neq j}^{N} r_{ij}\epsilon_i\epsilon_j \frac{\partial Q}{\partial x_i}\frac{\partial Q}{\partial x_j}. \tag{A5}$$

Similarly for Eq. (A4):

$$\epsilon_Q^2 = \sum_{i=1}^{J} Y_{Qi}^2 \epsilon_i^2 + \sum_{i\neq j}^{J} r_{ij}\epsilon_i\epsilon_j Y_{Qi} Y_{Qj}, \tag{A6}$$

where the ϵ_i are to be expressed in parts per million. Clearly, if $r_{ij}=0$ for $i\neq j$ (i.e., no correlation), then Eqs. (A5) and (A6) reduce to the usual law of error propagation for uncorrelated quantities.

In Table AI we give the variance–covariance matrix and correlation coefficients for the least-squares adjustment carried out in Sec. III.C to obtain the WQED values of the constants. In Table AII we give the similar matrix for the final adjustment carried out in Sec. V.B to obtain our recommended set of constants. The variances and covariances are on and above the major diagonal and are given in (parts per million)^2. The correlation coefficients are in boldface type (recall $r_{ii}=1$). For the convenience of the reader, we have expanded this matrix to include Λ, F, h, and m_e in addition to α^{-1}, e, K, and N, the constants actually used in our adjustment. Such an expansion follows from the fact that the covariance of two quantities Q and R, where Q is as given in Eq. (A1) and R is given by $R=R(x_1, x_2 \cdots x_N)$, is simply

$$v_{QR} = \sum_{i=1}^{N} \sum_{j=1}^{N} \frac{\partial Q}{\partial x_i}\frac{\partial R}{\partial x_j} v_{ij}. \tag{A7}$$

TABLE AII. Expanded variance–covariance matrix for the final adjustment from which the recommended set of constants was obtained. The variances and covariances, which are on and above the main diagonal, are in (parts per million)². The correlation coefficients are in boldface type.

	α^{-1}	e	K	N	Λ	F	h	m_e
α^{-1}	2.352	−5.141	−0.940	3.834	−1.278	−1.299	−7.930	−3.227
e	**−0.768**	19.049	7.518	−15.721	5.241	3.327	32.956	22.674
K	**−0.232**	**0.651**	6.992	−4.922	1.641	2.596	14.096	12.215
N	**0.381**	**−0.548**	**−0.283**	43.173	−14.391	27.451	−27.609	−19.940
Λ	**−0.157**	**0.227**	**0.117**	**−0.413**	28.114	−9.150	9.203	6.647
F	**−0.153**	**0.137**	**0.177**	**0.753**	**−0.311**	30.778	5.355	2.734
h	**−0.679**	**0.992**	**0.700**	**−0.552**	**0.228**	**0.127**	57.982	42.121
m_e	**−0.352**	**0.870**	**0.774**	**−0.508**	**0.210**	**0.083**	**0.926**	35.668

If Q is as given in Eq. (A3) and R by

$$R = r\prod_{j=1}^{J} Z_j^{Y_{Rj}},$$

Eq. (A7) becomes

$$v_{QR} = \sum_{i=1}^{J}\sum_{j=1}^{J} Y_{Qi}Y_{Rj}v_{ij},$$

where the v_{ij} are to be expressed in (parts per million)^2.

As an example of the use of these matrices, we compute the quantity h/e and its uncertainty. From the definition $\alpha = (\mu_0 c^2/4\pi)e^2(\hbar c)^{-1}$, we obtain

$$h/e = (2/\mu_0 c)(\alpha^{-1})^{+1}e^{+1}. \qquad \text{(A8)}$$

Using the final adjusted values for α^{-1} and e, Table XXXII, and our adopted value of c, Table XI, we find $h/e = 4.135708\times 10^{-15}$ J·sec C^{-1}. We calculate the uncertainty in h/e from Eq. (A4) and Table AII (c is assumed to be exactly known). Letting α^{-1} correspond to $j=1$ and e to $j=2$ gives

$$\epsilon_{h/e}^2 = Y_1Y_1v_{11} + Y_2Y_2v_{22} + 2Y_1Y_2v_{12}. \qquad \text{(A9)}$$

Comparing Eq. (A8) with (A3) yields $Y_1 = +1$ and $Y_2 = +1$. Thus we obtain from Eq. (A9) and Table AII

$$\epsilon_{h/e}^2 = [(+1)(+1)(2.352) + (+1)(+1)(19.049) + 2(+1)(+1)(-5.141)]\ \text{ppm}^2, \qquad \text{(A10)}$$

or $\epsilon_{h/e} = 3.33$ ppm. An alternate procedure would be to write $h/e = e^{-1}h^{+1}$, let h correspond to $j=3$, and find

$$\epsilon_{h/e}^2 = Y_2Y_2v_{22} + Y_3Y_3v_{33} + 2Y_2Y_3v_{23}$$

$$= [(-1)(-1)(19.049) + (+1)(+1)(57.982) + 2(-1)(+1)(32.956)]\ \text{ppm}^2,$$

which of course also yields $\epsilon_{h/e} = 3.33$ ppm.

References

Abramowitz, M., and I. A. Stegun, 1964, *Handbook of Mathematical Functions* (U. S. Government Printing Office, Washington, D. C.), Appl. Math. Ser. AMS-55, pp. 978–985.

Agaletskii, P. N., and K. N. Egorov, 1956, Izmeritel. Tekh. i Poverochnoe Delo No. 6, 29.

Anderson, H. L., C. K. Hargrove, E. P. Hincks, J. D. McAndrew, R. J. McKee, and D. Kessler, 1969, Phys. Rev. Letters **22,** 221.

Arnowitt, R., 1953, Phys. Rev. **92,** 1002.

Ayrton, W. E., T. Mather, and F. E. Smith, 1908, Phil. Trans. Roy. Soc. (London) **A207,** 463.

Bäcklin, E., 1935, Z. Physik **93,** 450.

Bailey, J., W. Bartl, G. von Bochmann, R. C. A. Brown, F. J. M. Farley, H. Jöstlein, E. Picasso, and R. W. Williams, 1968, Phys. Letters **28B,** 287.

——, W. Bartl, R. C. A. Brown, H. Jöstlein, S. van der Meer, E. Picasso, and F. J. M. Farley, 1967, Proceedings of the International Conference on Electron and Photon Interactions at High Energies, Stanford Linear Accelerator Center, Stanford, Calif., p. 48.

Barker, W. A., and F. N. Glover, 1955, Phys. Rev. **99,** 317.

Barrett, R. C., S. J. Brodsky, G. W. Erickson, and M. H. Goldhaber, 1968, Phys. Rev. **166,** 1589.

Bartlett, D. F., and E. A. Phillips, 1969, Bull. Am. Phys. Soc. **14,** 17.

Bearden, J. A., 1967, Rev. Mod. Phys. **39,** 78.

——, 1965a, Phys. Rev. **137,** B455.

——, 1965b, Phys. Rev. **137,** B181.

——, 1935, Phys. Rev. **48,** 385.

——, 1931, Phys. Rev. **37,** 1210.

——, and A. F. Burr, 1967, Rev. Mod. Phys. **39,** 125.

——, A. Henins, J. G. Marzolf, W. C. Sauder, and J. S. Thomsen, 1964, Phys. Rev. **135,** A899.

——, F. N. Huffman, and J. J. Spijkerman, 1964, Rev. Sci. Instr. **35,** 1681.

——, and J. S. Thomsen, 1959, Am. J. Phys. **27,** 569.

——, and J. S. Thomsen, 1957, Nuovo Cimento Suppl. **5,** 267.

——, and H. M. Watts, 1951, Phys. Rev. **81,** 73.

Bender, P. L., and R. L. Driscoll, 1958, IRE Trans. Instr. **I-7,** 176.

Bergstrand, E., 1957, Ann. Franc. Chronom. **2,** 97.

——, 1956, *Handbuch der Physik,* S. Flügge, Ed. (Springer–Verlag, Berlin), Vol. 24, p. 1.

——, 1950, Nature **165,** 405.

——, 1949, Nature **163,** 338.

Bethe, H. A., and E. E. Salpeter, 1957, *Quantum Mechanics of One- and Two-Electron Atoms* (Springer–Verlag, Berlin).
Bingham, G. M., 1963, Nuovo Cimento **27,** 1352.
Birge, R. T., 1932, Phys. Rev. **40,** 207.
Bjorken, J. D., 1966, Phys. Rev. **148,** 1467.
Bloch, F., and C. D. Jeffries, 1950, Phys. Rev. **80,** 305.
Bowcock, J. E., 1968, Z. Physik **211,** 400.
Boyne, H. S., and P. A. Franken, 1961, Phys. Rev. **123,** 242.
Brandenberger, J., 1968, Ph.D. thesis, Brown University (unpublished).
Brodsky, S. J., private communication.
——, and E. de Rafael, 1968, Phys. Rev. **168,** 1620.
——, and G. W. Erickson, 1966, Phys. Rev. **148,** 26.
——, and R. G. Parsons, 1967, Phys. Rev. **163,** 134.
——, and R. G. Parsons, 1968, Phys. Rev. **176,** 423, Erratum.
——, and J. R. Primack, 1968, Phys. Rev. **174,** 2071.
——, and J. R. Primack, Ann. Phys. (to be published).
——, and J. D. Sullivan, 1967, Phys. Rev. **156,** 1644.
Brown, R. A., and F. M. Pipkin, 1968, Phys. Rev. **174,** 48.
Burnett, T., and M. J. Levine, 1967, Phys. Letters **24B,** 467.
Cameron, A. E. and E. Wichers, 1962, J. Am. Chem. Soc. **84,** 4175.
Capptuller, H., 1964, *Proceedings of the Second International Conference on Nuclidic Masses,* W. H. Johnson, Jr., Ed. (Springer–Verlag, Vienna) p. 105.
——, 1961a, Z. Instrumentenk. **69,** 191.
——, 1961b, Z. Instrumentenk. **69,** 133.
——, private communication.
Carrico, J. P., E. Lipworth, P. G. Sandars, T. S. Stein, and M. C. Weisskopf, 1968, Phys. Rev. **174,** 125.
Charpak, G., F. J. M. Farley, R. L. Garwin, T. Muller, J. C. Sens, and A. Zichichi, 1965, Nuovo Cimento **37,** 1241.
Chu, D. Y., 1939, Phys. Rev. **55,** 175.
Clark, J. S., 1939, Phil. Trans. Roy Soc. (London) **A238,** 65.
Clarke, J., 1968, Phys. Rev. Letters **21,** 1566.
Cleland, W. E., J. M. Bailey, M. Eckhause, V. W. Hughes, R. M. Mobley, R. Prepost, and J. E. Rothberg, 1964, Phys. Rev. Letters **13,** 202.
Clothier, W. K., 1965, Metrologia **1,** 36.
Cochran, G. D., and P. A. Franken, 1968, Bull. Am. Phys. Soc. **13,** 1379.
Cohen, E. R., 1968, *Proceedings of the Third International Conference on Atomic Masses,* R. C. Barber, Ed. (University of Manitoba Press, Winnipeg), p. 478.
——, 1969, *Physics of the One- and Two-Electron Atoms,* F. Bopp and H. Kleinpoppen, Eds. (North-Holland Publishing Co., Amsterdam).
——, 1966, Nuovo Cimento Suppl. **I-4,** 839.
——, 1953, Rev. Mod. Phys. **25,** 709.
——, 1952, Phys. Rev. **88,** 353.
——, K. M. Crowe, and J. W. M. DuMond, 1957, *The Fundamental Constants of Physics* (Interscience Publishers, Inc., New York), p. 222.
——, and J. W. M. DuMond, 1965, Rev. Mod. Phys. **37,** 537.

——, J. W. M. DuMond, T. W. Layton, and J. S. Rollett, 1955, Rev. Mod. Phys. **27,** 363.
Coleman, C. D., W. R. Bozman, and W. F. Meggers, 1960, *Table of Wavenumbers Vol. I, 2000 Å to 7000 Å* (Natl. Bur. Std. Monograph 3).
Collington, D. J., A. N. Dellis, J. H. Sanders, and K. C. Turberfield, 1955, Phys. Rev. **99,** 1622.
Condon, E. U., and G. H. Shortley, 1957, *The Theory of Atomic Spectra* (Cambridge University Press, London), p. 134.
Cook, A. H., 1968, Contemp. Phys. **9,** 227.
——, 1967a, Phil. Trans. Roy. Soc. (London) **A261,** 211.
——, 1967b, Contemp. Phys. **8,** 251.
——, 1965, Metrologia **1,** 84.
——, private communication.
Cooper, A. S., 1965, Acta. Cryst. **18,** 1078.
Cosens, B. L., 1968, Phys. Rev. **173,** 49.
Craig, D. N., J. I. Hoffman, C. A. Law, and W. J. Hamer, 1960, J. Res. Natl. Bur. Std. **64A,** 381.
Crampton, S. B., D. Kleppner, and N. F. Ramsey, 1963, Phys. Rev. Letters **11,** 338.
——, H. G. Robinson, D. Kleppner, and N. F. Ramsey, 1966, Phys. Rev. **141,** 55.
Crane, P., and P. A. Thompson (private communication).
Crouch, E. A. C., and A. H. Turnbull, 1962, J. Chem. Soc. (London) 161.
Csillag, L., 1968, Acta Phys. Acad. Sci. Hung. **24,** 1.
——, 1966, Phys. Letters **20,** 645.
——, private communication.
Curtis, R. W., R. L. Driscoll, and C. L. Critchfield, 1942, J. Res. Natl. Bur. Std. **28,** 133.
Cutkosky, R. D., 1961, J. Res. Natl. Bur. Std. **65A,** 147.
——, private communication.
Dayhoff, E. S., S. Triebwasser, and W. E. Lamb, Jr., 1953, Phys. Rev. **89,** 106.
Deaver, Jr., B. S., and W. M. Fairbank, 1961, Phys. Rev. Letters **7,** 43.
Deutsch, M., and S. C. Brown, 1952, Phys. Rev. **85,** 1047.
Dickinson, W. C., 1951, Phys. Rev. **81,** 717.
Doll, R., and M. Näbauer, 1961, Phys. Rev. Letters **7,** 51.
Drell, S. D., 1969, *Proceedings of the First International Conference on Atomic Physics*, B. Bederson, V. W. Cohen, and F. M. J. Pichanick, Eds. (Plenum Press, Inc., New York), p. 53.
——, and H. R. Pagels, 1965, Phys. Rev. **140,** B397.
——, and J. D. Sullivan, 1967, Phys. Rev. **154,** 1477.
——, and J. D. Sullivan, 1965, Phys. Letters **19,** 516.
——, and J. S. Trefil (unpublished); see S. D. Drell, 1967, *Proceedings of the XIII International Conference on High Energy Physics* (University of California Press, Berkeley, Calif.), p. 85.
Drinkwater, J. W., O. Richardson, and W. E. Williams, 1940, Proc. Roy. Soc. (London) **A174,** 164.
Driscoll, R. L., 1964, Phys. Rev. **136,** A54.
——, 1958, J. Res. Natl. Bur. Std. **60,** 287.
——, private communication.
——, and P. L. Bender, 1958, Phys. Rev. Letters **1,** 413.

——, and R. D. Cutkosky, 1958, J. Res. Natl. Bur. Std. **60**, 297.
——, and P. T. Olsen, 1968, report to Comité Consultatif d'Electricité, Comité International des Poids et Mesures, 12th Session, October.
Dryden, H. L., 1942, J. Res. Natl. Bur. Std. **29**, 303.
DuMond, J. W. M., 1966, Z. Naturforsch. **21a**, 70.
——, and E. R. Cohen, 1956, Phys. Rev. **103**, 1583.
——, and E. R. Cohen, 1953, Rev. Mod. Phys. **25**, 691.
——, and E. R. Cohen, 1952, Am. Sci. **40**, 447.
——, and E. R. Cohen, 1951, Phys. Rev. **82**, 555.
Durand, L., 1962, Phys. Rev. **128**, 441.
Edlén, B., 1966, Arkiv Fysik **31**, 509.
——, and L. A. Svensson, 1965, Arkiv Fysik **28**, 427.
Elend, H. H., 1966a, Phys. Letters **20**, 682.
——, 1966b, Phys. Letters **21**, 720, Erratum.
Ensberg, E. S., and C. L. Morgan, 1968, Phys. Letters **28A**, 106.
Erickson, G. W., 1969, *Physics of the One- and Two-Electron Atoms*, F. Bopp and H. Kleinpoppen, Eds. (North-Holland Publishing Co., Amsterdam).
——, 1967, in 15 September Annual Report to the U. S. Atomic Energy Commission (UCD-CNL 88 AEC Research and Development Report); prepared under contract AT(11-1) Gen. 10, P.A. #15.
——, private communication.
——, and Henry H. T. Liu, 1968, AEC Research and Development Report UCD-CNL 81, July.
——, and D. R. Yennie, 1965a, Ann. Phys. (N. Y.) **35**, 271.
——, and D. R. Yennie, 1965b, Ann. Phys. (N. Y.) **35**, 447.
Faller, J. E., 1967, Science **158**, 60 (6 October).
——, 1963, Ph.D. thesis, Princeton University (unpublished).
——, and J. Hammond, private communication.
Farago, P. S., R. B. Gardiner, J. Muir, and A. G. A. Rae, 1963, Proc. Phys. Soc. (London) **82**, 493.
Farley, F. J. M., 1968a, Bull Am. Phys. Soc. **13**, 577.
——, 1968b, *Cargèse Lectures in Physics*, M. Levy, Ed. (Gordon and Breach, Science Publishers, Inc., New York) Vol. 2, p. 74.
——, private communication.
——, J. Bailey, R. C. A. Brown, M. Giesch, H. Jöstlein, S. van der Meer, E. Picasso, and M. Tannenbaum, 1966, Nuovo Cimento **45A**, 281.
Feinberg, G., and L. M. Lederman, 1963, Ann. Rev. Nucl. Sci. **13**, 431.
Fenster, S., R. Köberle, and Y. Nambu, 1965, Phys. Letters **19**, 513.
——, and Y. Nambu, 1965, Progr. Theoret. Phys. (Kyoto), Suppl. **1965**, 250.
Fil'kov, L. V., 1968, ZhETF Pis. Red. **7**, 352 [JETP Letters **7**, 275 (1968)].
Fisher, R. A., 1925, *Statistical Methods for Research Workers* (Oliver and Boyd, London) p. 55.
Fortson, E. N., F. G. Major, and H. G. Dehmelt, 1966, Phys. Rev. Letters **16**, 221.
Fried, H. M., and D. R. Yennie, 1960, Phys. Rev. Letters **4**, 583.
Froome, K. D., 1958, Proc. Roy. Soc. (London) **A247**, 109.

——, 1956, J. Brit. Inst. Radio Engrs. **16,** 497.
——, 1954, Proc. Roy. Soc. (London) **A223,** 195.
——, 1952, Proc. Roy. Soc. (London) **A213,** 123.
Galbraith, I. A., and R. B. Gardiner, 1968, J. Phys. A **1,** 194.
Gamow, G., 1968, Proc. Natl. Acad. Sci. U.S. **59,** 313.
Garcia, J. D., and J. E. Mack, 1965, J. Opt. Soc. Am. **55,** 654.
Goitein, M., J. R. Dunning, Jr., and R. Wilson, 1967, Phys. Rev. Letters **18,** 1018.
Gourdin, M., and E. de Rafael, 1969, Nucl. Phys. **10B,** 667.
Gräff, G., F. G. Major, R. W. H. Roeder, and G. Werth, 1968, Phys. Rev. Letters **21,** 340.
Grosse, H., 1967, Nachr. Karten-und-Vermessungwesen Ser. I **35,** 93.
Grotch, H., and D. R. Yennie, 1969, Rev. Mod. Phys. **41,** 350.
——, and D. R. Yennie, 1967, Z. Physik **202,** 425.
Guérin, F., 1967a, Nuovo Cimento **50A,** 1.
——, 1967b, Nuovo Cimento **50A,** 211.
Gutowsky, H. S., and R. E. McClure, 1951, Phys. Rev. **81,** 276.
Hagström, S., O. Hörnfeldt, C. Nordling, and K. Siegbahn, 1962, Arkiv Fysik **23,** 145.
Hamer, W. J., 1968, J. Res. Natl. Bur. Std. **72A,** 435.
——, 1967, Instrument Society of America Preprint No. M2-1-MESTIND-67, September; this gives an interesting history of the NBS as-maintained volt and its relationship to the Faraday.
——, private communication.
Hammond, J. A., and J. E. Faller, 1967, IEEE Trans. Quantum Electronics **QE-3,** 597.
Handbook of Physics and Chemistry, 1968, (Chemical Rubber Publ. Co., Cleveland, Ohio), 48th ed., p. E117.
Hara, K., N. Koizumi, H. Nakamura, and H. Imaizumi, 1965, report to the Comité Consultatif d'Electricite, Comité International des Poids et Mesures, 11th Session.
——, private communication.
——, H. Nakamura, T. Sakai, and N. Koizumi, 1968, report to the Comité Consultatif d'Electricite, Comité International des Poids et Mesures, 12th Session, October.
Harrick, N. J., R. G. Barnes, P. J. Bray, and N. F. Ramsey, 1953, Phys. Rev. **90,** 260.
Harriman, J. M., 1956, Phys. Rev. **101,** 594.
Harris, F. K., 1966, IEEE Spectrum **3,** 85, November.
——, 1964, "Electrical Units," Instrument Society of America Preprint No. 12.1-1-64.
——, private communication.
Henins, I., 1964, J. Res. Natl. Bur. Std. **68A,** 529.
——, and J. A. Bearden, 1964, Phys. Rev. **135,** A890.
Henry, G. R., and J. E. Silver, 1969, Phys. Rev. **180,** 1262.
Heyl, P. R., and P. Chrzanowski, 1942, J. Res. Natl. Bur. Std. **29,** 1.
——, and G. S. Cook, 1936, J. Res. Natl. Bur. Std. **17,** 805.
Himmell, L. C., and P. R. Fontana, 1967, Phys. Rev. **162,** 23.
Hipple, J. A., H. Sommer, and H. A. Thomas, 1949, Phys. Rev. **76,** 1877.
Houston, W. V., 1927, Phys. Rev. **30,** 608.

Huggins, R. W., and J. H. Sanders, 1965, Proc. Phys. Soc. (London) **86,** 53.
Hughes, V. W., 1969, *Proceedings of the First International Conference on Atomic Physics,* B. Bederson, V. W. Cohen, and F. M. J. Pichanick, Eds. (Plenum Press, Inc., New York), p. 15.
——, 1967, Phys. Today **20,** 29, December.
——, 1966, Ann. Rev. Nucl. Sci. **16,** 445.
——, S. Marder, and C. S. Wu, 1957, Phys. Rev. **106,** 934.
Hutchinson, D. P., J. Menes, G. Shapiro, and A. M. Patlach, 1963a, Phys. Rev. **131,** 1351.
——, J. Menes, G. Shapiro, and A. M. Patlach, 1963b, Phys. Rev. **131,** 1362.
Iddings, C. K., 1969, *Physics of the One- and Two-Electron Atoms,* F. Bopp and H. Kleinpoppen, Eds. (North-Holland Publishing Co., Amsterdam).
——, 1965, Phys. Rev. **138,** B446.
——, and P. M. Platzman, 1959a, Phys. Rev. **115,** 919.
——, and P. M. Platzman, 1959b, Phys. Rev. **113,** 192.
Jaklevic, R. C., J. Lambe, J. E. Mercereau, and A. H. Silver, 1965, Phys. Rev. **140,** A1628.
Jeffries, C. D., 1951, Phys. Rev. **81,** 1040.
Josephson, B. D., 1965, Advan. Phys. **14,** 419.
——, 1962, Phys. Letters **1,** 251.
——, private communication.
Karplus, R., and A. Klein, 1952, Phys. Rev. **87,** 848.
——, and N. M. Kroll, 1950, Phys. Rev. **77,** 536.
Kaufman, S. L., 1968, Ph.D. thesis, Yale University (unpublished).
——, W. E. Lamb, Jr., K. R. Lea, and M. Leventhal, 1969a, Phys. Rev. Letters **22,** 507.
——, W. E. Lamb, Jr., K. R. Lea, and M. Leventhal, 1969b, Phys. Rev. Letters **22,** 806, Erratum.
——, and K. R. Lea (private communication).
——, M. Leventhal, and K. R. Lea, 1968, International Conference on Atomic Physics, New York University, June.
Kendall, M. A., 1952, *Advanced Theory of Statistics* (Hafner Publ. Co., New York), Vol. 1, p. 292.
Khorana, B. M., and B. S. Chandrasekhar, 1967, Phys. Rev. Letters **18,** 230.
——, and D. H. Douglass, Jr., 1969, Bull. Am. Phys. Soc. **14,** 96.
Kinoshita, T., 1968, *Cargèse Lectures in Physics,* M. Levy, Ed. (Gordon and Breach, Science Publishers, Inc., New York), Vol. 2.
——, 1967, Nuovo Cimento **51B,** 140.
——, and R. J. Oakes, 1967, Phys. Letters **25B,** 143.
Kirchner, F., and W. Wilhelmy, 1955, Z. Naturforsch. **10a,** 657.
Kirkpatrick, H. A., J. W. M. DuMond, and E. R. Cohen, 1968, *Proceedings of the Third International Conference on Atomic Masses,* R. C. Barber, Ed. (University of Manitoba Press, Winnipeg), p. 347.
Klein, E., 1968, Z. Physik **208,** 28.
Kleppner, D., H. M. Goldenberg, and N. F. Ramsey, 1962, Appl. Opt. **1,** 55.
Knowles, J. W., 1964, *Proceedings of the Second International Con-*

ference on Nuclidic Masses, W. Johnson, Ed. (Springer–Verlag, Vienna), p. 113.
——, 1962a, Can. J. Phys. **40,** 257.
——, 1962b, Can. J. Phys. **40,** 237.
——, private communication.
Kolibayev, V. A., 1965, Geodesy and Aerophotography, No. 3, p. 228 (translated for the American Geophysical Union).
König, L. A., J. H. E. Mattauch, and A. H. Wapstra, 1962, Nucl. Phys. **31,** 1.
Lamb, Jr., W. E., 1952, Phys. Rev. **85,** 259.
——, 1941, Phys. Rev. **60,** 817.
——, and R. C. Retherford, 1952, Phys. Rev. **86,** 1014.
——, and R. C. Retherford, 1951, Phys. Rev. **81,** 222.
——, and R. C. Retherford, 1950, Phys. Rev. **79,** 549.
Lambe, E. B. D., 1959, Ph.D. thesis, Princeton University (unpublished).
——, 1969, *Polarisation, Matière et Rayonnement* (Societé Francaise de Physique, Paris), p. 441.
Langenberg, D. N., W. H. Parker, and B. N. Taylor, 1968, *Proceedings of the Third International Conference on Atomic Masses*, R. C. Barber, Ed. (University of Manitoba Press, Winnipeg), p. 439.
Lautrup, B. E., and E. de Rafael, 1968, Phys. Rev. **174,** 1835.
Layzer, A. J., 1960, Phys. Rev. Letters **4,** 580.
Leventhal, M., K. R. Lea, and W. E. Lamb, Jr., 1965, Phys. Rev. Letters **15,** 1013.
Liebes, Jr., S., and P. Franken, 1959, Phys. Rev. **116,** 633.
Lipworth, E., and R. Novick, 1957, Phys. Rev. **108,** 1434.
London, F., 1950, *Superfluids* (John Wiley & Sons, Inc., New York), Vol. 1.
——, 1935, Proc. Roy. Soc. (London) **A152,** 24.
Mader, D., and M. Leventhal, 1968a, International Conference on Atomic Physics, New York University, June.
——, and M. Leventhal, 1968b, Bull. Am. Phys. Soc. **13,** 575.
——, private communication.
Mamyrin, B. A., and A. A. Frantsuzov, 1968, *Proceedings of the Third International Conference on Atomic Masses*, R. C. Barber, Ed. (University of Manitoba Press, Winnipeg), p. 427.
——, and A. A. Frantsuzov, 1965, Zh. Eksp. i Teor. Fiz. **48,** 416 [Sov. Phys.—JETP **21,** 274 (1965)].
——, and A. A. Frantsuzov, 1964, Dokl. Akad. Nauk. USSR **159,** 777 [Sov. Phys.—Doklady **9,** 1082 (1965)].
Marion, J. B., 1968, *Proceedings of the Third International Conference on Atomic Masses*, R. C. Barber, Ed. (University of Manitoba Press, Winnipeg), p. 410.
——, private communication.
——, and H. Winkler, 1967, Phys. Rev. **156,** 1062.
Martin, W. C., 1959, Phys. Rev. **116,** 654.
Martsiniak, A. I., 1956, Izmeriteln. Tekh. i Poverochnoe Delo **5,** 11.
Mathur, B. S., S. B. Crampton, D. Kleppner, and N. F. Ramsey, 1967, Phys. Rev. **158,** 14.
Matsuda, H., and T. Matsuo, 1968, J. Phys. Soc. Japan **25,** 950.

Mattauch, J. H. E., W. Thiele, and A. H. Wapstra, 1965, Nucl. Phys. **67**, 1.

McNish, A. G., 1962, IRE Trans. Instr. **I-11**, 138.

Mercereau, J. E., 1969, *Tunneling Phenomena in Solids*, E. Burstein and S. Lundqvist, Eds. (Plenum Press, Inc., New York), p. 461.

Meservey, R., 1968, J. Appl. Phys. **39**, 2598.

Metcalf, H., J. R. Brandenberger, and J. C. Baird, 1968, Phys. Rev. Letters **21**, 165.

——, J. R. Brandenberger, and J. C. Baird (to be published).

——, J. R. Brandenberger, and J. C. Baird (private communication).

Mott, N. F., and H. S. W. Massey, 1965, *Theory of Atomic Collisions* (Clarendon Press, Oxford), 3rd ed., p. 224.

Myint, T., D. Kleppner, N. F. Ramsey, and H. G. Robinson, 1966, Phys. Rev. Letters **17**, 405.

Narasimham, M. A., 1968, Ph.D. thesis, University of Colorado, (unpublished).

——, private communication.

Newcomb, W. A., and E. E. Salpeter, 1955, Phys. Rev. **97**, 1146.

Newell, G. F., 1950, Phys. Rev. **80**, 476.

Nieto, M. M., 1968, Phys. Rev. Letters **21**, 488.

Nordfors, B., 1956, Arkiv Fysik **10**, 279.

Novick, R., E. Lipworth, and P. F. Yergin, 1955, Phys. Rev. **100**, 1153.

Page, C. H., 1968, report of Comité Consultatif d'Electricité, Comité International des Poids et Mesures, 12th Session, October.

——, private communication.

Parker, W. H., D. N. Langenberg, A. Denenstein, and B. N. Taylor, 1969, Phys. Rev. **177**, 639.

——, B. N. Taylor, and D. N. Langenberg, 1967, Phys. Rev. Letters **18**, 287.

Parsons, R. G., 1968, Phys. Rev. **168**, 1562.

Petermann, A., 1958a, Nucl. Phys. **5**, 677.

——, 1958b, Fortschr. Physik **6**, 505.

——, 1957a, Helv. Phys. Acta **30**, 407.

——, 1957b, Phys. Rev. **105**, 1931.

Petley, B. W. (private communication).

——, and K. Morris, 1968a, *Proceedings of the Third International Conference on Atomic Masses*, R. C. Barber, Ed. (University of Manitoba Press, Winnipeg), p. 461.

——, and K. Morris, 1968b, J. Phys. E **1**, 417.

——, and K. Morris, 1967, Nature **213**, 586.

——, and K. Morris, 1965, J. Sci. Instr. **42**, 492.

Pichanick, F. M. J., R. D. Swift, C. E. Johnson, and V. W. Hughes, 1968, Phys. Rev. **169**, 55.

Preston-Thomas, H., L. G. Turnbull, E. Green, T. M. Dauphinee, and S. N. Kalra, 1960, Can. J. Phys. **38**, 824.

Ramsey, N. F., 1950, Phys. Rev. **78**, 699.

Rayner, G. H., 1967, Metrologia **3**, 11.

Rich, A., 1968a, Phys. Rev. Letters **20**, 967.

——, 1968b, Phys. Rev. Letters **20**, 1221, Erratum.

——, 1968c, *Proceedings of the Third International Conference on Atomic Masses*, R. C. Barber, Ed. (University of Manitoba Press, Winnipeg), p. 383.

——, and H. R. Crane, 1966, Phys. Rev. Letters **17,** 271.
Richards, P. L., and P. W. Anderson, 1965, Phys. Rev. Letters **14,** 540.
Robiscoe, R. T., 1968, Phys. Rev. **168,** 4.
——, 1965, Phys. Rev. **138,** A22.
——, private communication.
——, and B. L. Cosens, 1966a, Phys. Rev. Letters **17,** 69.
——, and B. L. Cosens, 1966b, Bull. Am. Phys. Soc. **11,** 62.
Roesler, F. L., and J. E. Mack, 1964, Phys. Rev. **135,** A58.
Ruderman, M. A., 1966, Phys. Rev. Letters **17,** 794.
Sakuma, A., and J. Terrien, 1967, an excerpt (private communication) from a BIPM report to the International Union of Geophysics and Geodesy.
Sakurai, J. J., 1964, *Invariance Principles and Elementary Particles* (Princeton University Press, Princeton, N.J.), Chapt. 6.
Sanders, J. H., 1957, Nuovo Cimento Suppl. **6,** 242.
——, and K. C. Turberfield, 1963, Proc. Roy. Soc. (London) **A272,** 79.
Sandström, A. E., 1957, *Handbuch der Physik,* S. Flügge, Ed. (Springer–Verlag, Berlin), Vol. 30, p. 93.
Scully, M. O., and P. A. Lee, 1969, Phys. Rev. Letters **22,** 23.
Schupp, A. A., R. W. Pidd, and H. R. Crane, 1961, Phys. Rev. **121,** 1.
Schwartz, C., and J. J. Tiemann, 1959, Ann. Phys. (N.Y.) **6,** 178.
Schwinger, J., 1949, Phys. Rev. **76,** 790.
——, 1948, Phys. Rev. **73,** 416.
Shields, W. R. (private communication).
——, E. S. Garner, and V. H. Dibeler, 1962, J. Res. Natl. Bur. Std. **66A,** 1.
——, D. N. Craig, and V. H. Dibeler, 1960, J. Am. Chem. Soc. **82,** 5033.
Siegbahn, M., 1925, *The Spectroscopy of X-Rays,* (Oxford University Press, London).
Simmonds, M., W. H. Parker, and N. Nisenoff, 1968, Bull. Am. Phys. Soc. **13,** 1668.
Simkin, G. S., I. V. Lukin, S. V. Sikora, and V. E. Strelenskii, 1967, Izmeritel. Tekhn. **8,** 92; [Translation: Meas. Tech. **1967,** 1018].
Smakula, A., and J. Kalnajs, 1957, Nuovo Cimento Suppl. **X-6,** 214.
——, and J. Kalnajs, 1955, Phys. Rev. **99,** 1737.
——, J. Kalnajs, and V. Sils, 1955, Phys. Rev. **99,** 1747.
——, and V. Sils, 1955, Phys. Rev. **99,** 1744.
Smrž, P., and I. Úlehla, 1960, Czech. J. Phys. **10B,** 966.
Sommer, H., H. A. Thomas, and J. A. Hipple, 1951, Phys. Rev. **82,** 697.
Sommerfield, C. M., 1958, Ann. Phys. (N. Y.) **5,** 26.
——, 1957, Phys. Rev. **107,** 328.
Soto, Jr., M. F., 1966, Phys. Rev. Letters **17,** 1153.
Spijkerman, J. J., and J. A. Bearden, 1964, Phys. Rev. **134,** A871.
Stephen, M. J., 1968, Phys. Rev. Letters **21,** 1629.
Sternheim, M., 1963, Phys. Rev. **130,** 211.
Stevens, C. M., and P. E. Moreland, 1968, *Proceedings of the Third*

International Conference on Atomic Masses, R. C. Barber, Ed. (University of Manitoba Press, Winnipeg), p. 673.
Studentsov, N. V., T. N. Malyarevskaya, and V. I. Shifrins, 1968a, report to the Comité Consultatif d'Electricité, Comité International des Poids et Mesures, 12th Session, October.
——, T. N. Malyarevskaya, and V. I. Shifrin, 1968b, Izmeritel. Tekhn. **11,** 29 [English Transl.: Meas. Tech. **1968,** 1483].
Suura, H., and E. H. Wichmann, 1957, Phys. Rev. **105,** 1930.
Tate, D. R., 1968, J. Res. Natl. Bur. Std. **72C,** 1.
——, 1966, J. Res. Natl. Bur. Std. **70C,** 149.
Taylor, B. N., W. H. Parker, D. N. Langenberg, and A. Denenstein, 1967, Metrologia **3,** 89.
Terazawa, H., 1968, Progr. Theoret. Phys. (Kyoto) **39,** 1326.
Terent'ev, M. V., 1962, Zh. Eksp. i Teor. Fiz. **43,** 619 [Translation: Sov. Phys.—JETP **16,** 444 (1963)].
Terrien, J., 1960, Proceedings of the Symposium on Interferometry (National Physical Laboratory, Teddington, England) Suppl. 11, p. 103.
——, 1969, Metrologia **5,** 68.
——, private communication.
Theriot, Jr., E. D., R. H. Beers, and V. W. Hughes, 1967, Phys. Rev. Letters **18,** 767.
Thomas, H. A., 1950, Phys. Rev. **80,** 901.
——, R. L. Driscoll, and J. A. Hipple, 1950a, J. Res. Natl. Bur. Std. **44,** 569.
——, R. L. Driscoll, and J. A. Hipple, 1950b, Phys. Rev. **78,** 787.
Thompson, A. M., 1968, Metrologia **4,** 1.
——, 1959, Proc. Inst. Elec. Engrs. (London), Pt. B, **106,** 307.
——, private communication.
——, and D. G. Lampard, 1956, Nature **177,** 888.
Thompson, P. A., J. J. Amato, P. Crane, V. W. Hughes, R. M. Mobley, G. zu Putlitz, and J. E. Rothberg, 1969, Phys. Rev. Letters **22,** 163.
——, J. J. Amato, V. W. Hughes, R. M. Mobley, and J. E. Rothberg, 1967, Bull. Am. Phys. Soc. **12,** 75.
Thomsen, J. S., private communication.
——, and A. F. Burr, 1968, Am. J. Phys. **36,** 803.
Thulin, A., 1961, Trav. Bur. Int. Poids Mes. **22A,** 1.
Triebwasser, S., E. S. Dayhoff, and W. E. Lamb, Jr., 1953, Phys. Rev. **89,** 98.
Trigger, K. R., 1956, Bull. Am. Phys. Soc. **1,** 220.
——, 1955, Ph.D. thesis, Stanford University (unpublished).
——, private communication.
Tyrén, F., 1940, Nova Acta Regiae Soc. Sci. Upsaliensis **12,** No. 1.
——, 1938, Z. Physik **109,** 722.
Verganelakis, A., and D. Zwanziger, 1965, Nuovo Cimento **39,** 613.
Vessot, R., H. Peters, J. Vanier, R. Beehler, D. Halford, R. Harrach, D. Allan, D. Glaze, C. Snider, J. Barnes, L. Cutler, and L. Bodily, 1966, IEEE Trans. Instr. Meas. **IM-15,** 165.
Vigoureux, P., 1965, Metrologia **1,** 3.
——, 1962, Proc. Roy. Soc. (London) **A270,** 72.
——, 1938, Natl. Phys. Lab. Collected Researches **24,** 173.
——, 1936, Phil. Trans. Roy. Soc. (London) **A236,** 133.

——, private communication.
Weinstein, R., M. Deutsch, and S. Brown, 1955, Phys. Rev. **98,** 223.
——, M. Deutsch, and S. Brown, 1954, Phys. Rev. **94,** 758.
Weisskopf, M. C., J. P. Carrico, H. Gould, E. Lipworth, and T. S. Stein, 1968, Phys. Rev. Letters **21,** 1645.
Wells, T. E., 1956, Proc. Instr. Soc. Am. **11,** Paper 56-8-3.
Weneser, J., R. Bersohn, and N. M. Kroll, 1953, Phys. Rev. **91,** 1257.
Whittaker, E., and G. Robinson, 1944, *The Calculus of Observations* (Blackie & Son, Ltd., London), 4th ed., Chap. 8.
Wilhelmy, W., 1957, Ann. Phys. **19,** 329.
Wilkinson, D. T., and H. R. Crane, 1963, Phys. Rev. **130,** 852.
Williams, R. C., 1938, Phys. Rev. **54,** 568.
Wing, W., 1968, Ph.D. thesis, University of Michigan (unpublished).
——, private communication.
Woollard, G. P., and J. C. Rose, 1963, *International Gravity Measurements* (George Banta Co., Inc., Menasha, Wisc.) [Published for the Society of Exploration Geophysicists].
Yagola, G. K., V. I. Zingerman, and V. N. Sepetyi, 1966, Izmeritel. Tekhn. **7,** 44 [Translation: Meas. Tech. **1967,** 914].
——, V. I. Zingerman, and V. N. Sepetyi, 1962, Izmeritel. Tekhn. **5,** 24 [Translation: Meas. Tech. **1962,** 387].
Yanovskii, B. M., and N. V. Studentsov, 1962, Izmeritel. Tekhn. **6,** 28 [Translation: Meas. Tech. **1962,** 482].
——, N. V. Studentsov, and T. N. Tikhomirova, 1959, Izmeritel. Tekhn. **2,** 39 [Translation: Meas. Tech. **1959,** 126].
Yennie, D. R., 1967, Proceedings of the International Symposium on Electron and Photon Interactions at High Energies, Stanford Linear Accelerator Center, Stanford, Calif., p. 32.
Young, H. D., 1962, *Statistical Treatment of Experimental Data* (McGraw-Hill Book Co., New York).
Zemach, A. C., 1956, Phys. Rev. **104,** 1771.
Zimmerman, J. E., and J. E. Mercereau, 1965, Phys. Rev. Letters **14,** 887.
Zingerman, V., private communication.
Zwanziger, D. E., 1961, Phys. Rev. **121,** 1128.